Vermicomposting and Other Zoocomposting Systems for Solid Waste Valorization

This groundbreaking book explores the science and innovation behind zoocomposting – composting facilitated by animals such as earthworms, black soldier flies (BSF), and termites. It presents the high-rate vermicomposting (HRVC) paradigm, a revolutionary approach that enables the rapid conversion of phytowaste, even toxic and recalcitrant plant materials, into potent organic fertilizers without the need for pre-composting or manure supplementation. It details the design and performance of advanced vermireactors, the transformative effects of vermicomposting on plant growth, and the potential of BSF and termites in waste degradation. By tracing the history and mechanisms of both microbial and animal-assisted composting, this book offers valuable insights into sustainable waste management and soil fertility enhancement, making it an essential resource for researchers, environmentalists, and agricultural innovators. It serves as a practical guide for professionals and researchers working in agriculture, soil science, and sustainability in general.

Key Features:

- Explores HRVC, BSF composting, and termite-assisted waste degradation for efficient organic waste recycling.
- Presents cutting-edge reactor systems that accelerate the decomposition of phytowaste, including toxic and recalcitrant plant materials, without pre-composting or manure supplementation.
- Highlights the benefits of zoocompost products in improving soil fertility, enhancing plant growth, and supporting eco-friendly waste management practices.

Vermicomposting and Other Zoocomposting Systems for Solid Waste Valorization

Concepts, Designs, and Applications

Pratiksha Patnaik, Tabassum-Abbasi,
and S. A. Abbasi

CRC Press
Taylor & Francis Group
Boca Raton London New York

CRC Press is an imprint of the
Taylor & Francis Group, an **informa** business

Designed cover image: Shutterstock

First edition published 2026
by CRC Press
2385 NW Executive Center Drive, Suite 320, Boca Raton FL 33431

and by CRC Press
4 Park Square, Milton Park, Abingdon, Oxon, OX14 4RN

CRC Press is an imprint of Taylor & Francis Group, LLC

Library of Congress Cataloging-in-Publication Data
Names: Patnaik, Pratiksha author | Abbasi, Tasbassum, 1990- author |
Abbasi, S. A. (Shahid A.) author
Title: Vermicomposting and other zoocomposting systems for solid waste
valorization : concepts, designs, and applications / Pratiksha Patnaik, Tasbassum Abbasi, S.A. Abbasi.
Description: First edition. | Boca Raton, FL : CRC Press, 2026. |
Includes bibliographical references and index. |
Identifiers: LCCN 2025011482 | ISBN 9780367355135 hbk |
ISBN 9781041091318 pbk | ISBN 9780429331886 ebk
Subjects: LCSH: Vermicomposting | Organic wastes–Recycling
Classification: LCC TD796.5 .P37 2026
LC record available at https://lccn.loc.gov/2025011482

ISBN: 978-0-367-35513-5 (hbk)
ISBN: 978-1-041-09131-8 (pbk)
ISBN: 978-0-429-33188-6 (ebk)

DOI: 10.1201/9780429331886

Typeset in Times
by codeMantra

Contents

Preface xiv
Foreword xvi
The Authors xviii

PART I Composting: A General Introduction

Chapter 1 Composting as It Occurs in Nature in Comparison to Human-Controlled Composting 3

1.1 Introduction 3
1.2 How Composting Occurs in Nature and Under Human Control 3
1.3 Hot, Cold, and Hybrid Composting 9
1.4 Essential Difference Between Controlled Composting – Commonly Referred to as "Composting" – and The Most Common Animal-Mediated Composting Known as "Vermicomposting" 10
1.5 Introducing a New Term "Zoocomposting" 10
1.6 Distinctive Features of "Zoocomposting" 10
1.7 The Three Zoocomposting Processes Covered in This Book 13
1.7.1 Zoocomposting Based on EW and on BSF 13
1.7.2 Zoocomposting with Termites: Also Called "Termigradation" 13
References 15

Chapter 2 Distinguishing Features of Human-Controlled Composting 19

2.1 Introduction 19
2.2 The Composting Process 19
2.2.1 Distinguishing Features 26
2.3 The Vermicomposting Process 26
References 26

PART II Earthworm-Based Zoocomposting: Vermicomposting

Chapter 3 Distinguishing Features of Earthworm-Based Zoocomposting (Vermicomposting) 29

3.1 Introduction 29
3.2 Steps Involved in Vermicomposting and How They Differ from the Steps Associated with Human-Controlled Composting....... 29
3.3 Uniqueness of Vermicomposting as the First Animal–Mediated and Engineered Pollution Control Process....... 30
3.4 What Represents Vermicomposting? 31
3.5 Vermicomposting of Compost....... 32
3.6 Can Composting and Vermicomposting Be Done Simultaneously?....... 33
3.7 Pointers for Modelling, Design, and Operation of Both Processes 34
3.8 Engineering of the Vermicomposting Process....... 35
3.9 The Earthworm Bioreactor....... 35
3.9.1 The Peristomium....... 35
3.9.2 The Buccal Cavity 35
3.9.3 The Pharynx....... 36
3.9.4 The Oesophagus 36
3.9.5 The Gizzard....... 37
3.9.6 The Stomach....... 37
3.9.7 The Intestine....... 37
3.9.8 The Anus 37
3.10 Why Earthworm-Based Zoocomposting Is Much Quicker Than Composting? 38
3.11 Other Distinguishing Features of Earthworm-Based Zoocomposting....... 38
3.12 Why, Then, Zoocomposting Does Not Get Completed Within a Few Hours in an Earthworm Zooreactor?....... 38
References 40

Chapter 4 Earthworms and a Brief History of Appreciating Them 41

4.1 Introduction 41
4.2 A Brief History of Appreciating Earthworms....... 41
4.3 The Advent of Vermicomposting 45

4.4 Earthworms and the Species Considered Suitable for Vermicomposting 50
4.4.1 General 50
4.4.2 Morphology 52
4.4.3 Life Cycle 53
4.4.4 Other Habits 56
4.4.5 Ecological Classification 60
4.4.6 Species Suitable for Use in Vermicomposting 61
References 62

Chapter 5 Design and Operation of Conventional Vermireactors 63

5.1 Introduction 63
5.2 The Essence of a Vermireactor 63
5.3 The Reported Vermireactors (Vermicomposting Systems) 64
5.3.1 Static Pile Windrows 64
5.3.2 Top-Fed Windrows 64
5.3.3 Side-Fed Windrows 64
5.3.4 Top-Fed Beds 67
5.3.5 Stacked Bins 67
5.3.6 US3961603A: Habitat for Earthworm Cultivation 67
5.3.7 US4108625A: Method Of, and Apparatus For, Producing Fertilizer by Utilizing Earthworms 68
5.3.8 Wedges Vermicomposting 68
5.3.9 A Method and Apparatus for the Treatment of Organic Matter 70
5.3.10 Worm Reactor 71
5.3.11 Worm Wigwam 71
5.3.12 Earthworm Composting Device 72
5.3.13 Earthworm Breeding Plant 72
5.3.14 Waste Material Processing Apparatus 73
5.3.15 Vertically Stacked Vermihut and Horizontally Stacked Vermihut 74
5.3.16 Biolane™ 75
5.3.17 Biosafe™ 75
5.3.18 Worm Gin 75
5.3.19 High Efficiency Vermiculture Process and Apparatus 76
5.3.20 Vermiculture Composting Device 77
5.3.21 Apparatus for Upgrading Waste into Fertilizer 77
5.3.22 Multi-Layered Vermicomposting System 79
5.3.23 An Apparatus for Organic Waste Vermicomposting 79
5.3.24 Drawer Type Vermireactor 80

5.3.25 Method and Device for Processing Organic Waste for Vermicomposting 80
5.3.26 Worm Bin 81
5.3.27 Apparatus Used for Processing Organic Waste 82
5.3.28 Vermitech 200 83
5.3.29 Multiple-Tray Vermicomposter 83
5.3.30 Multiple-Tray Vermicomposter with Thermal Siphon Airflow 83
5.3.31 Digester Beds 84
5.3.32 Worm Culture Systems 84
5.3.33 Composting Apparatus and Method 86
5.3.34 Dual-Mode Vermicomposter 87
5.3.35 Vermicomposting Growing System 88
5.3.36 Vermicomposter Apparatus and Method 88
5.3.37 Vermicomposting Device 89
5.3.38 Vermicomposting Apparatus and System 89
5.3.39 Vermicomposting Method and Apparatus 89
5.3.40 Worm Composting Machine 90
5.3.41 Method for Manufacturing Vermicompost and Device for Its Implementation 92
5.3.42 A Housing for a Worm Farm 92
5.3.43 Worm Harvesting Apparatus 92
5.3.44 Vermicomposting Apparatus 96
5.3.45 Vermiculture Bioreactor System and Method of Use 96
5.3.46 Vermicompost Production Device 98
5.3.47 Installation for Complex Phytoremediation and Vermifiltration of Wastewater and Bottom Sediments 99
5.3.48 Vermicomposting Device, Method for Producing Vermicompost, Compost 100
5.3.49 Vermicompost Made of Aquatic Plants and Preparation Method and Use Thereof 100

References 100

Chapter 6 The Shortcomings of the Conventional Vermicomposting Technology, which Necessitated Development of the High-Rate Vermicomposting Paradigm and the Associated Technology 103

6.1 Introduction: The Key Challenges in the Path to Rational and Optimal Design of Vermicomposting Systems 103
6.2 Phytomass, Phytowaste, and the Imperative Need to Utilize It 106

References 123

Chapter 7 The Paradigm of High-Rate Vermicomposting and Its Validation 129

7.1 Introduction 129
7.2 Specifying the End Product, or "Aimed Product", of Vermicomposting 129
7.2.1 Vermicast as the Quantifiable Product of Vermireactors 130
7.3 The High-Rate Vermicomposting Paradigm (HRVP) 131
7.4 Validation of HRVP 134
References 137

Chapter 8 Vermireactors Designed on the Basis of the High-Rate Vermicomposting Paradigm 139

8.1 Introduction 139
8.2 A Flappable Units Vermireactors Train System (FLUVTS) 140
8.2.1 The Need for FLUVTS 140
8.2.2 Distinguishing Features of FLUVTS 140
8.2.3 Comparison with Conventional Designs 143
8.3 The High-Efficiency Vermicomposting System for Paper Waste and Other Refuse (HEVSPAR) 144
8.3.1 Novelty of the APSU 144
8.3.2 Novelty of the VSV Unit 144
8.3.3 Advantages Over Other Known Alternatives, National and Global 145
8.3.4 Detailed Description of the Invention 146
8.3.5 Distinguishing Features of the VSV Unit 151
8.3.6 Fabrication of a HEVSPAR Prototype 151
8.4 The Inclined Parallel Stack Continuously Operable (IPSCO) Vermireactor System 151
8.4.1 The IPSCO Vermireactor System 151
8.4.2 Distinguishing Features of the IPSCO Vermireactor Design 155
8.4.3 Novelty of the IPSCO Vermireactor Design 156
8.5 A Multiple Compartment Vermicomposting System (MCVS.) 159
8.5.1 Novelty of the Invention, in a Nutshell 159
8.5.2 Description of the Invention 160
8.5.3 Distinguishing Features of the Invention 161
8.6 Distinctive Features of the Four Designs 162
References 165

Chapter 9 Design and Performance Evaluation of Machines for Separating Earthworm-Vermicast-Substrate Mixtures 167

9.1 Introduction 167
9.2 The Separator of Earthworms, Vermicompost, and Substrate (SEVS) Machines 167
9.3 Testing of the SEVS 169
9.3.1 Earthworm–Vermicast–Paper Waste Mixtures 170
9.3.2 Earthworm–Vermicast–Salvinia Mixtures 170
9.3.3 Earthworms–Vermicast–Ipomoea Mixtures 170
9.3.4 Earthworm–Vermicast–Lantana Mixtures 171
9.4 Effect of Moisture on the Efficiency of Vermicast Recovery 172
9.5 Effect of the Batch Size 172

Chapter 10 Effect of Vermicomposts Derived from Phytomass and Waste Paper on Different Species of Commonly Used Plants 175

10.1 Introduction: The Pivotal Role of Manure as a Feedstock in Vermicomposting 175
10.2 But What about VC of Other Substrates? 176
10.3 Studies by S.A. Abbasi and Co-Workers 179
10.3.1 The Broad Scope 179
10.3.2 Key Steps of the Methodology 181
10.4 The Main Findings 187
10.4.1 Of the Studies Defined in Section 10.3.1.1 187
10.4.2 Studies Defined in Section 10.3.1.2 202
10.4.3 Studies Defined in Section 10.3.1.3 208
10.4.4 Studies Defined in Section 10.3.1.4 220
10.4.5 Studies Defined in Section 10.3.1.5 222
References 236

Chapter 11 Transformations Caused by Vermicomposting Which Convert Even Toxic and Allelopathic Weeds into Benign Organic Fertilizers 241

11.1 Introduction 241
11.2 Guiding Considerations 241
11.3 The Transformations 242
11.3.1 C:N Ratio 242
11.3.2 HI 243
11.3.3 The Fate of Allelochemicals and Other Harmful Biomolecules 244

11.3.4 Extent of Mineralization as Shown by TG and Differential Scanning Coulometric Studies 246
11.3.5 Molecular Rearrangements as Revealed by GC-MS Analysis 247
11.3.6 Surface Characteristics 258
11.4 Role of the Earthworm Species 259
11.4.1 C:N Ratio 259
11.4.2 The FTIR Spectra 260
11.4.3 TG Analysis and DC Scans 262
11.4.4 Scanning Electron Microscopy 264
11.5 Summary 266
References 266

Chapter 12 Long-Term Impact on the Growth and Fecundity of Successive Generations of Earthworms Fed Exclusively on One or the Other Species of Weeds 269

12.1 Introduction 269
12.2 Defining Attributes of the Methodology 270
12.3 Relative Performance of Three Earthworm Generations 271
12.4 Aspect to Ponder 276
References 276

Chapter 13 First-of-Its Kind Success in Rapidly Vermicomposting a Xerophyte and Utilizing the Product as an Organic Fertilizer 277

13.1 Introduction 277
13.2 Reaction of Earthworms to Prosopis Leaves 278
13.3 Demonstration of Uninterrupted and Sustained Vermicomposting of Prosopis Leaves for Close to Two Years 279
13.4 Prosopis Vermicompost (PRVC) as a Fertilizer 282
13.5 PRVC as Controller of Plant Pathogens 284
13.6 How Does PRVC Compare with CD and Cow Dung VC (CDVC)? 285
13.7 Factors Which Transform Highly Toxic and Insidious Prosopis Leaves into a Highly Benign and Beneficial Fertilizer-Cum-Pest-Repellent? 285
References 291

Chapter 14 Vermicomposting as an Ideal Foil to Upgrade Macrophyte/Algae-Based Sewage Treatment Systems into Close-Loop Biorefineries 292

14.1 Introduction: Problems Associated with Sewage and Its Treatment 292
14.2 The Inadequacy of Conventional Technologies and the Phytoremediation Systems 292
14.3 The SHEFROL Technology 294
14.4 Upgrading Macrophyte/Algae-Based Sewage Treatment Systems into Biorefineries 298
References 300

PART III Use of Black Soldier Flies in Zoocomposting

Chapter 15 The Unique Role of Black Soldier Flies in the Zoocomposting of Food Waste 305

15.1 Introduction 305
15.2 Food Waste and Abattoir Waste 307
15.3 Potential of Insect-Based Systems for the Management of Food Waste and Abattoir Waste 310
15.4 Why Is BSF So Called? 317
15.5 The Uniqueness of the BSF 317
15.6 Nativity and Distribution of BSF 321
15.7 The BSF Life Cycle 322
15.8 Designing BSF-Based Biowaste Processing Facilities at Different Scales: General Considerations 323
15.9 An Overview of the Design Approaches 324
References 340

Chapter 16 A Novel, Versatile, Effective, Yet Inexpensive System for Carrying out BSF-Mediated Zoocomposting of Difficult-to-Treat Food and Abattoir Waste: ZOCOFA 347

16.1 Introduction 347
16.2 Drawbacks in the Prior Art and How the Drawbacks Have Been Overcome by This 348
16.3 Summary of the Invention 348
16.4 Novelty and Distinguishing Features of the Invention 349
16.5 Detailed Description 350
16.5.1 Priming of the Reactor 355
16.5.2 Reactor Operation and Production of Larvae and Frass 355
References 356

PART IV Zoocomposting with Termites: Termigradation

Chapter 17 Termites and Their Unique Ability to Biodegrade Ligninous Waste and Fix Nitrogen 359

17.1 Introduction 359
17.2 The Challenges of Biocomposting Ligninous and "Hard" Biowaste 360
17.3 Termites: Some Relevant Facts 361
17.4 The Role and the Potential of Termites 365
17.5 Termites: Architects Extraordinary 367
17.6 Termites as Delicacies 371
References 372

Chapter 18 Opening a New Frontier in Bioprocess Technology: Direct Use of Termites in Biodegrading Highly Ligninous Waste 375

18.1 Introduction 375
18.2 Previous Attempts at Employing Termites for Solid Waste Treatment 377
18.3 The Work of These Authors 378
18.4 The Special Challenges Associated with the Use of Termites in Biocomposting 378
18.5 The Present Work 379
18.5.1 Identifying Termite Species to Be Worked With
18.6 Approaches to the Design of the Termireactors 380
18.7 The Design and the Performance Evaluation of the ISTs 381
18.7.1 The Design Approach 381
18.7.2 The Three Embodiments Designed Based on the Above Approach 381
18.7.3 Attachment of Trails 383
18.8 The Design and Performance Evaluation of the BBTs 386
18.8.1 Initiation of Termite Colonies 386
18.8.2 Present Work on Formation of Captive Termite Colonies 388
18.8.3 Controlled Use of BBTs in Processing Biowaste 389
References 393

Index 397

Preface

Among the authors of this book, the eldest – SAA – was the first to start exploring the field of vermicomposting. He set up his first vermireactor in 1986 on the basis of design and operation criteria that had been recommended by the accredited governmental agency – Centre for Advancement of People's Action and Rural Technology (CAPART). CAPART asked for an elaborate "vermibed" to be provided in all vermireactors to facilitate the earthworms in their feeding and mating efforts, as well as make available to them space to rest. The vermibed consisted of successive layers of sawdust, sand, and soil, covering more than half of the reactor space. CAPART also recommended that seven adult earthworms should be provided per kilogram of the reactor feed. Other purveyors of vermicomposting were also using the same or similar reactor designs and earthworm density. With this, vermicomposting took 3–4 months for completion.

When the reactors set by us went into operation and the periodic harvesting of vermicast began, it was realized that almost always, all earthworms were present in the feed layer itself. A few were found in the upper parts of the soil layer, but none in the sand or sawdust layers lying further down. It was also found that only a small portion of the feed was converted to vermicast every month, indicating that, for the most part, more feed was present in the vermireactor than could be consumed by the earthworm population of seven animals per kg of the feed. These observations raised serious doubts on the rationale behind the prevailing design and operation criteria of the vermireactors. By and by, other basic questions also emerged. The most haunting was the following question: Why are large vermireactors operated principally with animal manure instead of plant biomass (phytomass) when, in nature, earthworms feed much more on phytomass than they do on animal manure?

In the course of subsequent decades, the authors have tried to find answers to these questions. The resulting efforts – theoretical as well as experimental – have led to the development of the HRVC paradigm and associated technology. It has enabled rapid vermicomposting of phytowaste without any pre-composting and/or supplementation of animal manure. It has shown that vermicompost derived from not only horticultural waste but even toxic weeds like lantana, ipomoea, and parthenium is as potent an organic fertilizer as manure-based vermicompost is known to be. It has made possible to rapidly vermicompost even a xerophyte like *prosopis*– which stubbornly resistant to biodegradation while also being full of toxic biomolecules into a totally benign organic fertilizer. It has led to the development of several vermireactor designs aimed at processing different substrates at different scales of operation. Most importantly, this journey has enabled us to trace the history of composting and develop an appreciation of the basic differences between animal-mediated composting – which we have given the name *zoocomposting*, and the conventional "hot" composting, which is predominantly carried out directly by mesophilic and thermophilic microorganisms. We have also got an opportunity to foray into composting of food/abattoir and ligninous waste by black soldier flies (BSF) and termites, respectively. The challenges faced in the design, operation, and scaling up of each of these processes were also identified and reflected upon. This book is a distillate of all that exploration and churning. Authors thank the management of UPES, Dehradun, for supporting the book's writing. They express gratitude towards Mrs Naseema Abbasi for moral and logistic support, and Ms P. Anandhi for her prompt, intense and imaginative secretarial assistance.

Foreword

Very extensive research has been done, and is continuing to be done with ever greater intensity, on solving pollution control problems by nature-based means. Among the processes that have proved commercially viable, vermicomposting is arguably the simplest and the cleanest. It is simple because it needs no sophisticated reactors or controls and no significant external energy input to run, and it is among the easiest processes to operate and monitor. It is exceedingly clean because it generates no waste of its own, gives off no harmful emissions, and causes no nuisance like malodours. Also, vermicomposting is the least expensive of all bioprocesses because it only needs initial seeding of earthworms and periodic sprinkling of water. Above all, vermicomposting leads to a product – vermicast – which is an excellent organic fertilizer with almost limitless demand. Why, then, is the reach of vermicomposting limited in the world? Why is its enormous potential underutilized? Why is it so even in developing countries where massive quantities of biodegradable solid waste are generated but are disposed untreated, leading to many serious environmental problems? And where demand for organic fertilizers hugely outstrips the supply?

This book addresses these questions. It not only identifies the causes behind the very substantial underutilization of vermicomposting but also provides recently emerging solutions that have been introduced by the authors after extensive trials and tests. Many of the consequent innovations have received patents and have been published in reputed international journals. The resulting corpus has generated a new paradigm named high-rate vermicomposting (HRVC) by the authors. This book is a distillate of all those essences. In addition, chapters on the use of black soldier flies and termites to treat food/abattoir and ligneous wastes, respectively, which cannot be handled by vermicomposting, confer additional utility value upon this book. Of equal significance is the chapter that shows the potential of vermicomposting systems as sequels to aquatic plant-based wastewater treatment systems. Together, these two nature-based clean technologies lead to biorefineries that essentially run on wastewater but produce clean energy and organic fertilizers.

The seniormost author, Prof. S.A. Abbasi, is ranked among the top three environmental experts in India and the top 0.5% researchers worldwide. He has been researching the field of bioprocess engineering for 55 years and has pioneered many innovations, including HRVC. The co-authors, Dr Pratiksha Patnaik and Dr Tabassum Abbasi, have had brilliant academic careers and a proven research record. The team of youth and experience together has produced this book, which, I hope, will have a transformative impact on the field of bioprocess engineering in general and nature-based pollution control systems in particular.

I feel particularly delighted to mention that Uttarakhand, which is aspiring to be one of the main organic farming states of India, will greatly benefit from the innovations presented in this book, just as will other regions of India and, indeed, the world. In the process, the beneficiaries will also achieve control of noxious weeds, thereby arresting the ongoing harm to natural resources and biodiversity. This book makes this possible by providing know-how for utilizing large quantities of noxious weeds for generating organic fertilizers, that too in a very clean-green manner. The processes that the authors have developed are two to three times faster than the pre-existing ones, achieving a major benefit-cost advantage. Additionally, there are concomitant benefits associated with weed control. All these features, together, make this book an invaluable contribution.

(Prof. Durgesh Pant)
Director General
Uttarakhand State Council for Science and Technology,
Government of Uttarakhand,
Dehradun-248015, India

The Authors

Dr. Pratiksha Patnaik received her BSc and MSc degrees in ecology and environmental sciences, topping in both programmes and receiving gold medals. She later obtained her PhD in 2018 for her path-breaking research on zoocomposting with a focus on valorizing phytowaste and food waste. Besides several research papers in prestigious journals, Dr Pratiksha's work has resulted in a book, *Prosopis juliflora: Attributes, Impact, Utilization*, published by *Elsevier*, and four patents.

Dr. Tabassum-Abbasi has received her BTech, MTech, and PhD degrees in instrumentation and control engineering, environmental engineering, and environmental technology, respectively – all from Pondicherry University. She too has been a class topper throughout, receiving gold medals. She has been working as an Assistant Professor at the University of Petroleum & Energy Studies, Dehradun, since January 2019, earning promotion to senior grade in 2022, and Selection Grade in 2025. She has also spent a year (2023–2024) in postdoctoral research at the University of Edinburgh as a British Council Women in Science Fellow. Besides 87 research papers on bioprocess engineering and sustainability in highly cited journals and books, she has received three patents and has registered another two. She has authored a book and has another two in press.

Prof Shahid Abbas Abbasi is presently Distinguished Professor at the School of Advanced Engineering, UPES, Dehradun. He has been a full university Professor and Director (1987–2015) and Emeritus Professor (2015–2024) at the Centre for Pollution Control and Environmental Engineering, Pondicherry University. Prior to that, he was the founder and head of the Water Quality & Environment Division at the Centre for Water Resources, Kozhikode (1979–1987), and a Visiting-cum-adjunct Professor at California State University (1984–1987). He has also been a Visiting Professor at the universities of Minnesota, Florida, Nevada, California–Berkeley, Malaya, and Al-Ahsa.

With 50 books, close to 400 papers in indexed journals, 12 patents, over 25,000 citations, and a Hirsch Index currently at 84, Prof Abbasi is among the world's foremost experts in the fields of bioprocess engineering, process safety, and environmental engineering. In the Stanford University and ScholarGPS assessments, he is ranked among the world's top 0.5% scientists, and the latest AD Scientific rankings put him among India's top two in environmental science and engineering (https://www.thehindu.com/news/cities/puducherry/pondicherry-university-professor-felicitated/article33349572.ece).

Among numerous coveted honours and awards received by Prof Abbasi are the National Design Award in Environmental Engineering, the IPCL Award for Safety and Hazard Management in Petroleum Industries, the National Hydrology Award, the International Desalination Association's prize, The Trever Kletz Award, and Fellowships of the National Academy of Sciences and the Indian Institute of Chemical Engineers. Prof Abbasi is also well-known for his poetry (*ghazal* and *geet*), fiction, and his work on fostering inter-faith harmony. Details of his *oeuvre* can be found at *https://vidwan.inflibnet.ac.in/profile/61966.*

Part I

Composting

A General Introduction

1 Composting as It Occurs in Nature in Comparison to Human-Controlled Composting

1.1 INTRODUCTION

This chapter presents the essence of composting as it occurs in nature and in human-controlled composting systems. It has also elaborated upon how vermicomposting differs sharply from human-controlled composting, even as the end objective of both is to biodegrade organic waste and mineralize as much of it as possible. This chapter introduces the term "zoocomposting" to represent processes – of which vermicomposting is one of the major examples – in which one or other multi-cellular animal plays a leading, controlling role.

1.2 HOW COMPOSTING OCCURS IN NATURE AND UNDER HUMAN CONTROL

If biodegradable organic waste is left to itself in nature, it begins to be decomposed either by the microorganisms already present in it in the form of bacteria, fungi, and actinomycetes (endophytes), or by these microorganisms that come to it from air, water, or soil. Small invertebrates, especially insects, are also attracted to the degrading waste. There are also large animals who are either obligate scavengers – i.e., ones whose principal diet is carrion – or facultative scavengers who only scavenge occasionally to supplement the food they obtain by predation. The most prominent obligate scavengers are vultures (Figure 1.1), while hyenas, foxes, bears, raccoons, lions, jackals, coyotes, hens, crows, and crocodiles are examples of facultative scavengers. Among all the facultative scavengers, the spotted hyena (*Crocuta crocuta*) (Figure 1.2) is considered the most efficient. It is able to crush even the bones present in the carcasses of large animals and digest the feed.

Be it the large animals or small, or the microorganisms, by the acts of feeding upon organic waste and converting it into excreta, they hasten the biodegradation and mineralization of the waste. As the concentration of the microorganisms and invertebrates builds up in a waste pile, the rate of the waste's biodegradation gradually increases till all of it is converted to minerals. If the waste happens to have complex organics that resist biodegradation, those components are left behind as humus. Also generated in the process are some enzymes, hormones, and other organics, which,

DOI: 10.1201/9780429331886-2

FIGURE 1.1 Vultures are obligate scavengers, with carrion as their principal diet. (https://commons.wikimedia.org/wiki/File:Hyena_vs_Vultures.jpg.)

together with the mineralized major, minor, and trace elements, bestow fertilizer value upon the end product of this biodegradation.

When the abovementioned process is done in a controlled fashion by humankind, directing it towards the generation of fertilizer, and with an objective to achieve maximum biodegradation in minimum time, it is termed "composting". Moreover, it is not only the organic waste that gets composted in nature. All the food that is eaten by animals, including humans, also undergoes decomposition as it is digested, eventually getting mineralized in the form of excreta which has attributes similar to those of a compost.

It can be said that in its broadest sense, composting is a process for transforming organic matter into fertilizer – essentially by naturally available microorganisms

FIGURE 1.2 Spotted hyena is regarded as the most efficient among facultative scavengers. (https://commons.wikimedia.org/wiki/File:Sitting_Spotted_Hyena_Luangwa_Jul23_A7C_06239.jpg.)

but aided and abetted by different species of animals, principally insects. In that sense, all animals, including humans, ranging from single-cell protozoan to large mammals (Figure 1.3) are composters because they all ingest organic matter, of plant and/or animal origin, as food. The microorganisms carried by the food or present in the animal gut metabolize the food, supported by enzymes, and the animal leaves behind excreta which has the attributes of a compost. Besides microorganisms, coprophagous organisms such as dung beetles (Figures 1.4 and 1.5), several species of flies or earwigs, termites, etc., contribute to further stabilization and mineralization of animal excreta, including human faeces. By the continued action in this manner, of microorganisms and of other small animals, especially insects, the excreta is further processed in nature, increasing its fertilizer value. Due to the differences in the characteristics of the excreta generated by different species of animals, their ease-of-use and fertilizer values also differ. But, in nature, all animal droppings eventually get utilized as fertilizers by enriching the soil/water on which they fall with nutrients. Hence, every animal, including every human being, is a compost generator throughout its life.

FIGURE 1.3 All animals are composters – from the tiniest single-cell protozoan to the biggest land mammal. (https://commons.wikimedia.org/wiki/File:An_elephant_in_Queen_Elizabeth_National_Park_in_Uganda_01.jpg https://commons.wikimedia.org/wiki/File:Paramecium_-_Trichocysten_-_Opalblau_(8149944429).jpg.)

FIGURE 1.4 A dung beetle rolling human excreta into an easily transportable sphere. (https://commons.wikimedia.org/wiki/File:Flightless_Dung_Beetle_Circellium_Bachuss, _Addo_Elephant_National_Park, _South_Africa.JPG.)

FIGURE 1.5 Dung beetles colonizing a heap of dung. (https://commons.wikimedia.org/wiki/File:Namibia-dung-beetle-feast.jpg.)

When composting occurs in nature, it can be predominantly done by the microorganisms present in the environment, or predominantly by microorganisms present in the animal gut, or by combinations of both – depending on the situation. But when composting is done in a controlled fashion by humans, it is in two distinct forms. The first form, which is generally and simply referred to as composting, is carried out predominantly by microorganisms – mesophilic and thermophilic – with invertebrates playing a minor role towards the final phase of composting. Under this broad term of composting come "hot", "cold", and "hybrid" composting, as explained later. The second form is called vermicomposting, in which earthworms (EW) are introduced into organic waste and play the main role in the waste's degradation from start to finish. EW do it essentially by fragmenting, agitating, and aerating the waste, thereby helping microflora to proliferate and carry out biodegradation.

In recent years, another form of animal-mediated composting is gaining currency, which involves black soldier fly (BSF; *Hermetia illucens*). BSF (Figure 1.6) has been found particularly effective in handling food waste, especially the one rich in proteins, fats, and spices. Such substances are not easy to compost, nor do EW usually prefer them.

FIGURE 1.6 A BSF. It enables rapid zoocomposting of food and abattoir waste. (https://www.inaturalist.org/observations/81464966.)

1.3 HOT, COLD, AND HYBRID COMPOSTING

As detailed in Chapter 2, when composting is done with waste containing adequate nutrients, especially nitrogen, and is subjected to aeration and periodic turning, the microorganisms with which the waste has been inoculated multiply and feed upon the waste, and the resulting metabolic activity lifts the temperature above mesophilic and into thermophilic stage, reaching 60°C and beyond. These happenings speed up the rate of composting and also kill most pathogens and undesirable seeds, especially of weedy species of plants. At the same time, control is exercised not to let the temperature go much beyond 68°C (155°F) as it would then begin causing charring.

When such operational protocols and process controls are not exercised, the resulting composting is called "cold composting". In it, the waste is just piled up and left to slowly biodegrade.

"Hybrid composting" is something intermediate between hot and cold composting. It is controlled to some extent, but not as rigorously as hot composting is. Composting in bins is often of this type (Figure 1.7),

The closer a composting process is to proper hot composting, the quicker and better the end result, the fertilizer.

FIGURE 1.7 Composting in bins like these is generally of the hybrid type. (https://commons.wikimedia.org/wiki/File:Bacs_%C3%A0_compost_de_la_m%C3%A9tropole_grenobloise_dans_une_copropri%C3%A9t%C3%A9.jpg.)

1.4 ESSENTIAL DIFFERENCE BETWEEN CONTROLLED COMPOSTING – COMMONLY REFERRED TO AS "COMPOSTING" – AND THE MOST COMMON ANIMAL-MEDIATED COMPOSTING KNOWN AS "VERMICOMPOSTING"

Controlled composting, especially at larger-than-household scale, is mostly "hot composting" because of its much faster speed and its ability to kill pathogens and seeds of weedy species. As indicated above, hot composting is directly carried out by microorganisms – mesophilic as well as thermophilic – and is consciously driven in the direction of converting the waste into a relatively pathogen-free fertilizer. Even though carried out predominantly by microorganisms, small invertebrates in the form of nematodes, annelids, crustaceans, insects, etc., also play a part in hot composting in its post-thermophilic stage when the pile temperature has fallen to the levels tolerable to the animals. These organisms move in after the thermophilic phase is over and the temperature of the waste pile has become near ambient, from areas surrounding the compost piles. They help in further stabilizing the product by feeding upon, and turning to their excreta, the portions of the compost which had not yet been significantly stabilized.

1.5 INTRODUCING A NEW TERM "ZOOCOMPOSTING"

What, then, is special about vermicomposting? Or about composting carried out by other animals explored so far – about whom we will be discussing in some detail later?

Vermicomposting is different from composting essentially in the fact, and its far-reaching consequences, that in it one or the other chosen species of EW is involved right from the start. Due to this, EW control the rate as well as the efficacy of the biodegradation occurring in a vermicomposting system from start to finish. However, in composting of a general nature, multiple species of small invertebrates play a supporting role towards the final phase of biodegradation; the prime movers in vermicomposting are EW at all stages of the vermicomposting process. In that respect, vermicomposting should be regarded as a branch of "zoocomposting". Zoocomposting is a new term we are introducing to cover those processes in which one or the other species of animals play a decisive role from the beginning to the end (as EW do in vermicomposting). A prominent example of a zoocomposting process other than vermicomposting is the one controlled by BSF. Some authors have, in the past, used the term "biocomposting" to denote such processes, but it does not appear to be a correct term because all composting of all forms and variations is a biological process – there are no chemical-composting and physical-composting processes in existence. Hence, all forms of composting are biocomposting.

1.6 DISTINCTIVE FEATURES OF "ZOOCOMPOSTING"

As mentioned earlier in this chapter, all animals contribute to composting in one form or the other, in large measure or small. Among the animals, arguably, the most important and multi-faceted role in this aspect is played by insects. A large number of insect species have become specialized zoocomposters of different types of organic waste, ranging from plant debris to rotting animal flesh (Mondor et al., 2012; Mealey, 2022). In all cases, the zoocomposting is carried out by microorganisms, supported by naturally occurring biocatalysts (enzymes, other biomolecules contained in, or generated by, the microflora) present in the body of the concerned animal, while activities like ingesting, comminuting, munching, and turning of the animal facilitate the microflora.

But despite a bewildering variety of animals, carrying out zoocomposting, in ever more bewildering ways in nature, it has not been possible so far to use animals in large-scale and controlled systems in getting specific wastes zoocomposted in an economically viable manner. Only two exceptions exist to this general rule: EW and BSF.

As for other animals, 24 species, all invertebrates, have been explored so far in the zoocomposting of different types of biodegradable wastes (Table 1.1). But none of them have been found promising enough to be utilized at higher-than-bench scales, nor has anyone been exploited commercially. The list also does not include any species known to efficiently biodegrade ligninous waste.

TABLE 1.1
Insects and Other Invertebrates Explored so Far Towards Degradation of Solid Waste: All Available Reports are on Studies Done at Laboratory Scale

Species	Common Name	Targeted Solid Waste	Reference
Calliphora vicina	Blue blow fly	Slaughter waste	Barroso et al. (2014) and Pastor et al. (2015)
Chrysomya megacephala	Oriental blue fly	Restaurant waste, manure, slaughter wastes	Li et al. (2012), Barroso et al. (2014), Yang and Liu (2014), ČiČková et al. (2015), Pastor et al. (2015)
Conicera tibialis	Phorid fly or scuttle fly	Animal carcasses	Sharanowski et al. (2008)
Coproica ferruginata	Lesser dung fly	Solid urban waste	Morales and Wolf (2010)
Curtonotum helvum	Curtonotid fly	Solid urban waste	Morales and Wolf (2010)
Cydia pomonella	Codling moth	Starch and cheese waste, wastewater sludge	Meats et al. (2004), Brar et al. (2008), Fowles and Nansen (2019)
Desmometopa varipalpis	Free loader fly	Solid urban waste	Morales and Wolf (2010)
Drosophila melanogaster	Fruit fly	Urban waste, wastewater	Wheeler (1987) and Clark et al., (2021)
Galleria mellonella	Greater wax moth larvae	Polyethylene polystyrene	Billen et al. (2020)
Lucilia sericata	Common green bottle fly	Carrion, meat, manure, slaughter waste	Nuov et al. (1995), Marchaim et al. (2003), Yehuda et al. (2011), Barroso et al. (2014), ČiČková et al. (2015), Pastor et al. (2015)

(Continued)

TABLE 1.1 (*Continued*)
Insects and Other Invertebrates Explored so Far Towards Degradation of Solid Waste: All Available Reports are on Studies Done at Laboratory Scale

Species	Common Name	Targeted Solid Waste	Reference
Musca autumnalis	Face fly	Cattle manure	Koo et al. (1980), Arends and Wright (1981), ČiČková et al. (2015)
Musca domestica	House fly	Manure, garbage, food wastes, restaurant waste, corn silage, sawdust, gibberellin fermentation residue, agro-processing waste products	Dordević et al. (2008), Pastor et al. (2011), Wang et al. (2013), ČiČková et al. (2015), Pastor et al. (2015), Roffeis et al. (2015), van Zanten et al. (2015), Yang et al. (2015), Jiang et al. (2017), Niu et al. (2017), Bloukounon-Goubalan et al. (2020), Ganda et al. (2019), Fowles and Nansen (2019), Miranda et al. (2021)
Ornidia obesa	Syrphid fly	Solid urban waste	Morales and Wolf (2010)
Orthaga aenescens	Shout moth	Solid urban waste	Morales and Wolf (2010)
Protaetia brevitarsis	Oriental edible beetle larvae	plant matter	Li et al. (2019)
Protophormia terraenovae	Blue-bottlefly	Slaughterhouse waste	Barroso et al. (2014) and Pastor et al. (2015)
Sarcophaga carnaria	Common flesh flies	Decaying meat	Pape (1987), Yehuda et al. (2011), ČiČková et al. (2015), Pastor et al. (2015)
Sarcophaga peregrina	Flesh fly	Organic waste, defatted restaurant waste, faeces, decaying meat	Yang et al. (2012), ČiČková et al. (2015), Pastor et al. (2015)
Stomoxys calcitrans	Stable fly	Manure	Mramba (2007) and Dehghani et al. (2018)
Teleogryllus testaceus	Cambodian field cricket	Agriculture waste, plants, spent grain, weeds	Melillo (2014), Miech et al. (2016), Fowles and Nansen (2019)
Tenebrio molitor	yellow mealworm	Wheat straw, vegetable waste, corn stove, crop waste, plastic waste (polyurethane, polyethylene, polylactic acid, polystyrene)	Lush (1943), Meyer et al. (2012), Li et al. (2013), Wang et al. (2017), Yang et al. (2018, 2019), Fowles and Nansen (2019), Billen et al. (2020), He et al. (2021), Bulak et al. (2021)
Themira annulipes	Black scavenger fly	Solid urban waste	Morales and Wolf (2010)
Zophobas atratus	Super worm	Plastics such as polystyrene and polyethylene	Yang et al. (2020)
Zophobas morio	Super worm	Spent grain, distillers' grain, baking remains, plastics such as polystyrene and polyethylene	Van Broekhoven et al. (2015), Harsányi et al. (2020), Peng et al. (2020), Yang et al. (2020), Rumbos and Athanassiou (2021)

1.7 THE THREE ZOOCOMPOSTING PROCESSES COVERED IN THIS BOOK

1.7.1 Zoocomposting Based on EW and on BSF

Between EW and BSF, the EW-mediated biocomposting has had a far longer history, and the process, commonly referred to as vermicomposting, is in much more widespread use. For long, its range of applicability had been confined to animal manure because of the engineering problems associated with the use of plant biomass as vermicomposting feedstock (Edwards et al., 2011). But in recent years, these authors have overcome the problems by developing the paradigm of high-rate vermicomposting and applying it successfully to the processing of plant biomass (phytomass) in specially designed systems (Tauseef et al., 2021). These authors have also developed criteria with which the vermicomposting process can be monitored as well as controlled. Chapters 3–13 cover these aspects in detail.

As for BSF, it has also attracted increasing attention, and its potential for large-scale processing of different forms of biowaste is now widely accepted (Chaklader et al., 2021; Franco et al., 2022). Part III of this book is devoted to the role of BSF in zoocomposting.

It may be emphasized that when human-controlled aerobic composting is carried out in a traditional manner by seeding biodegradable waste with microorganisms and by supporting bacterial action/multiplication by periodic aeration and turning, animals do not play any role. Only after the bulk of the substrate has been biodegraded and the mass is kept for a few days for "curing", small invertebrates, especially insects, move in and facilitate completion of the composting process. But this happens in a more or less spontaneous manner, with no external controls. Hence, the animals involved in this manner in the curing have not been considered in the present book on zoocomposting, which focuses on specific classes of animals used in a controlled fashion for processing organic solid waste.

1.7.2 Zoocomposting with Termites: Also Called "Termigradation"

For long, termites (Figures 1.8 and 1.9) have been regarded as one among three groups of animals that have been termed "soil engineers" because of the dominant role they play in turning, shifting, fragmenting, aerating, nourishing, and rejuvenating the

(a) (b)

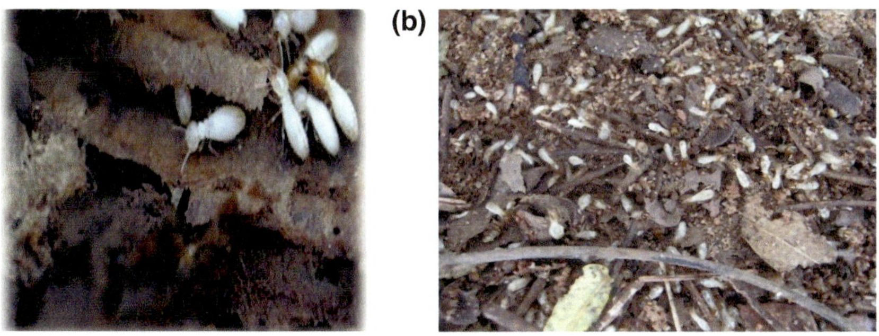

FIGURE 1.8 (a & b) Snow-white termite workers.

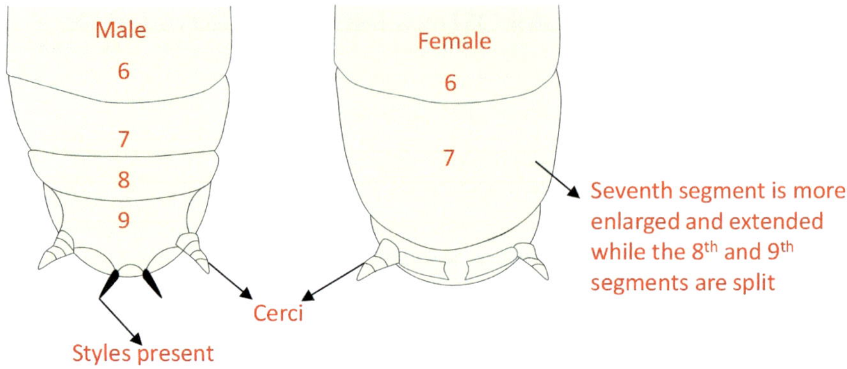

FIGURE 1.9 Distinction between a typical termite male and female.

world's soils. EW and ants are the other two groups given the sobriquet. It is not as if other animals do not contribute to the processing and rejuvenation of soils – almost all animals do in small measures. However, the sheer volumes of soil processed by the three "soil engineers" and the effectiveness of the processing make the three clearly distinct and dominant. Of the three, only termites have the additional, and crucial, ability to biodegrade and digest lignin. Only a very few other species of invertebrates have this ability, but none have the numbers and the efficiency to process the world's ligneous biomass to a fraction of the extent to which termites are able to. So crucial is the role of termites in the Earth's ecosystem that, without them, the world's soil surface will soon be covered by undecomposable ligninous biomass present in plant debris. And this pile-up will soon make all agriculture impossible.

But termites cannot be used to carry out controlled zoocomposting in the confines of reactor vessels, the way EWs can be used. This is because termites do not breed in captivity, barring very, very few and rare exceptions. When we operate a vermireactor with EW, the EW start breeding in them, generating cocoons and juveniles which grow to adulthood in the vermireactors and gradually replace the previous generation, which either dies in the vermireactor or is separated and utilized. But the same strategy cannot be used with termites because termites follow a very rigid, centralized, and hierarchical social structure. In it, only a queen breeds in the chamber of its nest, laying eggs in profusion, which give birth predominantly to workers who are blind and sterile.

When the workers are put in a vessel along with the feed, they soon die of the shock of being cut off from their nest. Even otherwise, being sterile, they cannot be cultured or farmed the way other animals can be.

Moreover, termites are a double-edged sword. While they can be very effective, if harnessed in processing ligninous waste, they can also be very destructive if allowed to function out of control, as they will begin devouring woody parts of buildings and other cellulosic materials such as clothes and paper. This risk, not associated with EW or BSF, makes it all the more challenging to use termites in solid waste management. As a consequence, there have been only a very few attempts to use termites directly, and most of the work has been aiming at isolating from termites the microflora and the biomolecules responsible for the termites' process in decomposing lignin, to be able to use them *ex situ*.

Given this reality, these authors had to explore novel ways to harness the "termite power". It led to the opening of a new frontier in bioprocess technology. The details are presented in Part IV of the book.

REFERENCES

Arends, J.J. and Wright, R.E., 1981. Mass rearing of face flies. *Journal of Economic Entomology*, *74*(3), pp.355–358.

Barroso, F.G., de Haro, C., Sánchez-Muros, M.J., Venegas, E., Martínez-Sánchez, A. and Pérez-Bañón, C., 2014. The potential of various insect species for use as food for fish. *Aquaculture*, *422*, pp.193–201.

Billen, P., Khalifa, L., Van Gerven, F., Tavernier, S. and Spatari, S., 2020. Technological application potential of polyethylene and polystyrene biodegradation by macro-organisms such as mealworms and wax moth larvae. *Science of the Total Environment*, *735*, p.139521.

Bloukounon-Goubalan, A.Y., Saïdou, A., Chrysostome, C.A.A.M., Kenis, M., Amadji, G.L., Igué, A.M. and Mensah, G.A., 2020. Physical and chemical properties of the agro-processing by-products decomposed by larvae of *Musca domestica* and *Hermetia illucens*. *Waste and Biomass Valorization*, *11*, pp.2735–2743.

Brar, S.K., Verma, M., Tyagi, R.D., Valéro, J.R. and Surampalli, R.Y., 2008. Wastewater sludges as novel growth substrates for rearing codling moth larvae. *World Journal of Microbiology and Biotechnology*, *24*, pp.2849–2857.

Bulak, P., Proc, K., Pytlak, A., Puszka, A., Gawdzik, B. and Bieganowski, A., 2021. Biodegradation of different types of plastics by tenebrio molitor insect. *Polymers*, *13*(20), p.3508.

Chaklader, M.R., Howieson, J., Foysal, M.J. and Fotedar, R., 2021. Transformation of fish waste protein to Hermetia illucens protein improves the efficacy of poultry by-products in the culture of juvenile barramundi, Lates calcarifer. *Science of the Total Environment*, *796*, p.149045.

ČiČková, H., Newton, G.L., Lacy, R.C. and Kozánek, M., 2015. The use of fly larvae for organic waste treatment. *Waste Management*, *35*, pp.68–80.

Clark, M., Tepper, K., Petroll, K., Kumar, S., Sunna, A. and Maselko, M., 2021. Bioremediation of industrial pollutants by insects expressing a fungal laccase. *ACS Synthetic Biology*, *11*(1), pp.308–316.

Dehghani, R., Mohammadi, M.J., Mostafaii, G., Gilasi, H., Shahrisvand, B., Charkhloo, E. and Hoseindoost, G., 2018. The frequency of stable fly larvae in the process of producing compost from municipal waste. *Fresenius Environmental Bulletin*, *27*, pp.5323–5328.

Dordević, M., Radenković-Damnjanović, B., VuČinić, M., Baltić, M., Teodorović, R., Janković, L., Vukašinović, M. and Rajković, M., 2008. Effects of substitution of fish meal with fresh and dehydrated larvae of the house fly (*Musca domestica L.*) on productive perfomrance and health of broilers. *Acta Veterinaria*, *58*, pp.357–368.

Edwards, C.A., Subler, S. and Arancon, N.Q., 2011. *Vermiculture Technology-Earthworms, Organic Wastes and Environmental Management*. CRC Press, Boca Raton.

Fowles, T.M. and Nansen, C., 2019. Artificial selection of insects to bioconvert pre-consumer organic wastes. A review. *Agronomy for Sustainable Development*, *39*, pp.1–14.

Franco, A., Scieuzo, C., Salvia, R., Mancini, I.M., Caniani, D., Masi, S. and Falabella, P., 2022. A mobile black soldier fly farm for on-site disposal of animal dairy manure. *Bulletin of Insectology*, *75*(1), pp.75–82.

Ganda, H., Zannou-Boukari, E.T., Kenis, M., Chrysostome, C.A.A.M. and Mensah, G.A., 2019. Potentials of animal, crop and agri-food wastes for the production of fly larvae. *Journal of Insects as Food and Feed*, *5*(2), pp.59–67.

Harsányi, E., Juhász, C., Kovács, E., Huzsvai, L., Pintér, R., Fekete, G., Varga, Z.I., Aleksza, L. and Gyuricza, C., 2020. Evaluation of organic wastes as substrates for rearing *Zophobas morio*, *Tenebrio molitor*, and *Acheta domesticus* larvae as alternative feed supplements. *Insects*, *11*(9), p.604.

He, L., Zhang, Y., Ding, M.Q., Li, M.X., Ding, J., Bai, S.W., Wu, Q.L., Zhao, L., Cao, G.L., Ren, N.Q. and Yang, S.S., 2021. Sustainable strategy for lignocellulosic crop wastes reduction by *Tenebrio molitor Linnaeus* (mealworm) and potential use of mealworm frass as a fertilizer. *Journal of Cleaner Production*, *325*, p.129301.

Jiang, C., Teng, C., Li, J., Zhu, J., Fen, D., Lou, L. and Zhang, Z., 2017. The effectiveness of bioconversion of food waste by housefly larvae. *Chinese Journal of Applied and Environmental Biology*, *23*, pp.1159–1165.

Koo, S.I., Currin, T.A., Johnson, M.G., King, E.W. and Turk, D.E., 1980. The nutritional value and microbial content of dried face fly pupae (Musca autumnalis (De Geer)) when fed to chicks. *Poultry Science*, *59*(11), pp.2514–2518.

Li, L., Zhao, Z. and Liu, H., 2013. Feasibility of feeding yellow mealworm (Tenebrio molitor L.) in bioregenerative life support systems as a source of animal protein for humans. *Acta Astronautica*, *92*(1), pp.103–109.

Li, R., Wang, J., Zhang, Z., Shen, F., Zhang, G., Qin, R., Li, X. and Xiao, R., 2012. Nutrient transformations during composting of pig manure with bentonite. *Bioresour. Technol. 121*, 362 –368. https://doi.org/10.1016/j.biortech.2012.06.065.

Li, Z., Meng, M., Li, S. and Deng, B., 2019. The transcriptome analysis of Protaetia brevitarsis Lewis larvae. *PLoS One*, *14*(3), p.e0214001.

Lush, J.L., 1943. *Animal Breeding Plans* (2nd ed.). The Iowa State College Press, Ames, Iowa, 658 p.

Marchaim, U., Gelman, A. and Braverman, Y., 2003. Reducing waste contamination from animal-processing plants by anaerobic thermophilic fermentation and by flesh fly digestion. *Applied Biochemistry and Biotechnology, 109*, 107–115.

Mealey, A., 2022. *Cooperative Decomposition: Recycling in Its Truest Form*. Montana Natural History Centre. https://www.montananaturalist.org/blog-post/cooperative-decomposition/. Last accessed April 2023.

Meats, A., Holmes, H.M. and Kelly, G.L., 2004. Laboratory adaptation of Bactrocera tryoni (Diptera: Tephritidae) decreases mating age and increases protein consumption and number of eggs produced per milligram of protein. *Bulletin of Entomological Research*, *94*(6), pp.517–524.

Melillo, E. D., 2014. Global entomologies: insects, empires, and the 'synthetic age'in world history. *Past & Present*, *223*(1), pp.233–270.

Meyer, R.S., DuVal, A.E. and Jensen, H.R., 2012. Patterns and processes in crop domestication: an historical review and quantitative analysis of 203 global food crops. *New Phytologist*, *196*(1), pp.29–48.

Miech, P., Berggren, Å., Lindberg, J.E., Chhay, T., Khieu, B. and Jansson, A., 2016. Growth and survival of reared Cambodian field crickets (Teleogryllus testaceus) fed weeds, agricultural and food industry by-products. *Journal of Insects as Food and Feed*, *2*(4), pp.285–292.

Miranda, C.D., Crippen, T.L., Cammack, J.A. and Tomberlin, J.K., 2021. Black soldier fly, *Hermetia illucens* (L.)(Diptera: Stratiomyidae), and house fly, *Musca domestica L.*(Diptera: Muscidae), larvae reduce livestock manure and possibly associated nutrients: an assessment at two scales. *Environmental Pollution*, *282*, p.116976.

Mondor, E.B., Tremblay, M.N., Tomberlin, J.K., Benbow, E.M., Tarone, A.M. and Crippen, T.L., 2012. The ecology of carrion decomposition. *National Education Knowledge*, *3*(10), p.21.

Morales, G.E. and Wolff, M., 2010. Insects associated with the composting process of solid urban waste separated at the source. *Revista Brasileira de Entomologia*, *54*, pp.645–653.

Mramba, F., Broce, A.B. and Zurek, L., 2007. Vector competence of stable flies, *Stomoxys calcitrans L.* (Diptera: Muscidae), *for Enterobacter sakazakii. Journal of vector Ecology*, *32*(1), pp.134–139.

Niu, Y., Zheng, D., Yao, B., Cai, Z., Zhao, Z., Wu, S., Cong, P. and Yang, D., 2017. A novel bioconversion for value-added products from food waste using *Musca domestica. Waste Management*, *61*, pp.455–460.

Nuov, S., Little, D.C. and Yakupitizage, A., 1995. Nutrient flows in an integrated pig, maggot and fish production system. *Aquaculture Research*, *26*, pp.601–606.

Pape, T 1987. *The Sarcophagidae (Diptera) of Fennoscandia and Denmark. Fauna entomologica scandinavica.* E.J. Brill/Scandinavian Science Press Ltd, Leiden, Copenhagen, *19.*

Pastor, B., ČiČková, H., Kozanek, M., Martinez-Sanchez, A., TakáČ, P. and Rojo, S., 2011. Effect of the size of the pupae, adult diet, oviposition substrate and adult population density on egg production in *Musca domestica* (Diptera: Muscidae). *European Journal of Entomology*, *108*(4).

Pastor, B., Velasquez, Y., Gobbi, P. and Rojo, S., 2015. Conversion of organic wastes into fly larval biomass: bottlenecks and challenges. *Journal of Insects as Food and Feed*, *1*(3), pp.179–193.

Peng, B.Y., Li, Y., Fan, R., Chen, Z., Chen, J., Brandon, A.M., Criddle, C.S., Zhang, Y. and Wu, W.M., 2020. Biodegradation of low-density polyethylene and polystyrene in superworms, larvae of *Zophobas atratus* (Coleoptera: Tenebrionidae): Broad and limited extent depolymerization. *Environmental Pollution*, *266*, p.115206.

Roffeis, M., Muys, B., Almeida, J., Mathijs, E., Achten W.M.J., W., Pastor, B., Velásquez, Y., Martinez-Sanchez, A.I. and Rojo, S., 2015. Pig manure treatment with housefly (*Musca domestica*) rearing–an environmental life cycle assessment. *Journal of Insects as Food and Feed*, *1*(3), pp.195–214.

Rumbos, C.I. and Athanassiou, C.G., 2021. The superworm, *Zophobas morio* (Coleoptera: Tenebrionidae): a ‘sleeping giant’ in nutrient sources. *Journal of Insect Science*, *21*(2), p.13.

Sharanowski, B.J., Walker, E.G. and Anderson, G.S., 2008. Insect succession and decomposition patterns on shaded and sunlit carrion in Saskatchewan in three different seasons. *Forensic Science International*, *179*(2–3), pp.219–240.

Tauseef, S.M., Tabassum-Abbasi, Patnaik, P., Abbasi, T. and Abbasi, S.A., 2021. A novel high-rate vermicomposting machine for downstream processing of biorefinery waste and other forms of organic solid waste. *Biomass Conversion and Biorefinery*, pp.1–10, https://doi.org/10.1007/s13399-021-01522-w.

Van Broekhoven, S., Oonincx, D.G., Van Huis, A. and Van Loon, J.J., 2015. Growth performance and feed conversion efficiency of three edible mealworm species (Coleoptera: Tenebrionidae) on diets composed of organic by-products. *Journal of Insect Physiology*, *73*, pp.1–10.

van Zanten, H.H., Mollenhorst, H., Oonincx, D.G., Bikker, P., Meerburg, B.G. and de Boer, I.J., 2015. From environmental nuisance to environmental opportunity: housefly larvae convert waste to livestock feed. *Journal of Cleaner Production*, *102*, pp.362–369.

Wang, H., Zhang, Z., Czapar, G.F., Winkler, M.K. and Zheng, J., 2013. A full-scale house fly (Diptera: Muscidae) larvae bioconversion system for value-added swine manure reduction. *Waste Management & Research*, *31*(2), pp.223–231.

Wang, H., Liu, X., Yang, Q., Zheng, L., Li, W., Cai, M., Li, Q., Zhang, J. and Yu, Z., 2017. Insect biorefinery: a green approach for conversion of crop residues into biodiesel and protein. *Biotechnol Biofuels*, 10(1), pp.1–13.

Wheeler, M.R. (1987) Drosophilidae. *Manual of Nearctic Diptera, 2*, pp.1011–1018.

Yang, S. and Liu, Z., 2014. Pilot-scale biodegradation of swine manure via Chrysomya megacephala (Fabricius) for biodiesel production. *Applied Energy, 113*, pp.385–391.

Yang, S., Li, Q., Zeng, Q., Zhang, J., Yu, Z. and Liu, Z., 2012. Conversion of solid organic wastes into oil via *Boettcherisca peregrine* (Diptera: Sarcophagidae) larvae and optimization of parameters for biodiesel production. *PLoS One, 7*, p.e45940.

Yang, S., Xie, J., Hu, N., Liu, Y., Zhang, J., Ye, X. and Liu, Z., 2015. Bioconversion of gibberellin fermentation residue into feed supplement and organic fertilizer employing housefly (*Musca domestica L.*) assisted by *Corynebacterium* variabile. *PLoS One, 10*(5), p.e0110809.

Yang, S.S., Brandon, A.M., Flanagan, J.C.A., Yang, J., Ning, D., Cai, S.Y., Fan, H.Q., Wang, Z.Y., Ren, J., Benbow, E. and Ren, N.Q., 2018. Biodegradation of polystyrene wastes in yellow mealworms (larvae of *Tenebrio molitor Linnaeus*): Factors affecting biodegradation rates and the ability of polystyrene-fed larvae to complete their life cycle. *Chemosphere, 191*, pp.979–989.

Yang, S.S., Zhang, Y., Zhou, H.M., Ji, X.Y., He, L., Xing, D.F., Ren, N.Q., Ho, S.H. and Wu, W.M., 2019. A novel clean production approach to utilize crop waste residues as co-diet for mealworm (*Tenebrio molitor*) biomass production with biochar as byproduct for heavy metal removal. *Environmental Pollution, 252*, pp.1142–1153.

Yang, Y., Wang, J. and Xia, M., 2020. Biodegradation and mineralization of polystyrene by plastic-eating superworms *Zophobas atratus*. *Science of the Total Environment, 708*, p.135233.

Yehuda, B., Marchaim, U., Glatman, L., Drabkin, V., Chizov-Ginzburg, A., Mumcuoglu, K.Y. and Gelman, A., 2011. Bioconversion of poultry and fish waste by *Lucilia sericata* and *Sarcophaga carnaria* larvae. *Asian Journal of Water, Environment and Pollution, 8*(4), pp.69–75.

2 Distinguishing Features of Human-Controlled Composting

2.1 INTRODUCTION

In the preceding chapter, we have explained that the process that is generally referred to as "composting" is basically "hot" composting in which thermophilic microorganisms play a major role. On the other hand, vermicomposting is one of the forms of zoocomposting in which animals play a decisive role from start to finish. As far as the goal of transforming organic waste into a highly mineralized product, of significant fertilizer value, is concerned, composting and vermicomposting appear to be quite similar bioprocesses, the only apparent difference between them being the presence or absence of a good number of earthworms. But when one sets out to engineer these bioprocesses, several significant differences are encountered, which need very different approaches to the design and operation of the two processes. In order to identify the differences, it will be worthwhile to recapitulate the steps and the nuances associated with composting as contrasted to vermicomposting.

2.2 THE COMPOSTING PROCESS

Composting occurs when aerobic microorganisms use organic waste as a food source. To this end, microorganisms decompose the food source, breaking its content of complex organic compounds into simpler organics and then still more simple ones (Ipek et al., 2002; Gajalakshmi & Abbasi, 2008):

$$\text{Organics} + O_2 \xrightarrow[\text{metabolism}]{\text{microbial}} \text{Compost} + CO_2 \uparrow + H_2O + \text{heat}$$

As would become clear from the steps involved in composting, described below (Figure 2.1), composting is a quintessential batch process and cannot be operated even in semi-batch mode, let alone in a continuous fashion. The steps described below have been illustrated with the example of the classical windrow-based systems but are, in essence, common to all the composting systems (Gajalakshmi & Abbasi, 2008; Abbasi & Abbasi, 2009):

i. *Setting up windrows*: In this step, the substrate to be composted is laid out in layers about 15 cm thick, alternated with thinner layers (5 cm) of microorganism-rich seed material such as cow dung or sewage sludge. After being laid out, the windrows have the appearance as typified in Figure 2.2.

DOI: 10.1201/9780429331886-3

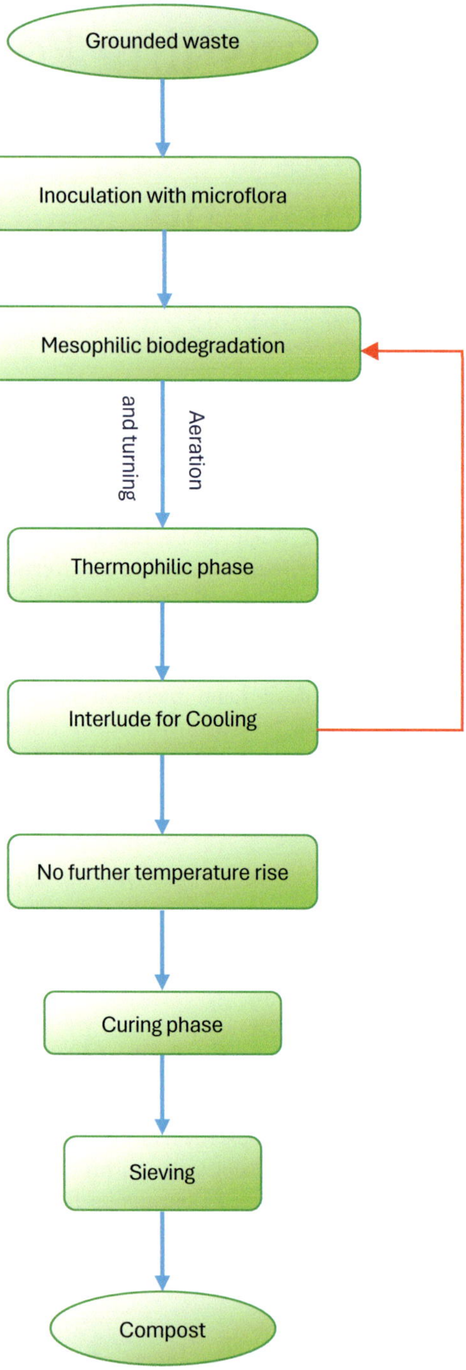

FIGURE 2.1 Broad steps associated with composting.

FIGURE 2.2 Typical windrow, front view. (https://commons.wikimedia.org/wiki/File:Composting_piles,_Digby_Farm_Compost_Site_-_geograph.org.uk_-_5834953.jpg.)

ii. *Providing adequate moisture (60% of the reactant mass)*: This is accomplished through sprinkling of water over the windrows, and a passive or active means of aeration. In the former, open-ended, perforated pipes are partially inserted into the windrows at different points. Air flows into the open ends of the pipes and through the windrows because of the chimney effect created as hot gases rise upward out of the windrow. In the latter, the outer ends of the pipes are connected to aeration devices with which air is forced into the composting pile with the help of pumps. Some of the equipment used in these steps is shown in Figures 2.3 and 2.4.

iii. *Covering of the windrow*: Cohesive clay or open plastic sheets are utilized for this purpose. The covering sets the aerobic decomposition of the substrate in motion. For the initial 2–4 days, the mesophilic bacteria dominate the process. Then, the heat released from the exothermic biodegradation starts lifting the temperature of the waste piles. As the temperature rises, thermophilic bacteria take over, gradually lifting the temperature of the pile to 55°C or higher. One can see steam and other hot gases coming from one of the windrows (Figures 2.5 and 2.6). In this thermophilic phase, the rate of biodegradation increases. The raised temperature also serves the purpose of killing many pathogens present in the waste. To ensure that the rapid rate of oxygen utilization within the waste pile does not lead to anoxic conditions, which would kill the aerobic bacteria involved in composting, the pile is provided with aeration. Apart from passive or active aeration mentioned above, some aeration is also achieved when the waste is turned,

FIGURE 2.3 Typical windrow, side view. (https://commons.wikimedia.org/wiki/File:Andains_de_compost,_sur_une_plateforme_de_compostage.JPG.)

FIGURE 2.4 Equipment required for watering, aerating, and turning the windrows. (https://commons.wikimedia.org/wiki/File:Outils_pour_compost.jpg.)

FIGURE 2.5 Windrows can be seen emitting gases when going through their thermophilic phase. (https://commons.wikimedia.org/wiki/File:Wheel-loader_working_with_compost_at_Sunnhordland_interkommunale_milj%C3%B8verk_(SIM)_02.jpg.)

FIGURE 2.6 Isolated compost pile in the thermophile phase. (https://commons.wikimedia.org/wiki/File:Compost_steaming_-_detail_3.jpg.)

as explained below. Then, as the availability of the substrate in the aerated zones declines, further decomposition is reduced, and the temperature begins to fall. In forced air composting systems, the aeration is shut off once the temperature goes beyond 55°C, and, instead, a blower is used to push off hot gases so that the pile temperature does not shoot up so high as to kill even the beneficial microorganisms. If that is allowed to happen, the composting would be incomplete, yielding an unsatisfactory product.

iv. *Turning the windrow*: This step aims to rejuvenate the process by bringing in previously undecomposed or partially decomposed substrate in contact with microorganisms and air. Depending on the scale of operation, this is either done manually with shovels or through turning machines. After a freshly turned windrow, in which the moisture has been replenished, is covered again, another wave of mesophilic-cum-thermophilic aerobic decomposition occurs. Again, the mesophilic phase causes the temperature to gradually rise, followed by a thermophilic phase. Another temperature peak results before the temperature begins to decline after a brief plateau. In a forced air composting system, the substrate isn't turned, but aeration is restarted once the blower cools the pile to near ambient temperature. This leads to the restart of composting and the generation of the next temperature wave. With the passage of time, during which the substrate gets more and more stabilized, the peak temperature attained by each wave gradually comes down. In 40–70 days, depending on the nature of the substrate and efficiency of process operation, the peak pile temperature falls to ~40°C. This is illustrated in the results of the composting of acacia leaf litter by the authors (Figure 2.7).

v. *Completion of the predominantly microbial decomposition phase*: This occurs after the steps ii to iv have been repeated 3–4 times or more, and the pile temperature becomes more or less constant.

vi. *The "curing" phase*: In this, the concluding phase of a typical composting process, the compost pile is left undisturbed for 2–4 weeks. This enables (a) degradation of the more refractory organics which had resisted previous steps; (b) decomposing some of the waste which had escaped degradation because of the slowing of the process imposed by kinetic rate limitations, and (c) re-establishing lower temperature microbial populations in the compost so that the "maturing" of the compost continues, albeit very slowly, metabolizing phytotoxic compounds, and suppressing plant pathogens. Finally, after sieving, the compost is ready for utilization (Figure 2.8).

As reviewed elsewhere (Gajalakshmi & Abbasi, 2008), many modifications and advancements have been made in the windrow-based composting systems over several decades. These pertain mainly to the mechanization of the composting process and development of waste-specific machines suited to different particle sizes and chemical characteristics of different substrates. Some pre-composting steps have also been introduced for some substrates, such as the "drum" system, in which the substrate is fragmented and anaerobically digested to make it more "compostable". But, in essence, the composting process remains the same as described in steps i to vi above.

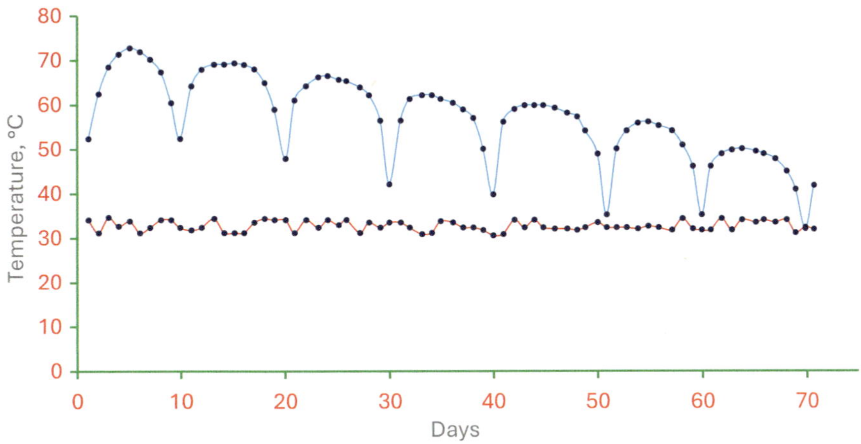

FIGURE 2.7 Illustrative example of temperature waves encountered during composting.

FIGURE 2.8 Finished compost ready for use. (https://commons.wikimedia.org/wiki/File:Compost_after_sieving,_ready_for_application_(8916542948).jpg.)

2.2.1 Distinguishing Features

The following aspects constitute the distinguishing features of the composting process:

a. It is a strictly batch process and cannot be made even semi-continuous. Once composting has begun in a pile, no fresh substrate can be added to it as it would hinder and retard the ongoing process. Likewise, no new substrate can be added at any of the steps without destabilizing the ongoing composting process.
b. It is absolutely essential to provide for mechanical turning of the reactor contents at periodic intervals. Without it, the composting will be incomplete.
c. The process has several temperature waves during which there is a sharp rise in temperature. But for it, the composting will be incomplete, and pathogens will not be destroyed significantly.

2.3 THE VERMICOMPOSTING PROCESS

As elaborated in Chapter 3, in contrast to composting, the vermicomposting process does not require auxiliary accretion, nor mechanical turning is needed because both these functions are naturally achieved by the breathing (through skin) of the earthworms, as also their movements within the vermireactor. The third most important difference is that there is no thermophilic phase. Indeed, no perceptible increase in temperature relative to the ambient has ever observed in the vermireactors deployed by these authors over the past 30 years. Once appropriate feed is put in a vermireactor and earthworms of appropriate species are released into the feed, vermicomposting begins and proceeds with a steady pace to completion in a few weeks. The only external control necessary is maintaining a moisture content of $60\% \pm 10\%$ in the vermireactor. There are no separate and distinct phases as they occur in controlled composting, wherein each phase is controlled by different consortia of microflora, which cannot function together at the same time.

These differences between composting and vermicomposting have a far-reaching influence on the way one of the two processes is engineered – designed, optimized, and operated – relative to the other.

The question that immediately arises is: if it takes an earthworm just a few hours to convert most of the waste it ingests into vermicompost, why do most vermicomposting systems take 4–6 weeks or more to process their substrate? The answer, as we will explain and demonstrate in the following chapters, is in the engineering problem associated with (a) enhancing the earthworm–substrate contact so that the quantity of the substrate consumed per worm is maximized, and (b) harvesting of the vermicast as soon as it is generated.

REFERENCES

Abbasi, T. and Abbasi, S.A., 2009. Water quality management of rooftop rainwater harvesting systems. *Journal of Environmental Science & Engineering*, *51*(4), pp.325–330.

Gajalakshmi, S. and Abbasi, S.A., 2008. Solid waste management by composting: state of the art. *Critical Reviews in Environmental Science and Technology*, *38*(5), pp.311–400.

Ipek, U., Öbek, E., Akca, L., Arslan, E.I., Hasar, H., Doğru, M. and Baykara, O., 2002. Determination of degradation of radioactivity and its kinetics in aerobic composting. *Bioresource Technology*, *84*(3), pp.283–286.

Part II

Earthworm-Based Zoocomposting

Vermicomposting

3 Distinguishing Features of Earthworm-Based Zoocomposting (Vermicomposting)

3.1 INTRODUCTION

This chapter elaborates upon the distinguishing features of earthworm-based zoocomposting, or vermicomposting, which bestow upon vermicomposting a uniqueness that is very distinct from the features of human-controlled composting. These aspects need to be made the basis for the design and operation of vermireactors in the most optimum manner.

3.2 STEPS INVOLVED IN VERMICOMPOSTING AND HOW THEY DIFFER FROM THE STEPS ASSOCIATED WITH HUMAN-CONTROLLED COMPOSTING

The broad steps involved in vermicomposting are shown in Figure 3.1. Once earthworms are released into the substrate, the ingestion of the substrate particles by the earthworms begins. It is followed by physical size reduction of the ingested particles by the action of the earthworm's gizzard, which is located next to the worm's mouth.

There is digestion of the substrate as it passes through the earthworm's body and is acted upon by the microorganisms and enzymes present in the earthworm's gut.

Finally, there is the exit of the substrate as "vermicast" a few hours after the ingestion. The number of hours depends on the nature of the substrate, the worm species, and the length of the worm's body. In general, earthworms of shorter body length take less time to deliver the vermicast than longer-bodied earthworms, and epigeic earthworms process their feed faster than the anecics or the endogeics.

As briefly mentioned in the previous chapter, vermicomposting differs from human-controlled composting in several ways. The most significant difference is that, in contrast to composting, vermicomposting does not accompany exothermic reactions; hence, there is no measurable rise (or, for that matter, fall) in temperature in the vermireactors. Nor do vermireactors need supplementary aeration. In fact, most species of earthworms are not able to thrive at temperatures exceeding 45°C; hence, if put in an under-composting substrate pile, most worms would perish when the pile temperature rises (as it is a must to ensure proper composting) to greater than 45°C.

Also, in contrast to composting, there is no necessity for the periodic turning or mechanical mixing of the substrate. Rather, the natural moving and burrowing

DOI: 10.1201/9780429331886-5

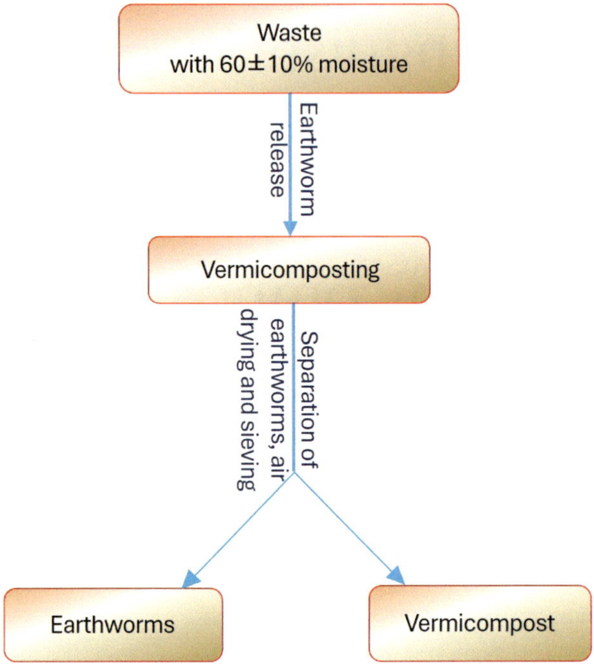

FIGURE 3.1 Broad steps involved in vermicomposting.

activities of the worms accomplish the task of keeping the substrate well-mixed. These activities of worms also keep the substrate adequately aerated and, hence, prevent anaerobic regions from developing in the vermireactors.

All things considered, composting requires labour and machinery to accomplish fragmentation, mixing, and aeration (Gajalakshmi & Abbasi, 2008), while in vermicomposting, these tasks are by and large accomplished by earthworms (Abbasi et al., 2009). The only human intervention needed is in keeping the vermireactors moist to the extent of 60% ± 10%.

3.3 UNIQUENESS OF VERMICOMPOSTING AS THE FIRST ANIMAL–MEDIATED AND ENGINEERED POLLUTION CONTROL PROCESS

Till the introduction of composting systems based on black soldier flies during the later half of the previous century (Chapter 14) and of termite-based composting systems introduced by these authors recently (Chapter 17), vermicomposting had a unique position in the domain of environmental engineering: it was the only pollution control bioprocess that had a multi-cellular animal as the main bioagent. Indeed, it was the only bioprocess – outside the ones involving one or other type of animal farming – in which a multi-cellular animal was used in reactor systems with the prime objective to generate a product other than the animal offspring.

All the other engineered bioprocesses, except a few which are based on plants (botanical species), revolve around the use of enzymes, bacteria, microfungi, or microalgae in mobile or immobilized modes. Such processes have been intensively studied, modelled, designed, and engineered. In contrast, no design or operation criteria based on bioprocess engineering principles had been developed for vermicomposting till as recently as 2005, when these authors took up the task. Admittedly, considerable work has been going on, which has addressed the science of vermicomposting, especially in terms of biotic and abiotic factors, which influence vermicast production, earthworm growth, and fecundity. Studies have also been increasingly reported on the "vermicompostability" of newer substrates and newer earthworm species. But the aspects of vermicomposting process design, control, operation, and optimization had by and large remained unaddressed till these authors carried out the work described at different places in this book.

3.4 WHAT REPRESENTS VERMICOMPOSTING?

The term "vermicomposting" is generally believed by many authors, including the present ones, to denote the process in which earthworms are made to feed upon

FIGURE 3.2 Blend of weeds and paper waste before their vermicomposting.

FIGURE 3.3 The vermicompost obtained from the material shown in Figure 3.2. (https://commons.wikimedia.org/wiki/File:Home_Vermicompost_Thrissur_DSCN2713.jpg.)

certain substrates in order to convert the substrates into the worm's excreta, or the vermicast. This is a product clearly distinct from the parent substrate and has consistent characteristics (Figures 3.2 and 3.3). However, several authors also use the term vermicomposting to denote processes in which earthworms are added to raw or partly decomposed waste and the act of natural degradation of the waste at ambient temperatures (usually ±°C), purportedly assisted by the worms, is deemed "vermicomposting". The end product, which is usually a mixture of still incompletely stabilized waste and vermicast, is dubbed "vermicompost".

3.5 VERMICOMPOSTING OF COMPOST

Composting of a substrate prior to vermicomposting can facilitate the subsequent vermireactor operations and improve product quality in several ways:

The thermophilic phase, which is an essential component of the composting process, kills most of the pathogenic organisms. This not only ensures the survival and growth of the earthworms when the compost is subjected to vermicomposting but also ensures a much more pathogen-free vermicompost. The thermophilic phase also kills seeds of weedy plants, an advantage that is carried over to the vermicomposting stage.

On the other hand, the mesophilic decomposition that takes place during the post-thermophilic "curing" phase bestows upon the compost a rich and diverse community of microorganisms. These microorganisms, which are known to be beneficial to soil and plants, have the ability to kill several pathogens due to the former's ability to generate pathogen inhibitors. Vermireactors can enhance these gains because the microorganisms may be taken as feed by the earthworms, digesting some of the species and making the more beneficial species proliferate before excretion as part of the vermicast.

Composting of any substrate is accompanied by a loss of carbon (which is converted to CO_2 by microbial action), but little or no nitrogen is lost; neither are other nutrients lost. This not only improves (by reducing) the C/N ratio of the substrate but also, effectively, enhances the concentration of nutrients in the compost relative to the uncomposted substrate. This situation favours the subsequent vermicomposting as greater concentration of nutrients now exists in the substrate and in better bioavailable forms.

The brisk microbial action, as well as the periodic turning that is done during composting, breaks down larger lumps in the substrate, making it more easily digestible by the earthworms.

Most significantly, composting stabilizes all the rapidly biodegradable parts of the substrate. When the substrate happens to be vegetable waste, poultry waste, or slaughterhouse waste, which have a large proportion undergoing rapid biodegradation, anaerobic zones can get set in, which can be hazardous to the earthworms. Moreover, rapidly biodegrading fruits, vegetables, and animal wastes generate volatile fatty acids and other decomposition products, which severely stress the worms, killing many (Nayeem Shah et al., 2013). By and large, the environment within any rapidly biodegrading substrate is hostile to earthworms. This situation gets rectified to a significant extent if such substrates are first stabilized through composting.

In several studies reported earlier, the authors and coworkers have used well-stabilized composts of municipal solid waste and different weeds as substrates for generating vermicompost. Earthworms not only feed voraciously on compost to produce vermicompost but also display good growth and fecundity. Moreover, during its passage through the earthworm's gut, the compost acquires additional microflora, enzymes, and hormones known to be beneficial to plants. Apparently, the complex organic matter that survives composting also gets "digested" to a significant extent during the vermicomposting phase. Due to all these reasons, vermicompost of a compost is a better fertilizer than the vermicompost of the uncomposted substrate. However, the extra time and effort needed to first compost a substrate and then subject it to vermicomposting more than offsets the gains. As a consequence, this option is not exercised frequently and is usually tried only on substrates that defy direct vermicomposting.

3.6 CAN COMPOSTING AND VERMICOMPOSTING BE DONE SIMULTANEOUSLY?

Composting not only entails totally different process control parameters and operational strategies from vermicomposting, but it also differs significantly in terms of the governing biological and physico-chemical aspects. If the two are attempted in a single reactor, both will proceed sub-optimally, interfering with each other instead of facilitating each other. Hence, composting and vermicomposting cannot be done together and simultaneously for the same substrate.

3.7 POINTERS FOR MODELLING, DESIGN, AND OPERATION OF BOTH PROCESSES

From the foregoing, it is clear that composting is to be modelled as a strictly batch process in which the contents are homogenized intermittently and the oxygen deficit caused in the course of the aerobic fermentation is made good by surface aeration effected during the turning of the reactants or/and by pumping in air.

Vermicomposting, on the other hand, approximates a continuously stirred tank reactor (CSTR) wherein the contents are mixed and kept aerobic by the tunnelling and burrowing action of the earthworms, as also the stirring of the substrate caused due to the movement of the worms towards substrate particles. The species of earthworms (epigeic and anecic) which are usually employed in vermireactors deposit their cast (which is the reactor output – the vermicompost) either at the top of the reactor contents (as *Eudrilus eugeniae* or *Eisenia fetida*) or at the bottom. In either case, the product is fairly well separated from the substrate–worm zone. This behaviour contributes to the CSTR-like operation of the vermireactor. Indeed, the vermicomposting process can be operated not only in pseudo-continuous mode but also in batch and semi-batch modes.

There is another facet unique to the vermireactors, i.e., in a vermireactor, the earthworms function as "reactors within a reactor" in the sense that the entire processing of the substrate from the reactant to the product occurs within each worm. It follows that the boundary condition for the minimum time a vermireactor will require to produce vermicast is the time taken by the concerned species of earthworm to "digest" a substrate. A number of medium- and large-scale vermicomposting units are in operation across the world. Of these, some are run in batch mode, but several are supposed to be "continuous flow" units. Yet all of these are based on solid retention times (SRTs) of at least 6 weeks, more commonly 3 months.

This is surprising because the longest duration that the species of earthworms normally used in vermicomposting take to generate vermicompost from any substrate is less than 10 hours. Moreover, if the vermicast is harvested early and air dried, it becomes fairly stable with little subsequent biodegradation. This being the case, it may not be in the interest of product quality and process efficiency to leave the vermicast (which is produced daily) unharvested for long durations, as is being done in the existing large-scale systems at present.

The question that immediately arises is: if it takes an earthworm just a few hours to convert most of the waste it ingests into vermicompost, why do most vermicomposting systems take 4–6 weeks or more to process their substrate? The answer, as we will explain later, is in the engineering problem associated with a) enhancing the earthworm–substrate contact so that the quantity of the substrate consumed per worm is maximized, and b) harvesting of the vermicast as soon as it is generated.

Other forcing factors being equal, the vermireactor efficiency will be controlled by the access to food of the earthworms. Lesser the distance each earthworm must travel before reaching the food and lesser the competition it faces each time it sets out to ingest a food particle, closer will be the time taken for vermicast production to the minimum time it takes for the given species to process the given substrate (time taken from ingestion to excretion). But the rate of vermicast production, i.e., mass fraction of the substrate converted to vermicast per day, is a function of earthworm density because the greater the number of worms simultaneously feeding upon the substrate, the greater the rate of vermicast production. But the earthworm density can be increased only up to the level below which competition for food does not become a restrictive factor. Hence, it is important to optimize the earthworm density and the surface-to-volume ratio of vermireactors to ensure the easiest possible access to the substrate by the largest possible number of animals. Studies carried out to test this concept have revealed that vermireactors of a high surface-to-volume ratio yield more vermicast per unit time than reactors of a lower surface-to-volume ratio, evidently because shallower reactors prevent crowding and also reduce the vertical movement needed by each worm to deposit its vermicast.

3.8 ENGINEERING OF THE VERMICOMPOSTING PROCESS

As brought out earlier as well, unlike composting, the vermicomposting process can be operated in batch, semi-batch, and even pseudo-continuous modes. This is because vermicomposting of a substrate, representing most of the substrate's biodegradation, takes only as much time as it takes an earthworm to release its excreta (vermicompost) after ingesting the waste. This time is of the order of 6–12 hours, even as composting needs several weeks. Hybrid and cold composting can take much longer than hot composting.

But before we address this issue in detail, let us have an overview of the steps that are involved in the conversion of waste into its vermicompost. As this is predominantly accomplished by the action of individual earthworms, each of the earthworms functions as a stand-alone bioreactor. The journey of waste from the point of its ingestion by an earthworm to its exit from the earthworm's body comprises the key steps involved in vermicomposting.

3.9 THE EARTHWORM BIOREACTOR

A brief description of the earthworm body parts (Figure 3.4) that contribute to vermicomposting is presented below:

3.9.1 The Peristomium

It is the mouth of the earthworm by which the waste is ingested as its food. It is situated in the first body segment of the earthworm. It is forceful enough to enable an earthworm to detach soil particles in nature and make burrows for itself. It is a Crescent-shaped aperture opening into the buccal cavity.

3.9.2 The Buccal Cavity

It is a thin-walled chamber, occurring in the second and third segments of an earthworm's body. The expansion and contraction of muscles attached to the buccal cavity's body wall enable the cavity to either protrude out or retract. This helps it to hold

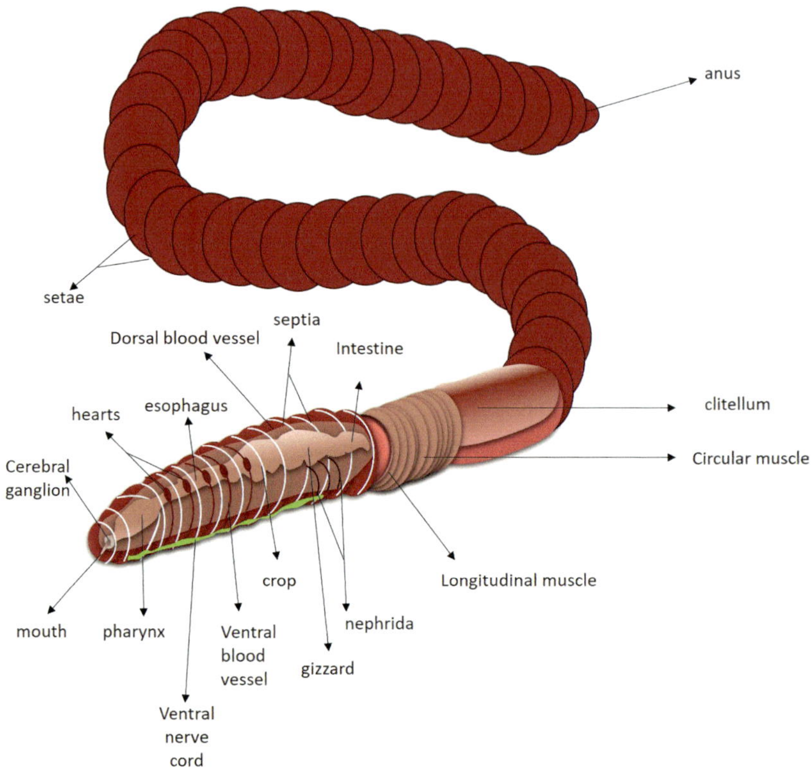

FIGURE 3.4 The earthworm body parts that make the "earthworm bioreactor".

the food during ingestion. The cavity then opens into a muscular chamber referred to as the pharynx.

3.9.3 The Pharynx

It's a thick-walled, pear-shaped chamber, which extends from the third to the fourth segments of the earthworm's body. It is made distinct from the buccal cavity by a dorsal groove which houses the earthworm's brain. The salivary glands situated in the pharynx, which are unicellular and are called chromophil, secrete saliva containing proteolytic enzymes such as protease and mucin, which make the ingested waste soft and begin aiding its decomposition. The contraction or dilation of the pharyngeal wall, which is connected to the body wall through muscular strands, acts as a suction pump. It promotes the movement of the digested waste through the pharynx.

3.9.4 The Oesophagus

The waste is delivered into the oesophagus from the pharynx. It is a narrow and thin, tube-like structure present on the earthworm's five to seven segments. It passes the food on to the gizzard.

3.9.5 The Gizzard

The gizzard plays a major part in the comminution of waste. It functions like a sludge crusher. Occupying the eighth and the ninth segments of the earthworm's body, the gizzard is a highly muscular, thick-walled organ. The muscles of the gizzard are supported by a hard lining of cuticle, whose motion results in a grinding action. The organic waste, inert particulates like the soil particles, and decaying leaves that are delivered to it by the oesophagus are finely ground in the gizzard, preparing them for rapid biodegradation.

3.9.6 The Stomach

The stomach covers one of the larger parts of the earthworm's body, occupying its 9th to 14th segments. It is a highly vascular and tubular structure. In it, six calciferous glands exist at segments 10–12. These glands store calcium carbonate, which they release to help in the neutralization of the acids present in the now rapidly digesting waste. The glands present in the stomach also secrete proteolytic enzymes, which cause the decomposition of the otherwise bioresistant proteins.

3.9.7 The Intestine

This penultimate, and by far the longest, body part of relevance to the functioning of the "earthworm bioreactor" is the intestine. It is a long, thin-walled tube extending from the 15th segment of the earthworm's body to the last segment. It is teeming with diverse microflora that act upon the digesting waste to generate vitamins, hormones, and minerals. The intestinal caeca present in the 26th segment secrete enzymes encompassing amylase, sucrose, maltose, chitinase, lipase, and proteolytic enzymes. The enzymes act upon the waste to rapidly reduce its organic carbon content, mineralize it, and embellish it with microflora, enzymes, and hormones known to be soil-friendly, plant-friendly, and anti-pathogen. The inner lining of the intestine is ciliated, vascular, and glandular, helping the intestine to take positions which increase the effective area of absorption of the passing nutrients. The intestine comprises three parts:

a. The pre-typhlosolar region: It is present from the 15th segment to the 26th segment. It contains villi, which are short conical projections on the 26th segment called intestinal caeca.
b. The typhlosolar region: It constitutes the large internal median fold of the dorsal wall of the intestine, forming a longitudinal ridge. It increases the area of absorption in the intestine.
c. The post-typhlosolar region: It is also called as rectum. This segment stores faecal pallets and leads to the anus.

3.9.8 The Anus

It is a small round aperture present in the last segment. Through it, the earlier ingested waste, which by now has been biodegraded, mineralized, and stabilized into "vermicompost", is excreted.

3.10 WHY EARTHWORM-BASED ZOOCOMPOSTING IS MUCH QUICKER THAN COMPOSTING?

What are the factors that enable zoocomposting by earthworms to be completed within a few hours, in contrast to 6 weeks or more needed in composting?

From the details presented in the preceding section, it can be seen that in each earthworm, functioning as a bioreactor, the following processes occur at a rapid rate:

a. Comminution, and the consequent size reduction, of the waste particles of the waste.
b. Improvement in the pH of the waste towards neutral.
c. Action of several enzymes and microorganisms, present in the earthworm's gut, on the constituents of the waste, as the waste passes through different zones of the earthworm's body, causing its biodegradation.

There is no perceptible rise in temperature in or around the earthworm when it is engaged in biodegrading its feed. This is in sharp contrast to the thermophilic phase in composting, during which temperatures can reach up to 60°C and can even go higher. But even though no temperature rise occurs, during earthworm-based zoocomposting, pathogens are still killed to as significant an extent in earthworm-based zoocomposting as they do in composting. This is likely due to the microflora and other constituents present in the earthworm's gut destroying many of the pathogenic organisms present in the waste.

3.11 OTHER DISTINGUISHING FEATURES OF EARTHWORM-BASED ZOOCOMPOSTING

The movement of earthworms towards food, as well as their up-and-down movement in the course of their diurnal bio-rhythm, keeps the substrate in the zoocomposting systems well-agitated and aerated. Moreover, earthworms do not possess organs capable of breathing and obtain their oxygen needs by absorbing atmospheric oxygen. This process also facilitates the aeration of the waste. Due to the combined effect of all these happenings, the substrate undergoing zoocomposting by earthworms does not need periodic turning or aeration, which are necessary in the composting systems. However, as explained later, these advantages accrue only as long as the depth of the substrate in the zooreactor is not more than about 20 cm. At greater substrate depths, anaerobic pockets can form, as there may also be a sharper increase in the moisture levels as we go deeper. But conventional vermireactors (Figures 3.5 and 3.6) are several feet deep, which reduces their efficiency to a drastic extent.

3.12 WHY, THEN, ZOOCOMPOSTING DOES NOT GET COMPLETED WITHIN A FEW HOURS IN AN EARTHWORM ZOOREACTOR?

In reality, all earthworms present in a zooreactor convert what they eat into compost within a few hours. But neither is it possible to give precisely as much feed to an earthworm as it can process every day, nor is it possible to precisely separate

FIGURE 3.5 A conventional vermireactor. (https://commons.wikimedia.org/wiki/File:Rubber_worm_farm.jpg.)

FIGURE 3.6 A typical vermicomposting facility based on conventional vermireactors. (https://commons.wikimedia.org/wiki/File:Rubber_worm_farm.jpg.)

the compost formed each day in a zooreactor from the unutilized feed present in a reactor. Only when 75% or more of the feed has been converted to compost is it possible to separate the litter clearly. It is due to these factors that conventional vermicomposting systems take 6–12 weeks to deliver their products. We have significantly reduced this time span to 15–25 days by several innovations detailed in Chapters 5 and 6.

REFERENCES

Abbasi, T., Gajalakshmi, S. and Abbasi, S.A., 2009. Towards modeling and design of vermicomposting systems: mechanisms of composting/vermicomposting and their implications. *Indian Journal of Biotechnology*, *8*, pp.177–182.

Gajalakshmi, S. and Abbasi, S.A., 2008. Solid waste management by composting: state of the art. *Critical Reviews in Environmental Science and Technology*, *38*(5), pp.311–400.

Nayeem-Shah, M., Gajalakshmi, S. and Abbasi, S.A., 2014. Direct vermicomposting of vegetable waste using the concept of high-rate vermireactor operation. *International Journal of Environmental Research*, *4*, pp.59–65.

4 Earthworms and a Brief History of Appreciating Them

4.1 INTRODUCTION

This chapter presents an overview of the distribution, the classification, the life cycle, the anatomy, the physiology, and the ecology of earthworms in the context of their role in vermicomposting. It identifies the epigeics – the group of earthworms that are phytophagous, also called detritus-feeders or humus-feeders, and predominantly surface-dwelling – as the most suitable for vermicomposting.

4.2 A BRIEF HISTORY OF APPRECIATING EARTHWORMS

The history of the use of animal manure as a fertilizer is probably as old as the history of agriculture. The art of composting also has ancient origins – some of the engraved mentions of composting are found on clay tablets going back 3,000 B.C. The ancient Indian scripture – the Bhagvad Geeta – has reference to composting, so has the Bible, and so have the ancient Chinese writings. Composting continues to be an essential part of farming (Figures 4.1 and 4.2).

FIGURE 4.1 Livestock farmers preparing manure piles to carry out composting. (https://commons.wikimedia.org/wiki/File:106349_kibbutz_tel_yosef_PikiWiki_Israel.jpg.)

DOI: 10.1201/9780429331886-6

FIGURE 4.2 Horticultural waste being transported for composting. (https://commons.wikimedia.org/wiki/File:Farmers_of_forty_centuries_-_returning_from_genya_lands_with_herbage.jpg.)

Almost parallel to the appreciation of the use of manure as a fertilizer – in its composted form or direct form – has been the acknowledgement that earthworms have the ability to rejuvenate soil and make it more plant-friendly. Ancient Egyptians placed such high value upon earthworms that Cleopatra (Figure 4.3) had banned the removal of earthworms from the Egyptian soil, and capital punishment awaited the defaulters. Aristotle (Figure 4.4) was the first among great scientists who hailed the contribution of earthworms in decomposing fallen organic matter into "rich, compost-like humus" – the substance we now call vermicompost or vermicast. Thus, it can be said that interest in composting and in earthworms has evolved together. But it was only in the 1970s that vermicomposting was recognized as a separate process. About 150 years earlier, Charles Darwin (Figure 4.5) had brought earthworms to global attention with his research on the behavioural ecology of earthworms, which eventually led to his 1881 classic *The formation of vegetable mould through the action of worms with observations on their habits*. The book sold at a faster rate than his epoch-making *The origin of species*, which was published in 1859 (Lessiter, 2023) (Figures 4.6–4.8).

The laudatory tone of Darwin's description of earthworms tickled cartoonists, resulting in some very famous caricatures. Several quotes of Darwin on earthworms have also remained in great currency, for instance, "their sexual passion is strong enough to overcome for a time their dread of light". Interestingly,

FIGURE 4.3 Cleopatra, the Egyptian queen, who had outlawed all actions that harmed earthworms. (https://commons.wikimedia.org/wiki/File:Benedetto_Servolini_and_Frederick_Stibbert_-_The_toilette_of_Cleopatra.jpg.)

FIGURE 4.4 Aristotle was among the first globally acknowledged intellectuals who spoke about the great importance of earthworms. (https://commons.wikimedia.org/wiki/File:Philosophie-gr%C3%A8ce_Aristotle.jpg.)

FIGURE 4.5 Charles Darwin has been the single most significant influence in making the world conscious about the pivotal role of earthworms in soil processing and rejuvenation. (https://commons.wikimedia.org/wiki/File:Charles_Darwin_portrait.jpg.)

Darwin had published his first paper on earthworms in 1838, bringing out the ability of the earthworms to bioturbate soil and bury in it objects lying on its surface (Brown et al., 2003). This work places Darwin as one of the pioneers of soil science. After almost 44 years of this paper, and after a life-time spent on researching earthworms, Darwin had authored *The formation*. While his conclusions on the role of earthworms in soil engineering, on driving to biogeochemical cycles, and on "benevolence" have largely been accepted, there remains unsettled controversy on how accurate Darwin was in attributing intelligence to earthworms.

FIGURE 4.6 A cartoon depicting Darwin immersed in deep thought on earthworms. (https://commons.wikimedia.org/wiki/File:PSM_V20_D414_The_worm_turns.jpg.)

4.3 THE ADVENT OF VERMICOMPOSTING

The attention Darwin brought towards earthworms early in his scientific career was progressively intensified by him as reflected in his famous words,

> The plow is one of the most ancient and most valuable of man's inventions; but long before he existed the land was in fact regularly plowed, and still continues to be thus plowed by earthworms. It may be doubted whether there are many other animals which have played so important a part in the history of the world, as have these lowly organized creatures.

FIGURE 4.7 A caricature implying that Darwin had evolved from an earthworm. (https://commons.wikimedia.org/wiki/File:Caricature_of_Charles_Darwin_Wellcome_L0003921.jpg.)

It gradually became very well-recognized that earthworms feed upon a variety of substrates, turning them into a well-formed compost-like substance. It is also easy to imagine that whenever vermiculture was practiced, the formation of vermicast was noticed. Likewise, it must have been noticed that when either manure or its compost was applied to agricultural fields, earthworms tended to move in and leave behind their signature offering – the vermicast – before moving away. But neither the term

FIGURE 4.8 Darwin portrayed in animal form. (https://commons.wikimedia.org/wiki/File:Editorial_cartoon_depicting_Charles_Darwin_as_an_ape_(1871).jpg.)

vermicomposting, nor the use of worms to convert animal manure, its compost, or biodegradable solid wastes into useful fertilizer, emerged till almost 100 years after Darwin had passed away. The credit for bringing vermicomposting to global attention is given to American biologist Mary Arlene Appelhof (1936–2005).

The trend-setting 1972 Stockholm Conference for the Human Environment, which led to the formation of the UN Environment Programme (UNEP), was perhaps a major source of inspiration to Appelhof (Figure 4.9). It had generated in her the vision of "tonnes of worms eating tonnes of garbage". A school teacher in biology at the Kalamazoo Central High School, who had BS and MS degrees from Michigan State University, East Lansing, Appelhof had no access to well-equipped laboratories or research grants. But she had enormous zeal and drive. So she did the best she could and turned the basement of her dwelling into a kind of

FIGURE 4.9 Mary Appelholf: the pioneer of human-controlled vermicomposting.

laboratory-cum-workshop-cum-manufacturing-cum-vermiculture-cum-publishing house. Operating from there, she began designing vermicomposting bins, culturing worms, and marketing both. She acquired a barely functioning mimeographic machine from a local Democratic party office at a cost of US $ 5. Using it, she published her maiden work, a brochure, titled *Basement Worm Bins Produce Potting Soil and Reduce Garbage*. It is perhaps the first-ever monograph on vermicomposting. Interestingly, Appelhof was a lady of many talents. Besides being a biology teacher and self-guided researcher, she was also an expert swimmer and a prize-winning photographer. But she chose to channelize most of her spare time and energy in taking vermicomposting to the people, especially to the budding citizens – the schoolchildren (Figure 4.10). In 1976, she authored and self-published, under the banner Flower Press, a slim 100-page book titled *Worms Eat My Garbage*. The book became a runaway success and by 1997 had sold 250,000 copies (Figure 4.11). It educates the readers on the essentials of vermicomposting (also called *worm composting*), how to put together a vermicomposting bin, how to operate and maintain it, and what to do with the vermicast. The book continues to be very popular to this day. Appelhof also set up, in 1982, Flowerfield Enterprises, which continues to be active and growing.

By the end of the 1970s, Appelhof had sensitized millions of people across the world on the virtues of vermicomposting. In the bargain, she had acquired the

FIGURE 4.10 A caricature of Mary Appelholf tending earthworms and dreaming of a bumper harvest. (https://www.hachettebookgroup.com/contributor/mary-appelhof/; Mary's photos needed permission).

nickname *The Worm Woman* and had begun to be considered the "mother of modern vermiculture". She travelled around the world speaking on the virtues of worm composting. In 2000, she organized a seminar named *Vermillinieum,* which attracted people from all over the globe.

While Appelhof's forte was household-scale vermicomposting, and her most popular contraption was her household-scale worm bin, attempts to design larger-scale vermicomposting systems soon began as a follow-up to Appelhof's invention. Initial efforts revolved around adapting windrow composting to vermicomposting. The use of static pile windrows, the top-fed windrows, and the side-fed windrows, which were developed in 1960 for composting, was extended to vermicomposting. To prevent earthworm poaching by rodents, reptiles, poultry, and other predators, as well as preventing the earthworms from migrating, the Top-fed Bed was introduced by, which was similar to top-fed windrow but was provided with four walls. These systems have been described in detail later in this chapter. In 1982, the first-ever patent on a vermicomposting system was filed, which described a 30-ft tall silo-like machine that used a conveyor belt. Since then, a large number of systems have been developed, of which about half have been patented.

FIGURE 4.11 Mary Appelhof's *Worms Eat My Garbage* continues to be a bestseller. (https://en.wikipedia.org/wiki/Worms_Eat_My_Garbage#/media/File:Worms_Eat_My_Garbage.jpg.)

4.4 EARTHWORMS AND THE SPECIES CONSIDERED SUITABLE FOR VERMICOMPOSTING

4.4.1 General

In classical systematics, earthworms have been placed under the class *Oligochaeta* or *Clitella* and phylum *Annelida*. They are crawling and wriggling animals, with a serpent-like body shape of a very high length/width ratio.

Earthworms have been so named because they are predominantly terrestrial and dwell either in the organic matter lying on the soil surface or in the burrows they make underground (Abbasi & Ramasamy, 1999, 2001; Gajalakshmi & Abbasi, 2004). Earthworms are ubiquitous. They are found everywhere on land except in the deserts (where they do occur but are rare), permafrost areas, and totally barren soils.

Estimates of how many species of earthworms exist on Earth vary from 3,000 (ESB, 2024) to 7,000 Tabassum-Abbasi et al., 2019. The estimate, which is often quoted, is 6,000.

The lengths of earthworm bodies can range from a few millimetres to more than a metre, while the width can also range from minuscule to 250 mm or more. Within the same species, too, the size can vary greatly – for instance, the length of *Lumbricus rubellus* can be less than 250 mm to greater than 1,000 mm. The longest earthworm on record is a *Microchaetus rappi* specimen, found in South Africa in 1967, which was about 6.7 metres long (GBWR, 2024). Interestingly, the average length of *M. rappi* (Figure 4.12) is only about 1.8 m (GBWR, 2024).

The life expectancy of earthworms is 1–7 years, depending on the conditions under which they live. Earthworms dwelling in urban areas have shorter lifespans – 2 to 3 years – due to the stresses caused by pollution and the hustle and bustle of life. In good natural settings, earthworms can go on for 4–7 years. Earthworms have circular segments and grooves which distinguish them from snakes (Figure 4.13).

Three aspects of earthworm growth are special. The first is that they are homomorphous developers. When two or more species are forced to live together (as in a container used for culturing earthworms), they start tending towards sizes and growth rates that are similar across species (Gaddie & Douglas, 1977). To enable this, the individuals of the otherwise larger-sized species begin to gradually reduce

FIGURE 4.12 A *Microchaetus rappi* specimen on Australian soil. It being curled up, its large length gets masked. (https://www.inaturalist.org/observations/121628659.)

FIGURE 4.13 Ring-like segments and circular grooves that distinguish an earthworm from other snake-shaped animals. (https://en.wikipedia.org/wiki/Earthworm#/media/File:Staphylinus.olens.vs.lumbricus.terrestris.jpg.)

in size, while the individuals of the otherwise smaller-sized species begin to grow bigger. This tendency is prompted to make it easier for each species to survive. In other words, different species, when made to live together, orient themselves to make the most of the average availability of food and other conditions. This tendency of earthworms is mistaken for hybridization, but in reality, no hybrids result from the cohabitation of multiple earthworm species (Gaddie & Douglas, 1977).

The second earthworm specialty, which they share with a few other animals, is that in some of the earthworm species, the individuals keep growing all their lives, thus exhibiting "indeterminate growth".

Earthworms are also among the few animals that possess the ability to regenerate. For example, *Perionyx excavatus* can regenerate even its head if, by some accident, or in a tussle with a predator, it has lost it. The animal's central nervous system, which is housed in its head, also gets regenerated.

4.4.2 Morphology

Earthworms are slow-moving, and their soft bodies are distinguished by ring-like segments all along their lengths, made conspicuous by circular grooves called *annuli*.

On the ventral surface of each segment are short, stubby bristles called *setae*. The setae provide the earthworms with the traction they need to be able to generate their peristaltic motion, characterized by fits and starts. Setae also enable the earthworms to resist being pulled out of their burrows when predators seek them.

Earthworms do not possess a well-defined head. Instead, they have a pre-oval part named *prostomium*. Each individual has openings at either ends of their wire-like bodies which perform the roles of the mouth and the anus, respectively.

4.4.3 Life Cycle

The earthworm life cycle (Figure 4.14) involves five distinct stages: mating, cocoon formation, hatching of eggs encapsulated in cocoons, growth to the juvenile stage, and adulthood.

The Mating phase: Earthworms are hermaphrodites, possessing both male and female reproductive organs. But they cannot fertilize their own eggs because of being protandrous. The relative position of the earthworm's male and female genital apertures is such that self-fertilization cannot be accomplished. What occurs is reciprocal cross-fertilization. Earthworms prefer to mate on moist surfaces by exchanging sperm while clinging to each other in a 69 formation, giving the appearance of having been tied into a knot (Figure 4.15). As the mating – which lasts for an

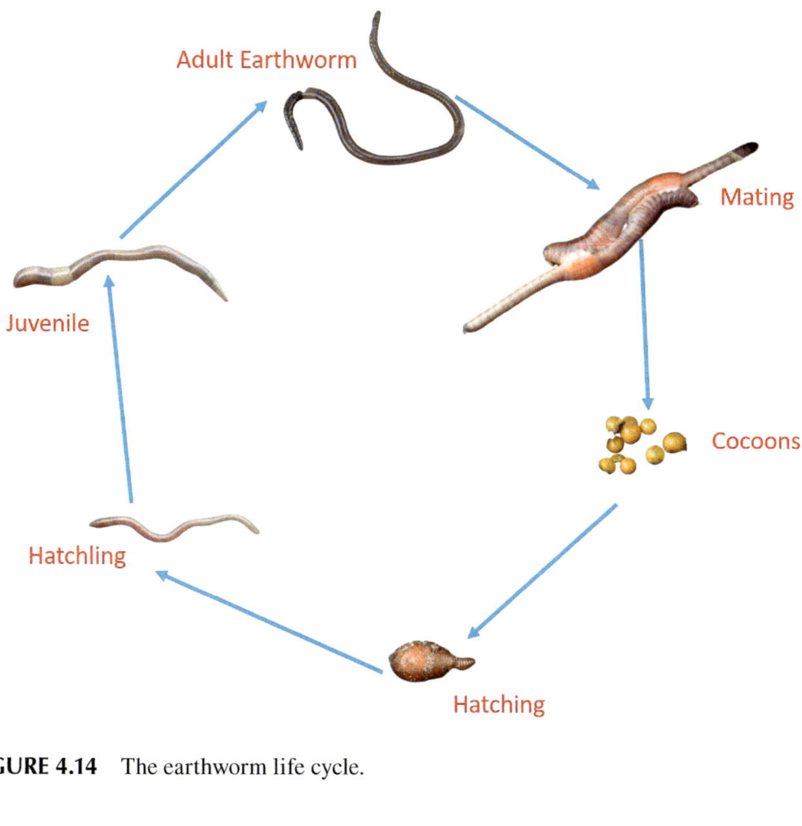

FIGURE 4.14 The earthworm life cycle.

FIGURE 4.15 Two earthworms engaged in mating. (https://en.wikipedia.org/wiki/Earthworm#/media/File:Mating_earthworms.jpg.)

hour – progresses, a mucus ring forms around the clitellum of each of the partners, and the sperm is transferred from each individual's male pores to the spermathecae of the partner. It leads to the formation of slime around the clitellum, which dries as it is filled with albumin. The tube passes over the female pore, depositing eggs into the capsule while releasing sperm as it passes over the male pore. After this mutual interchange of sperm, the copulating earthworms separate, and fertilization takes place in cocoons, which are egg capsules formed as the secretion from the clitellum hardens on exposure to air into bulb-shaped cocoons.

The cocoon phase: The cocoons are tiny, bulb-shaped encasements (Figure 4.16), built to withstand adversity. An earthworm can produce, on average, 20–30 cocoons per year, each cocoon containing 1–10 eggs.

In 2–3 weeks, the cocoons hatch, releasing hatchlings which resemble small bits of black thread. But the number of days needed for cocoons to hatch can be much more than 2–3 weeks – up to several months – if environmental conditions are not favourable, especially temperature and humidity.

The hatching phase. For the initial 1–2 days, the very soft and vulnerable hatchlings survive on the nutrients remaining in their egg sac. They then begin to feed on the organic matter available around them. During the following 2–3 weeks, the hatchlings steadily increase their mass and size as they move into the juvenile phase.

The juvenile phase. As the hatchlings-turned juveniles grow, they begin resembling adult earthworms (Figure 4.17). They are distinguished from their adults on the basis of the ability of the adults to procreate, which juveniles do not possess.

FIGURE 4.16 Typical bulb-shaped earthworm cocoons. They belong to *Lumbricus terrestris*. (https://en.wikipedia.org/wiki/Earthworm#/media/File:Earthworm_-_L._terrestris_cocoons.jpg.)

FIGURE 4.17 Earthworm juveniles (of *Perionyx excavatus*). (https://www.inaturalist.org/observations/112940685.)

FIGURE 4.18 An adult earthworm (typified by *Lumbricus rubellus*). (https://www.inaturalist.org/observations/244937735.)

The adult phase. Eventually, after several months of feeding and growing, juveniles acquire the wherewithal and the skills to mate, turning into adult earthworms. They also develop a tell-tale sign of earthworm adulthood – the clitellum, which is a thickened band near the earthworm's head (Figure 4.18) and which plays a significant role in the earthworm's reproduction.

4.4.4 Other Habits

Sensitivity to moisture: Earthworms need moist conditions to survive and can die quickly of dehydration when deprived of moisture. To minimize water loss from their bodies, earthworms cling to each other in ball-shaped formations (Figure 4.19). When they have a sufficient reservoir of moisture, their bodies are kept moist by the secretion of moisture to the inner and outer body surfaces.

Sensitivity to temperature: Earthworms are exceedingly temperature sensitive. Indeed, their cycles of activity are dictated by temperature, together with the availability of moisture and food. Temperature is the biggest limiting factor for earthworms, followed by moisture and food.

FIGURE 4.19 Earthworms tend to herd together and communicate by touch. (https://commons.wikimedia.org/wiki/Category:Lumbricidae#/media/File:Nightcrawlers_Worms_Nest.jpg.)

Sensory organs: Earthworms are bereft of eyesight, sense of smell (olfaction), and ability to hear. But special cells present along the length of their bodies enable them to pick up vibrations and be guided by them. At times, this ability proves lethal to them because some of the predator birds are aware of it. These predators stamp their feet on the ground under which earthworms may be lying to generate vibrations which the earthworms mistake for a mate or a moisture source. They come to the surface and get poached.

Social behaviour: In recent years, a lot of interest has emerged towards the social behaviour of earthworms after it became known that they are the only non-chordates to show social behaviour (Patnaik et al., 2022).

Earthworms have been observed to herd together (Figure 4.19) and appear to take collective decisions. They seem to communicate with each other through touch to guide the behaviour of each other. They are able to take collective decisions on which direction the herd should move that will enhance their prospects of survival, growth, and reproduction. This enables the earthworms to find and benefit from the ecological niches that would otherwise have been missed by them. As the number of individuals in a herd increases, indicative of safety and success associated with that herd, the probability of other individuals joining that herd increases in proportion. Conversely, if the sustainability of a herd is reduced for any reason, individuals begin deserting that herd. This knowledge can be of value when sizing up vermireactors and handling earthworms.

Scientists have recorded events in which a significant number of earthworms were seen to be migrating. It is suggested that social cues may be operating to make such collective movement possible. According to Edwards and Bohlen (1996), both chemical and/or physical "communication" may be occurring in the earthworms.

Soil-working: All except a few species, which dwell in the leaf litter lying on the soil surface or in the piles of animal manure, of earthworms make either shallow or deep burrows below ground (Figure 4.20). To accomplish it, they use their bodies to act as mini-drills by pushing their anteriors into crevices and expanding their segments to force apart the particles bound to the crevices. As they keep boring in this fashion, particle by particle, tunnels form (Figure 4.21). If the soil happens to be very compact, earthworms eat their way through it. The resulting tunnels can go up to several feet below ground. The diameters of these tunnels can range from 3 to 12 mm and can have a broader space at the end for earthworms to rest and turn.

FIGURE 4.20 A vertical tunnel dug by a geophagus (endogenic or soil-feeding) earthworm. (https://en.wikipedia.org/wiki/Earthworm#/media/File:Earthworm_-_L._terrestris_permanent_vertical_burrow.jpg.)

While the majority of earthworm species feed upon the decaying organic matter present in the soil, some species specialize in feeding on leaf litter and other plant debris that fall on the soil surface. The rest do not normally consume plant material directly. They first pull in the phytomass to a few centimetres into the entrance of their burrow. The phytomass is then moistened with an enzymatic secretion of alkaline pH, released by the earthworm. It decomposes some of the cellulose and starch present in the plant material, making it easier to ingest. Earthworms also tear the plant leaves by holding their edges between the animal's prostomium and the mouth and pushing the pharynx forward. Small portions are sucked up by first pressing the mouth against the pieces of the leaves and then withdrawing the pharynx, thus creating the suction that pulls the piece in.

The organic material ingested by the earthworms eventually comes out as vermicast (Figure 4.22). As detailed in the previous and subsequent chapters, vermicast is a very efficient and potent organic fertilizer due to its content of bioavailable minerals and other nutrients. Vermicast also improves soil quality in various ways (Gajalakshmi and Abbasi, 2008).

These burrowing and feeding activities of the earthworms go on almost continuously. In the process, earthworms turn, till, aerate, and move enormous masses of soil and enrich it with organic material as well as minerals. Their actions shift the

FIGURE 4.21 Typically branched burrows of endogeic (geophagus or soil-eating) earth worms. (https://commons.wikimedia.org/wiki/File:Upper-30cm-showing-view-through-the-glass-side-of-an-80cm-deep-Evans-box-used-to-examine-burrowing-behaviour-of-a-single.jpg.)

FIGURE 4.22 A pile of vermicast resulting from the digestion of organic material by earthworms. (https://en.wikipedia.org/wiki/Earthworm#/media/File:Earthworm_faeces.jpg.)

soil pH from the acidic range to the more plant-friendly alkaline range. The earthworm burrows facilitate fragmentation, aeration, drainage, and porosity of the soil, making it porous and preventing it from becoming too compact for plant growth.

Earthworm activity supports mound formation as the earthworms tend to return to the same spot to deposit casts and breathe. In the course of it, the soil around the mound is moved to the mound, adding to the latter's size. As such mounds made by different individual earthworms grow, they coalesce into larger mounds.

4.4.5 Ecological Classification

Attempts have been made from time to time to classify earthworms on the basis of their colour, texture, and habits.

- Evans and Guild (1947) classified earthworms on the basis of their burrowing ability into surface-dwelling and deep-dwelling types.
- Graff (1953) noted that surface-dwelling and pigmented varieties of earthworms generally occur in habitats rich in organic matter.
- Byzova (1965) suggested that surface-dwelling smaller worms can be distinguished as a separate type due to their higher metabolic rate, as distinct from the much slower metabolism of deep-dwelling larger worms.
- Bouché (1977) proposed an ecological classification of earthworms that is widely followed. He slotted earthworms as (a) epigeics, (b) anecics, and (c) endogeics. The characteristics with which Bouché distinguished these categories are summarized in Table 4.1.

TABLE 4.1
Relative Features on the Basis of Which Banche (1977) Has Classified Earthworms Into Three Main Types

Characteristic	Relative Extent in		
	Epigeic	Anecic	Endogeic
Average body size	Smallest	Medium	Largest
Nature of burrows	Very shallow or absent	Deepest	Deep but less than anecic burrows
Mobility	Most rapid	Moderate	Feeble
Photosensitivity	Least	Intermediate	Most
Fecundity	Highest	Intermediate	Lowest
Maturation	Most rapid	Intermediate	Slowest
Metabolic rate	Highest	Intermediate	Lowest
Diet	Predominately biomass and zoomass	Biomass, zoomass, and soil	Predominantly soil

i. *Epigeics.* This constitutes a group of species that dwell above the mineral soil surface. They are predominantly plant and humus-feeders – hence called phytophagous. While epigeics successfully process massive quantities of phytomass, returning their nutrients to soil in a more easily bioavailable form as their castings, they generally have no effect on the soil structure. This is because epigeics cannot dig into the soil as anecics and endogeics can. Epigeics are typically smaller in size than anecics and endogeics, with uniform pigmentation.

ii. *Anecics.* Anecics are among the species that live below ground in vertical burrows but come to the surface to feed on plant debris, which they drag into their burrows to process and ingest. They play an important role in taking sources of nutrients available at the soil surface in the form of plant debris to deeper soil layers, thus improving the soil's nutrient content and physical structure across depths of several feet. Since anecics ingest phytomass as well as soil, they are called geophytophagous.

iii. *Endogeics.* These are the group of species that form horizontal branching burrows and mainly feed upon soil, which happens to be enriched with decomposing organic matter. As they ingest such organic matter while principally ingesting soil, they are called "soil eaters" or geophagous.

4.4.6 Species Suitable for Use in Vermicomposting

The faster an earthworm species consumes a substrate, the better it is for the vermicomposting of that substrate, provided that the species can sustain such a rate indefinitely.

Since epigeics have a higher metabolic rate than anecics and endogeics, they generally consume more food per gram of body mass per day than anecics and endogeics do. For some substrates, anecics like *Drawida willsi* and *Lampito mauritii* can run close to the epigeics in rate of feeding, but so far these authors have not found any anecic or endogeic overtaking an epigeic in this respect. Epigeics also have, in general, much higher fecundity than the other two groups. Above everything else, epigeics are – as found by these authors and described in the subsequent chapters – as much at home with phytowaste, even toxic weeds, as they are with animal manure.

For all these reasons, and not having any disadvantage in other respects in comparison to anecics and endogeics, epigeics make the most ideal among other earthworm species for use in vermicomposting.

REFERENCES

Abbasi, S.A. and Ramasamy, E., 1999. *Biotechnological Methods of Pollution Control.* Universities Press, Hyderabad, viii + 168 pages.

Abbasi, S.A. and Ramasamy, E., 2001. *Solid Waste Management with Vermitechnology.* Discovery Publishing House, New Delhi, xiv + 178 pages.

Bouché, M.B., 1977. Strategies lombriciennes. In 'Soil organisms as components of ecosystems'. (Eds U Lohm, T Persson). *Ecological Bulletins, 25,* pp.122–132.

Brown, G.G., Feller, C., Blanchart, E., Deleporte, P. and Chernyanskii, S.S., 2003. With Darwin, earthworms turn intelligent and become human friends: The 7th international symposium on earthworm ecology·Cardiff·Wales 2002. *Pedobiologia*, *47*(5–6), pp.924–933.

Byzova, Y.B., 1965. Comparative rate of respiration in some earthworms (*Lumbricidae, Oligochaeia*). *Review of Soil Ecology and Biology*, 2, 207–216.

ESB (Earthworm Society of Britain), 2024. Earthworm diversity. https://www.earthwormsoc.org.uk/earthworm-diversity. Last accessed 10 September 2024

Evans, A.C. and Guild, W.M., 1947. Studies on the relationships between earthworms and soil fertility: I. Biological studies in the field. *Annals of Applied Biology*, *34*(3), pp.307–330.

Edwards, C.A. and Bohlen, P.J., 1996. *Biology and Ecology of Earthworms*, Vol. 3, Springer Science & Business Media, Chapman & Hall, London, 426pp.

Gaddie, R.E., and Douglas, D.E., 1977, *Earthworms for Ecology and Profit*, Bookworm Publishing Company, Inc, New York, 213 pages.

GBWR (Guinness Book of World Records), 2024. Longest earthworm. https://www.guinnessworldrecords.com/world-records/70873-longest-earthworm#:~:text=The%20longest%20earthworm%20is%20Microchaetus,Alice%20and%20King%20William's%20Town; last accessed September 10, 2024.

Graff, O., 1953. Investigations in soil zoology with special reference to the terricole Oligochaeta. *Zeitschrift für Pflanzenernährung, Düngung und Bodenkunde*, *61*, pp.12–22.

Gajalakshmi, S. and Abbasi, S.A., 2004. Earthworms and vermicomposting. *Indian Journal of Biotechnology*, *3*, pp.486–494.

Gajalakshmi, S. and Abbasi, S.A., 2008. Solid waste management by composting: state of the art. *Critical Reviews in Environmental Science and Technology*, *38*, pp.311–400.

Lessiter, M., 2023. Darwin's 'lowly earthworm' gets its due. https://www.no-tillfarmer.com/articles/12203-darwins-lowly-earthworm-gets-its-due. Last accessed August 2024.

Patnaik, P., Ghazala, D., Abbasi, S.A., Tabassum-Abbasi, and Abbasi, T., 2022. Social behaviour of earthworms in the context of their use in bioremediation. In: Siddiqui N., Tauseef S., Abbasi S., Dobhal, R., Kansal, A. (eds) *Advances in Sustainable Development.* Springer Nature Singapore Pte Ltd. Springer, Singapore. Chapter 1, pp. 1–14. https://doi.org/10.1007/978-981-16-4400-9_1

5 Design and Operation of Conventional Vermireactors

5.1 INTRODUCTION

This chapter covers efforts made towards vermireactor design based on the conventional vermicomposting technology. Beginning with the simplest form of reactors, it covers a large number of representative devices which were developed across the world, especially North America, to enhance the capacity, the convenience, the catchment, and the economic viability of the vermicomposting technology.

5.2 THE ESSENCE OF A VERMIREACTOR

In its most rudimentary form, a vermicomposting system just needs a container that can hold the waste matter and cow dung, horse dung, or other forms of animal manure, and earthworm species that tend to feed upon the animal manure.

Only two other controls are required: the ambient temperature should be such that it remains within the temperature limits tolerable to the earthworm species being used, and the moisture content of the reactor feed should be within $60\% \pm 10\%$. Both controls are easy to exercise because several earthworm species have a wide range of temperature as well as moisture tolerance. If the conditions are met, earthworms feed upon the substrate and convert it into vermicast in 3–4 months. In this period, earthworms also reproduce, generating cocoons and juveniles, so the net zoomass contained in a vermireactor at the culmination of vermicomposting is generally higher than it was at the start.

When it is aimed to enhance the efficiency of vermicomposting systems and operate them at larger scales, several challenges appear. Besides finding a way to make earthworms convert more substrate to vermicast per unit time, it also becomes a major challenge to efficiently separate vermicast, earthworms, and unreacted substrate in a way that vermicast can be utilized or sold, while the earthworms and the unconsumed substrate can be recycled.

To make the process operate in a continuous fashion is also a challenge because it needs to synchronize the rate of feeding with the very slow rate of vermicomposting and the complex operations necessary to simultaneously (and clearly) separate vermicast away from the process cycle.

Attempts to meet these challenges have led to a large number of designs which differ from each other principally in the following terms:

DOI: 10.1201/9780429331886-7

1. Nature of feeding: continuous, semi-continuous, or in purely batch mode
2. Manner of feed input: top fed or side fed
3. Manner of harvesting of earthworms and their vermicast: manual or mechanized

Within the permutations–combinations of these basic functions, a large variety of vermicomposting systems have been designed, which further differ from each other in terms of the contraptions and modalities used in supporting the three basic functions listed above. The variety available is summarized in Figure 5.1.

5.3 THE REPORTED VERMIREACTORS (VERMICOMPOSTING SYSTEMS)

We present below, in chronological order, vermireactor systems which have been patented or the designs of which have been described in accessible prior art. It is an exceedingly representative list, though not exhaustive.

5.3.1 Static Pile Windrows

Next to the simplest form of a vermireactor – a container with substrates and earthworms – comes a static pile windrow, which is the next least mechanized of vermireactor systems. In this system (Figure 5.2), which is inspired by a similar composting system (Gajalakshmi and Abbasi, 2008), elongated piles of the substrate are laid on a ground layered with some lining to avoid leakage. The piles are laid in either rectangular- or square-shaped rows. Organic material is mixed and added to the bedding along with earthworms. Each windrow is left for 2–3 months until the conversion of the substrate to vermicast is complete. Harvesting is done manually or using a worm harvester to separate vermicast. Besides being slow, this system is not amenable to easy control of moisture and temperature. Being unconfined, it can have earthworms migrating out.

5.3.2 Top-Fed Windrows

These (Figure 5.3) are similar to static pile windrows except that bedding and feed are not mixed. The bedding is first done. Then, worms are introduced, followed by a thin (about 10 cm thick) layer of feed. As the worms consume the feed, they move up. The feed is added from the top periodically so that the worms keep feeding upon the substrate and keep moving up, leaving the vermicast behind. Harvesting is done by first removing the top 10–20 cm with a front-end loader fitted with a bucket. The top layer can be removed easily, which contains most of the worms.

The top layer can serve as seed for the next windrow. The leftover bed has mostly vermicast and some unconsumed material, which are either used directly as organic manure or are screened to separate the undigested material for processing, while the vermicast is packed and sold.

5.3.3 Side-Fed Windrows

Side-fed windrows are similar to the windrows described above except that fresh feed is loaded as a new windrow along the side of an existing windrow (Figure 5.4).

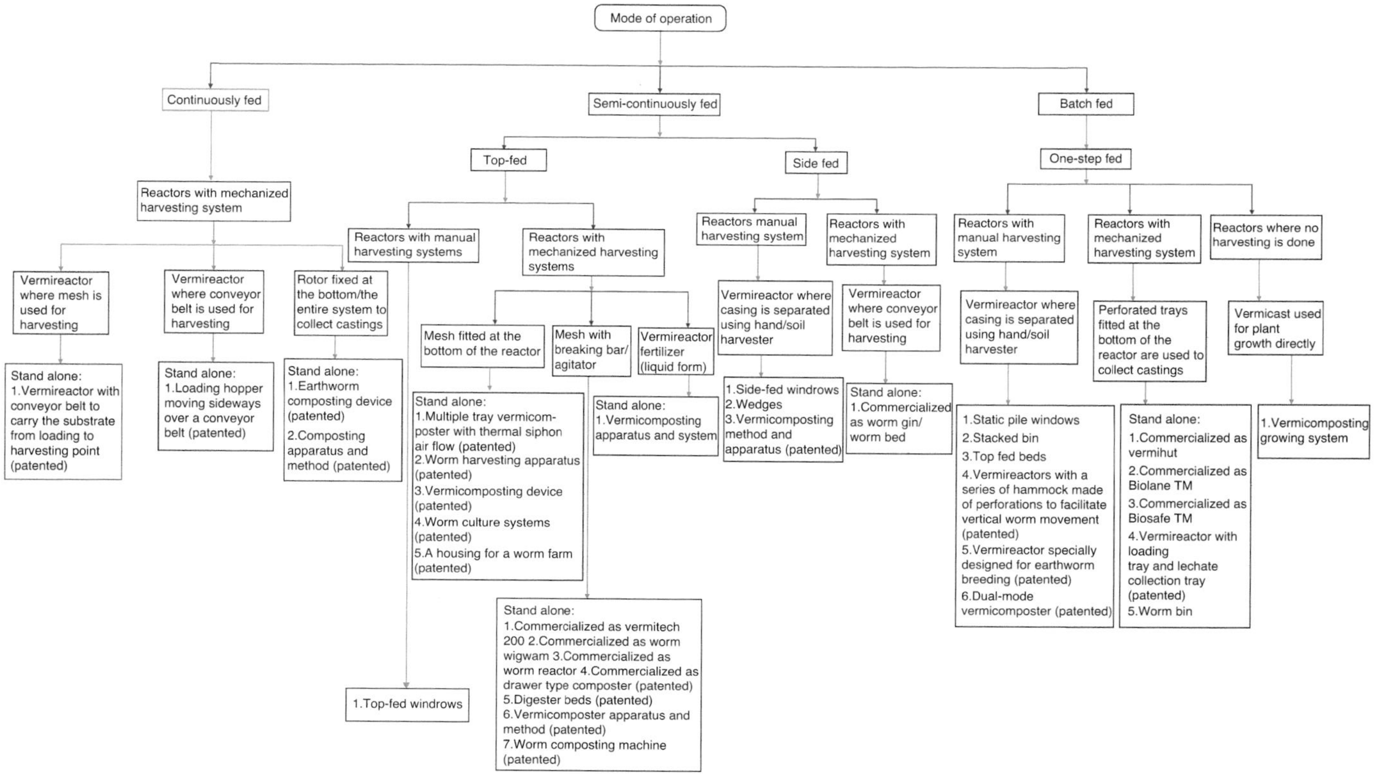

FIGURE 5.1 Different types of feeding, operating, and harvesting strategies employed in the reported vermireactors.

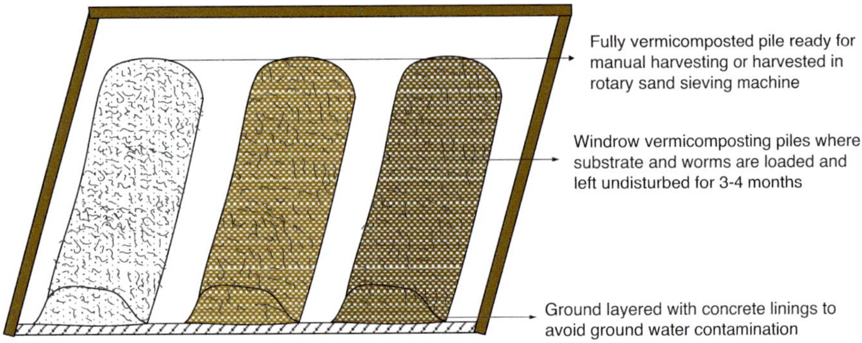

FIGURE 5.2 Static pile windrows.

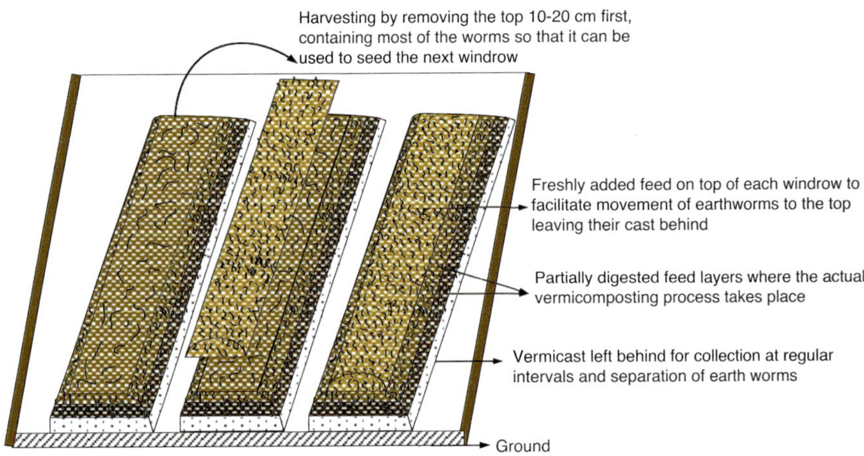

FIGURE 5.3 Typical top-fed windrows.

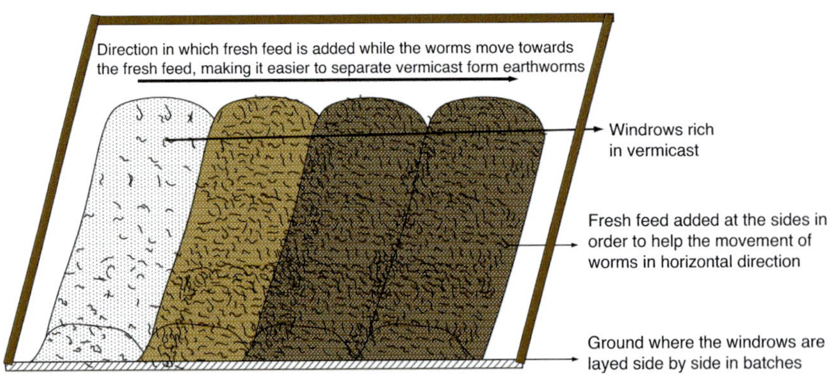

FIGURE 5.4 Typical side-fed windrows.

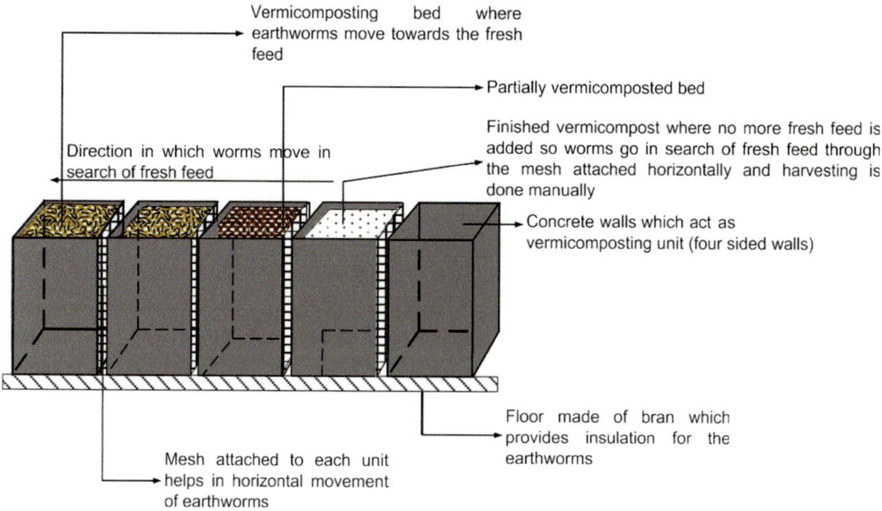

FIGURE 5.5 Typical top-fed vermicomposting beds.

This enables the earthworms to move sidewise from the existing (substrate-lean) windrow to another (substrate-rich) windrow. Even though it is operated in batches, it can be deemed a semi-continuously fed system.

5.3.4 Top-Fed Beds

Top-fed beds (Figure 5.5) are similar to top-fed windrows except that they are confined within four walls. These beds are laid on the floor, protected by concrete or bran, which provides sufficient heat during winters. These are separated using a metal screen to aid earthworm movement. The organic material is mixed and added to the beds along with earthworms. Each bed is left until the conversion of the substrate to vermicast is complete. During harvesting, the operator stops feeding the digested bed, so that the worms migrate horizontally from the existing (substrate-lean) bed to another (substrate-rich) bed. Besides being confined, the system is slow and is classified as a batch-fed system.

5.3.5 Stacked Bins (Flowerfield Enterprises, 1973)

These are confined, vertically stacked bins (Figure 5.6), arranged in plurality covering an area about 120 cm^2 up to a depth of 30–45 cm. Organic material is mixed and added to the bins along with about 2.5 kg of earthworms. These bins are left undisturbed until the vermicomposting process is complete. Harvesting is done manually. The bins can also be lifted manually, or the step can be mechanized. The bins need much less space per unit mass of vermicompost than the more conventional systems described earlier.

5.3.6 US3961603A: Habitat for Earthworm Cultivation (Gaddie, 1976)

It consists of stackable containers with inward projections at their upper flanges and perforated bottoms. The projections sustain the weight of the upper containers,

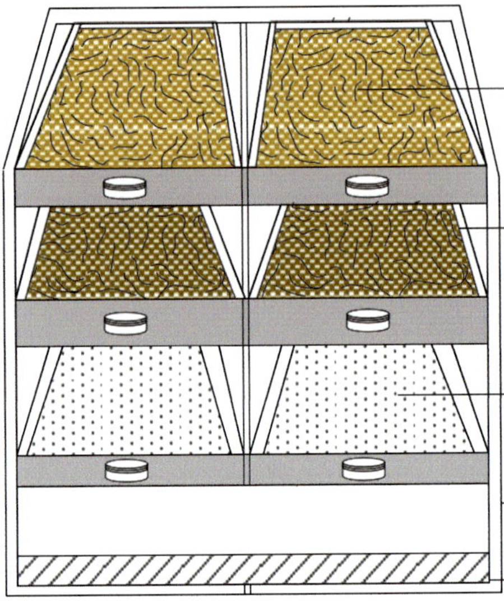

FIGURE 5.6 The Stacked Bins system.

while admitting air to individual containers and permitting feeding and breeding of the earthworms throughout the assembly. The perforations at the bottoms of the containers enable the movement of the earthworms from one level to another. The container at the bottom has no perforations but carries a provision to hold water in a pile of pebbles for the control of the moisture reaching from the beddings above (Figure 5.7).

5.3.7 US4108625A: Method Of, and Apparatus For, Producing Fertilizer by Utilizing Earthworms (Okada, 1978)

Earthworms are bred in a layer of fermented vegetable matter, such as waste cotton and straw, which is maintained in an adequately moistened condition. The earthworm casting is gathered and refined into an organic fertilizer through a series of steps, which include the removal or killing of earthworm eggs (Figure 5.8).

5.3.8 Wedges Vermicomposting (Edwards, 1988)

These systems (Figure 5.9) are similar to top-fed windrows except that removable barriers are placed between each layer. The bedding is first done, where organic material is mixed and added along with earthworms. Each layer is laid at an angle of 45 degrees on the concrete floor. Thereafter, successive thin layers of 5–10 cm of organic material are added by placing a removable barrier.

The wedges can be of any width or length, but the height should be 1.2–1.5 m for easy addition of organic waste to the leading edge. As the worms consume the feed in 3–4 months, they keep moving upwards, towards the removable barrier.

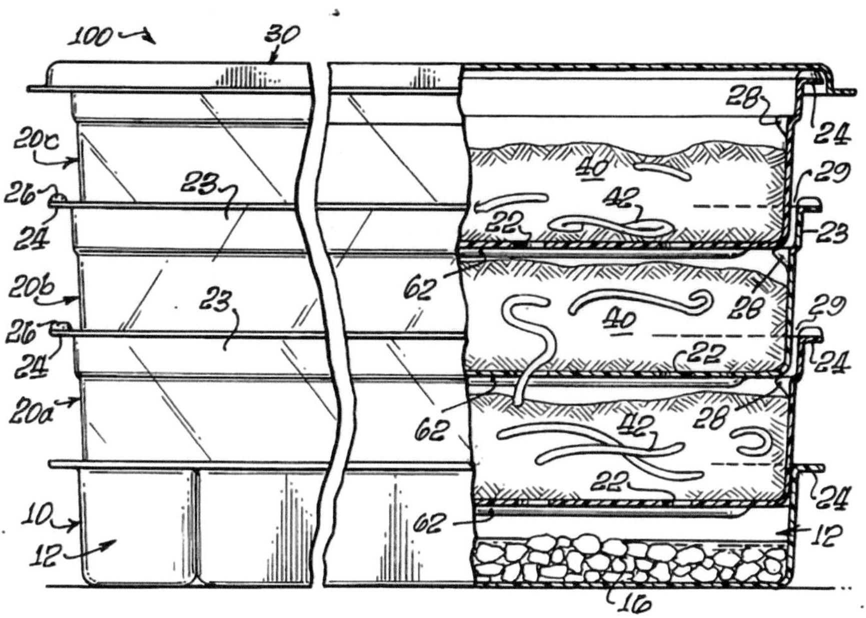

FIGURE 5.7 Habitat for earthworm cultivation. 10 base pan; 12 support structures; 16 Pan; 20a, 20b and 20c breeding pans; 22 orifices; 23 upper rim; 24 horizontal flange; 26 convex buttons; 28 gussets; 29 ventilating gap; 30 cover; 40 bedding; 42 earthworms; 62 ribs; 100 habitat.

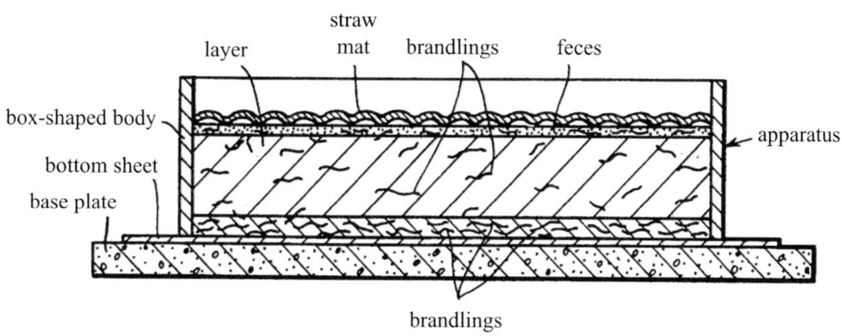

FIGURE 5.8 Method of, and apparatus for, producing fertilizer by utilizing earthworms.

Harvesting is done by first removing barriers placed above these piles. This enables earthworms to move upwards from the existing (substrate-lean) pile to another (substrate-rich) pile. The leftover pile has mostly vermicast and some unconsumed material. This is then screened to separate undigested material for further processing, and the vermicast is packed and sold. Even though it is operated in batches, it can be classified as a semi-continuous fed system. These are less labour-intensive than windrows and need less space.

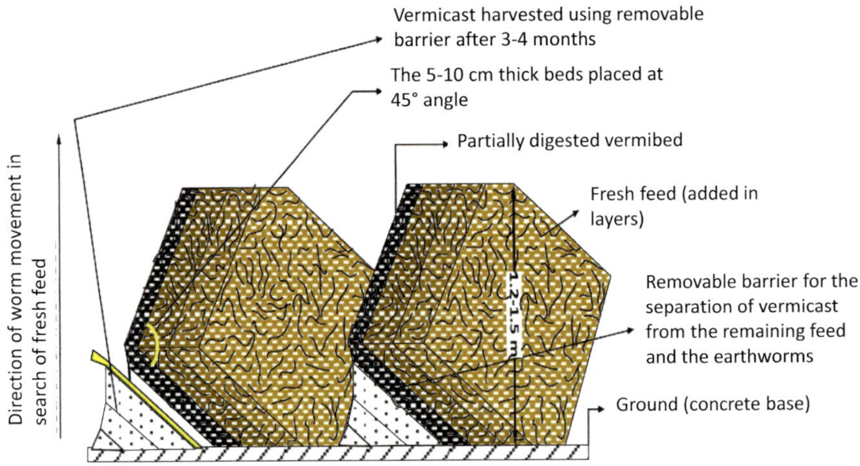

FIGURE 5.9 The Wedges Vermicomposting System.

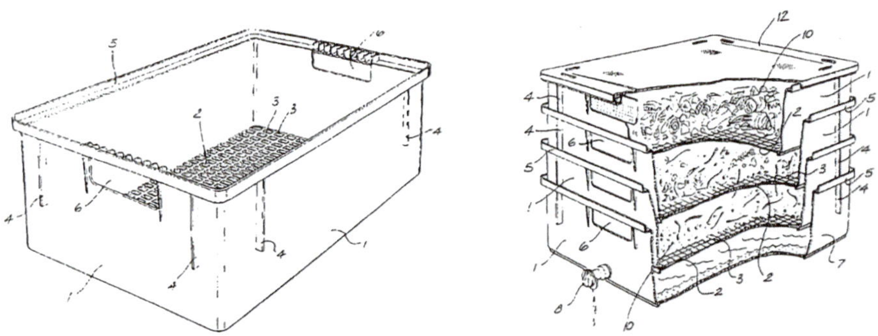

FIGURE 5.10 A method and apparatus for the treatment of organic matter. 1. Stack of containers; 2. a perforated base; 3. plurality of perforations; 4. projecting portions; 5. rim portions; 6. handles; 7. the base container; 8. liquid run-off tap; 10. organic matter; 12. Lid.

5.3.9 A Method and Apparatus for the Treatment of Organic Matter (Nattrass, 1994)

In this contraption, organic matter is treated in a stack of containers in which the lowermost container is given a solid base and has a liquid run-off tap. All the remaining containers have perforated bases to provide access to earthworms. Initially, the worms are placed in the next-to-lowest container in the stack, and after they have digested the organic matter in that container, they migrate upwards through the perforations in the base to continue feeding upon the substrate and also generating offspring. Eventually, the next-to-lowest container is removed from the stack, the worm casting is harvested, and the empty container is replaced at the top of the stack. The procedure is repeated, resulting in the cycle of steps involving the digestion of the organic material, the breeding of worms, and the collection of the liquid run-off (Figure 5.10).

5.3.10 Worm Reactor (Holocombe, 1995)

These (Figure 5.11) are modular units of dimensions $40 \times 2 \times 1\,m^3$. They are constructed under a greenhouse in order to protect them from winter rains. The working depth is kept at 0.7 m, while the remaining void provides adequate space for the feed. Organic material is shovelled into the gantry, which runs along the rails set on top of the bin side walls, as a thin layer. Earthworms move upwards as they consume feed from the bottom (substrate-lean) to top (substrate-rich) zones. Harvesting is carried out by meshes placed underneath the modular units. Those units are fitted with breaker bars or scrapers.

5.3.11 Worm Wigwam (Sustainable Agricultural Technologies, 1995)

In Worm Wigwam (Figure 5.12), organic material is mixed and added to the bedding along with 7–10 kg of earthworms. The movement of earthworms provides sufficient aeration inside these systems. After the bulk of the substrate is converted to the

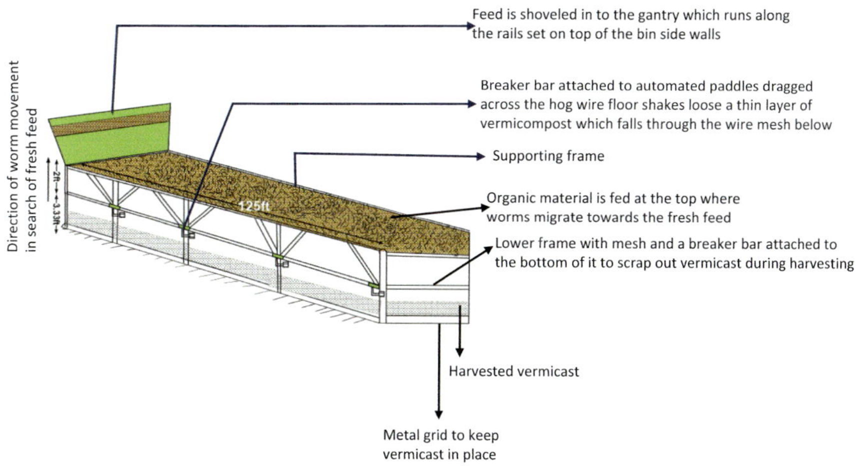

FIGURE 5.11 A two-dimensional representation of the Worm Reactor Systems.

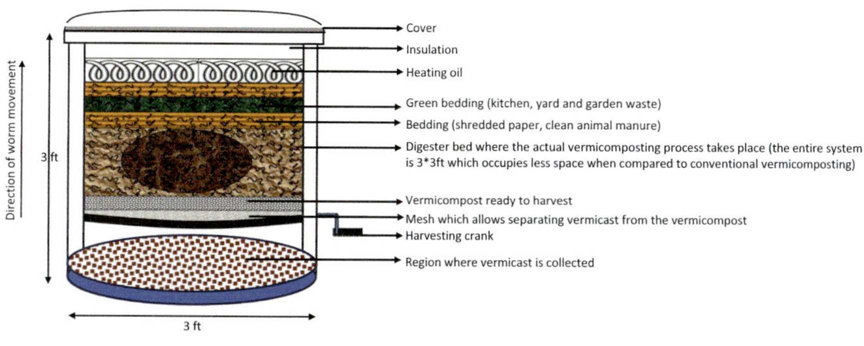

FIGURE 5.12 The Worm Wigwam.

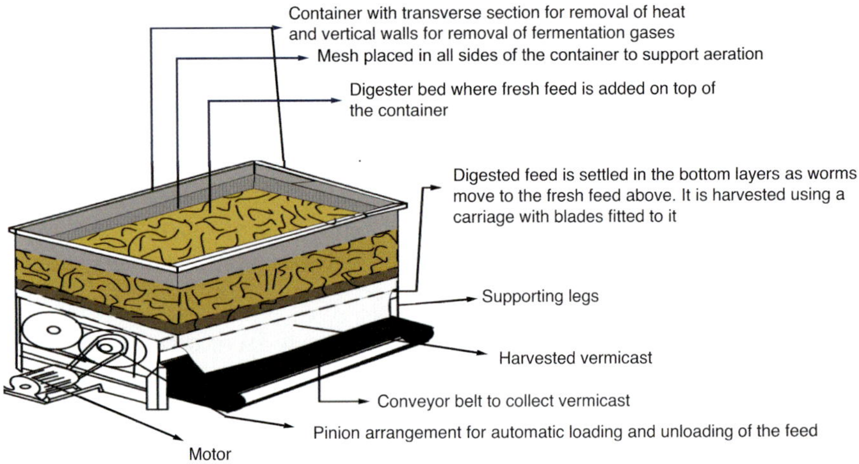

FIGURE 5.13 The Earthworm Composting Device.

vermicast, the cast is harvested using a mechanical crank, fitted to the mesh placed below the vermicomposting unit. A typical commercial-sale Worm Wigwam can process 35 kg of vermicast/week, up to 2 tonnes/year.

5.3.12 Earthworm Composting Device (Chareyre and Naturba, 1996)

In this device (Figure 5.13), loading and unloading are done at a predetermined frequency. Containers are provided with transverse and vertical walls, which facilitate retention of heat and removal of fermentation gas, respectively. A mesh provides scope for the aeration of the digester beds. A rack and pinion arrangement helps in loading the organic material mixed with earthworms at the top of the container. Harvesting is done by a carriage fitted with blades. A conveyor belt receives the harvested vermicast. Hence, the system needs less manpower compared to conventional systems, even as the commissioning and operation of the conveyor belt increases the cost of the system considerably.

5.3.13 Earthworm Breeding Plant (Facchi and Ferdinando, 1998)

The Earthworm Breeding Plant (Figure 5.14) has multiple frames to receive the organic material. Each frame is protected in order to avoid infiltration of leachate. A micro-slotted drainage pipeline is provided, which travels the whole length of the digester bed, finally draining into the leachate collection tray. The beds are covered with a green roof in order to protect the earthworms from the winter cold. A gravel bed is placed under the digester bed. Probes are attached for the control of temperature, moisture, and humidity in the bedding. The organic material is mixed and added to the bedding along with the earthworms. Once the bed is fully digested, the vermicast and the worms are harvested manually.

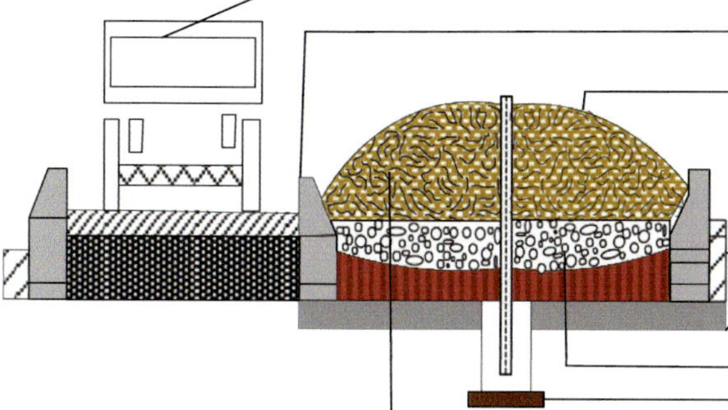

FIGURE 5.14 The Earthworm Breeding Plant.

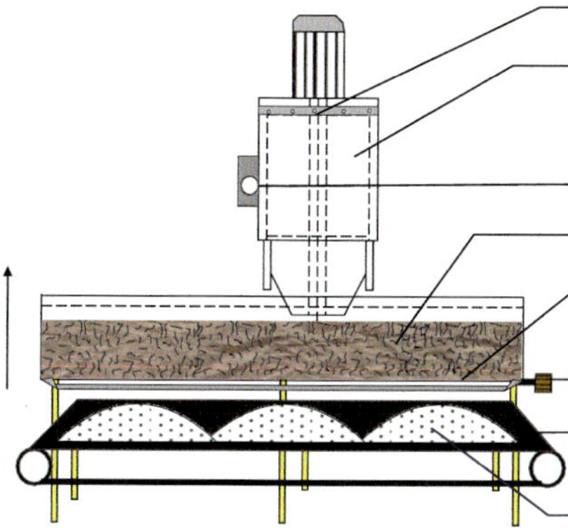

FIGURE 5.15 The Waste Material Processing Apparatus.

5.3.14 Waste Material Processing Apparatus (Taylor, 1998)

In this apparatus (Figure 5.15), organic material is mixed with earthworms and loaded by a hopper moving sidewise along the digester bed. Moisture content is maintained using a misting system, which is attached to the hopper. Heaters and ventilators help in maintaining temperature and aeration along the vermibed. The lower end of the vermibed has two rows of struts which move relative to each other to facilitate harvesting. The harvested vermicast is separated by the actuation of the retrieval mechanism. Vermicast falling on the conveyor belt is collected manually and sent for storage and packing.

5.3.15 Vertically Stacked Vermihut and Horizontally Stacked Vermihut (Rodgers, 2000)

These contraptions (Figures 5.16 and 5.17) can be designed either vertically or horizontally according to the user's needs. These are usually stackable, and each stack can be removed easily.

Organic material is loaded onto the wooden box along with the earthworms. Once the earthworms complete feeding on one tray, they migrate towards the fresh feed.

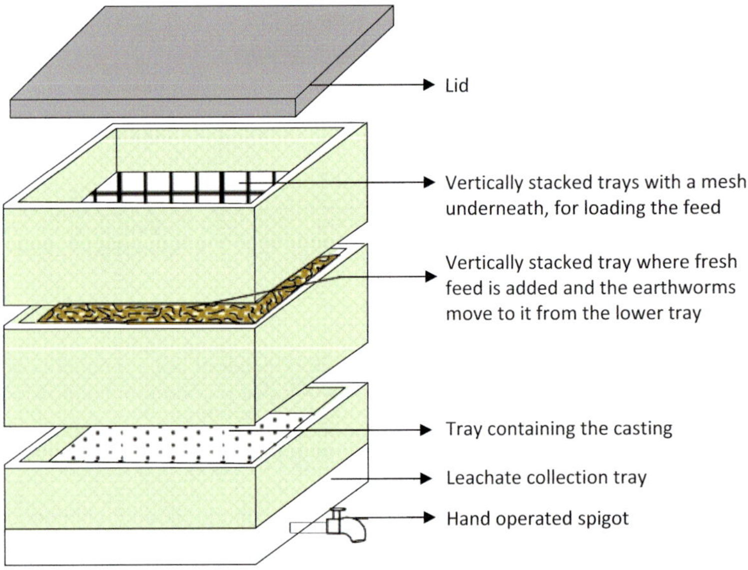

FIGURE 5.16 The Vertically Stacked Vermihut.

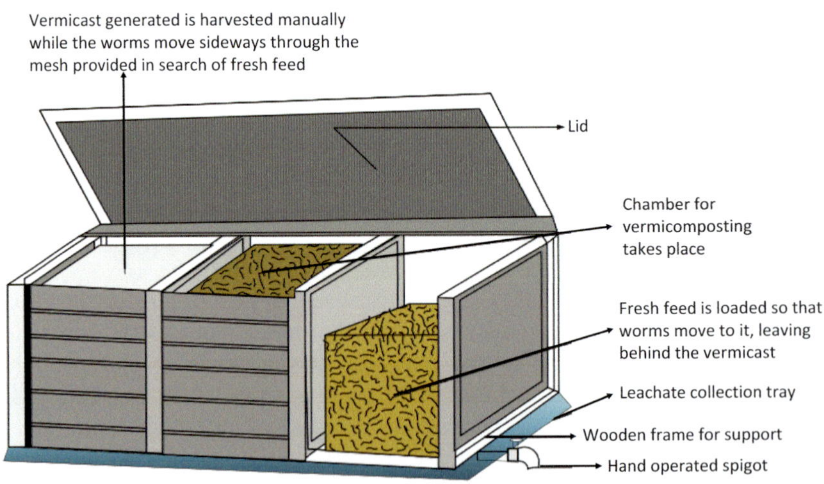

FIGURE 5.17 The Horizontally Stacked Vermihut.

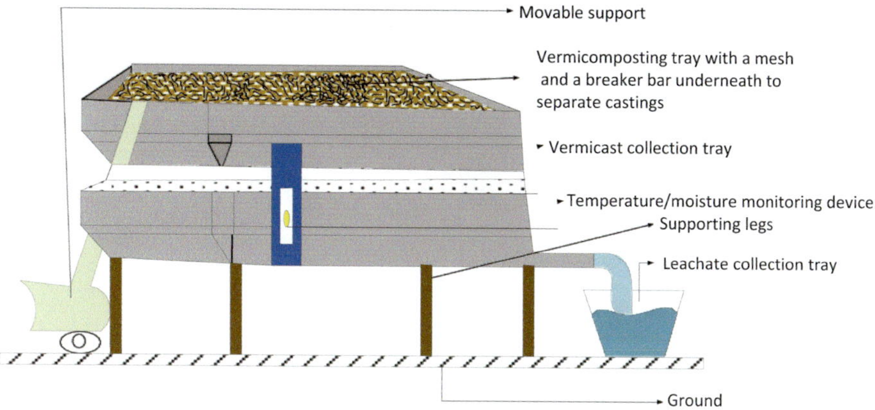

FIGURE 5.18 The Biolane^{TM}.

Harvesting is done by removing the digested substrate. The leachate is collected at the bottom of the system and can be used as a foliar spray. The system can be deemed a semi-continuously fed vermireactor and can be used indoors as it controls odours to a great extent.

5.3.16 Biolane^{TM} (Kingmoor, 2001)

Biolane^{TM} (Figure 5.18) is claimed to be the only automated and scalable in-vessel vermicomposting system. In it, vermicomposting is carried out in stackable trays onto which organic material is loaded along with earthworms. The trays are fixed with mesh with a breaker bar underneath to support harvesting the vermicast. Temperature and moisture are controlled. Leachate is collected at the bottom of the system, which is fitted with wheels for easy movement.

5.3.17 Biosafe^{TM} (Kingmoor, 2001)

Biosafe^{TM} (Figure 5.19) is a small-volume, in-vessel vermicomposting system. The system is provided with stackable drawers ($0.7 \times 1 \times 1.25\,m^3$) where earthworm and organic material can be loaded, and the entire system is enclosed with a double-layered recyclable plastic material. A lock is provided to keep off pests in a manner that does not interfere with the aeration. Each drawer is fitted with a mesh and a scraper bar for easy harvesting. A leachate collection tray is also provided at the bottom. The system carries arrangements for maintaining warmth in case ambient temperatures drop below the tolerable levels for the earthworms.

5.3.18 Worm Gin (Windle, 2001)

In "Worm Gin" (Figure 5.20), a conveyor belt acts as the vermibed, with the top of each belt providing the base for the subsequent beds. Two stacks of such vermibeds are placed back to back, and organic material is loaded in the middle. In the

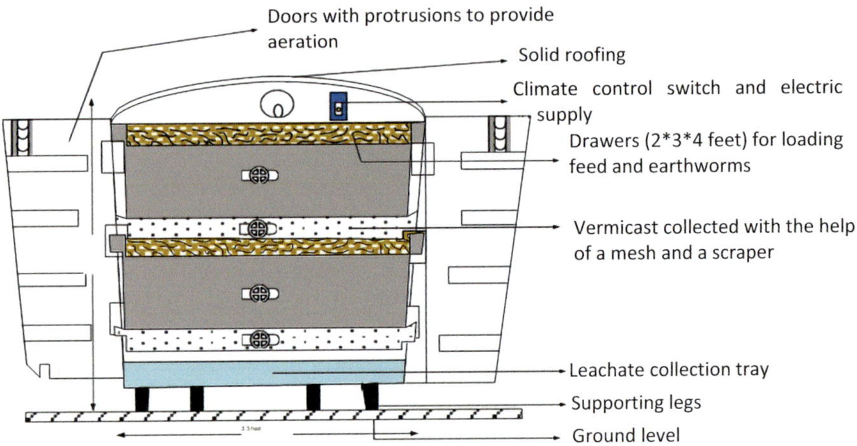

FIGURE 5.19 The Biosafe™.

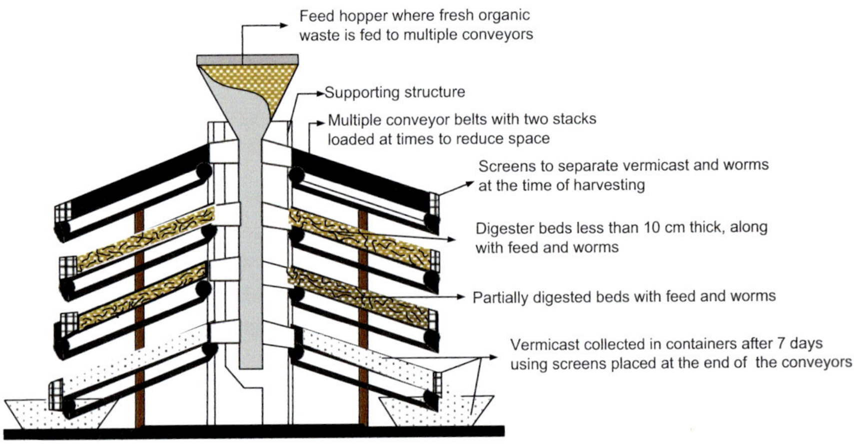

FIGURE 5.20 The Worm Gin.

marketed Worm Gin model, the feed loading rate is 30 tonnes/day. As the beds move automatically, the material falls off the ends and is conveyed to a screen. Vermicast is removed, and the leftover is mixed with the fresh feedstock. As the vermibed is vertically stacked, the system occupies less space per unit feed processed than horizontally stacked vermibeds.

5.3.19 High Efficiency Vermiculture Process and Apparatus (Windle, 2001)

This system (Figure 5.21) uses a hopper to load organic material along a moving conveyor belt. As the earthworms consume the substrate, they move towards fresh

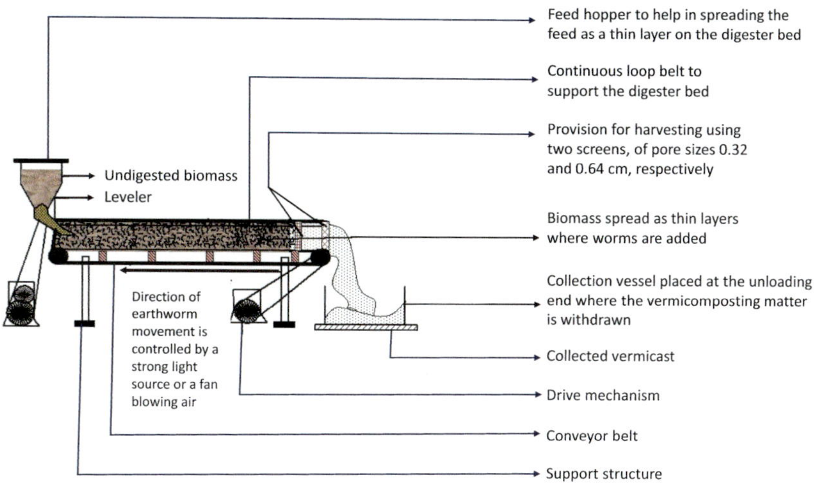

FIGURE 5.21 The High Efficiency Vermiculture Apparatus.

substrate. The conveyor belt movement is in the direction opposite to that of the earthworm movement. A strong light source also forces the earthworms to move away from the harvesting end towards the freshly added feed. The vermicast crumbles and falls down by gravity as it reaches the unloading point. Harvesting is done by placing two screens of 0.32 and 0.64 cm^2 mesh sizes at the unloading point. The material passing through the 0.32 cm^2 mesh, along with earthworms, is returned to the undigested biomass. This helps to maintain worm mass and supports the worm population. The system can be scaled up by arranging parallel reactor lines with vertical stacks.

5.3.20 Vermiculture Composting Device (Thompson, 2003)

The device (Figure 5.22) includes a container with four rectangular sides, a base, and a removable, reversible cover. A sealable compost extraction opening is provided on the front side of the container near the base, while ventilation openings are positioned on the sides. One side of the cover and the front side of the container are painted black to absorb sunlight and heat, even as the opposing side is painted white to facilitate reflection of the sunlight. The waste is added through the top for worms and microflora to act upon it. The resulting vermicompost is removed through the opening adjacent to the base for recycling. The temperature in the compost is adjusted by increasing or decreasing the flux of sunlight falling over it. Due to the temperature gradient thus formed in the container, air is drawn through the ventilation openings and into the container. This provides aeration for the biodegrading waste.

5.3.21 Apparatus for Upgrading Waste into Fertilizer (Lepeltier and Pierre, 2004)

This apparatus (Figure 5.23) consists of three containers placed one above another. Each container is of $60 \times 40 \times 17.5\,cm^3$ dimensions. The organic material is mixed

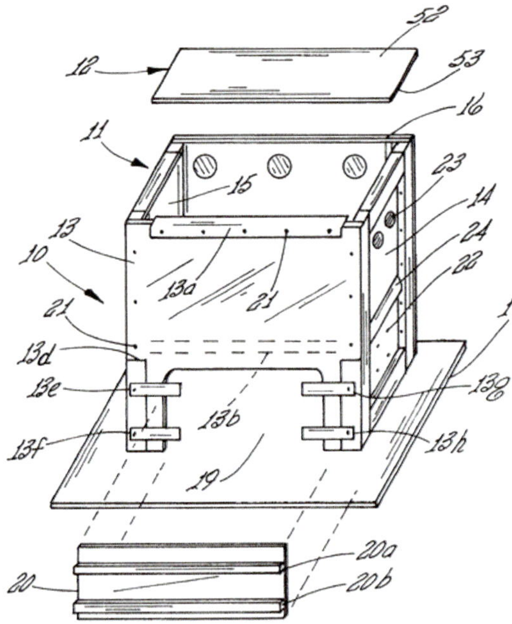

FIGURE 5.22 Vermiculture composting device. 10 - composting device; 11 - rectangular container; 12 - removable cover; 13 - front panel ; 13a - galvanized steel angle iron; 13b - 2 inch reinforcing stringer; 13d - spacers; 13e, 13f, 13g, 13h - pivoting locking bars; 14 - right side panel; 15 - left side panel; 16 - rear panel; 17 - optional base; 19 - extraction opening; 20 - removable extraction door; 20a, 20b - two reinforcing stringers; 21 - galvanized steel bolts; 22 - ventilation orifices; 23 - ventilation apertures; 24 - hot dipped galvanized steel bolts; 52 - top surface; 53 - bottom surface.

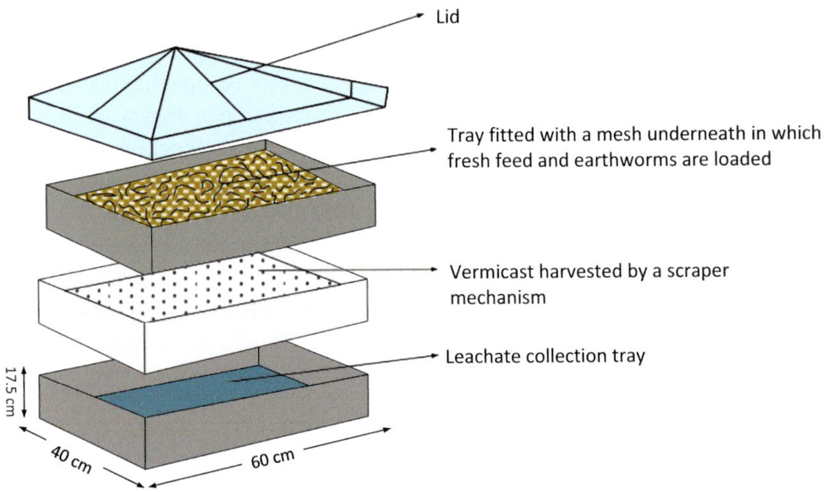

FIGURE 5.23 An Apparatus for Upgrading Waste Into Fertilizer.

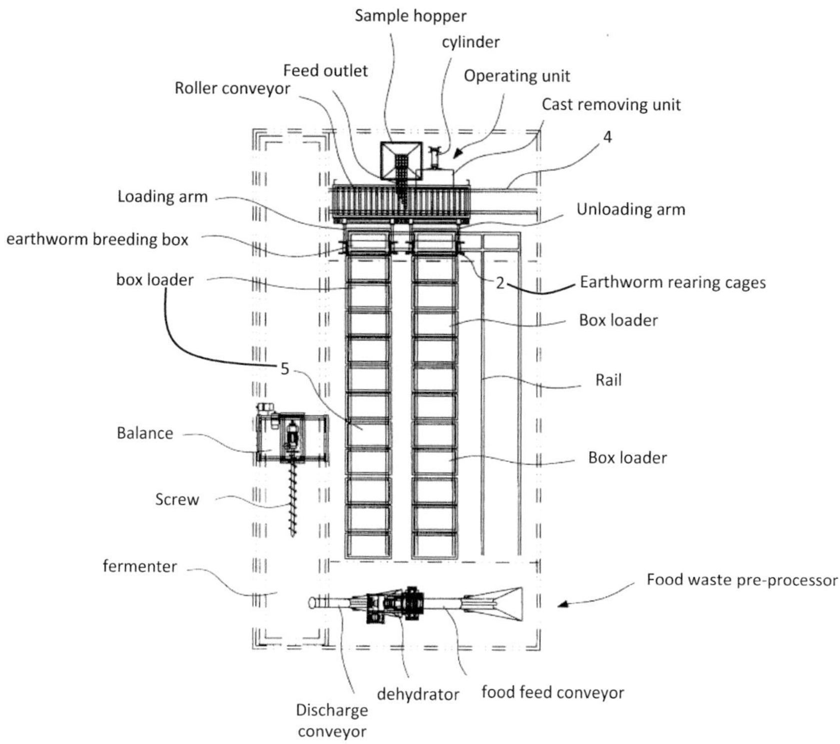

FIGURE 5.24 Multi-layered Vermicomposting System.

and loaded onto the upper tray along with the earthworms. The system is closed and left undisturbed until the organic material is converted to vermicast. Harvesting is done by the scraper mechanism; the vermicast falls onto the second tray, and the leachate is collected in the third. The system is completely opaque and doesn't facilitate gas exchange from the surroundings.

5.3.22 Multi-Layered Vermicomposting System (Hwan, 2005)

The system (Figure 5.24) consists of an operating unit to which multiple earthworm rearing cages are attached, along with a food waste pre-processor. A box loader moves on a rail and, besides loading the waste, assists in removing the vermicast.

5.3.23 An Apparatus for Organic Waste Vermicomposting (Neklyudov et al.), 2006

It is a two-compartment system (Figure 5.25) in which the left compartment is used to keep the waste for a few days to enable some natural composting. The slightly stabilized waste is then sent to the right-hand compartment, where earthworms convert it into vermicompost.

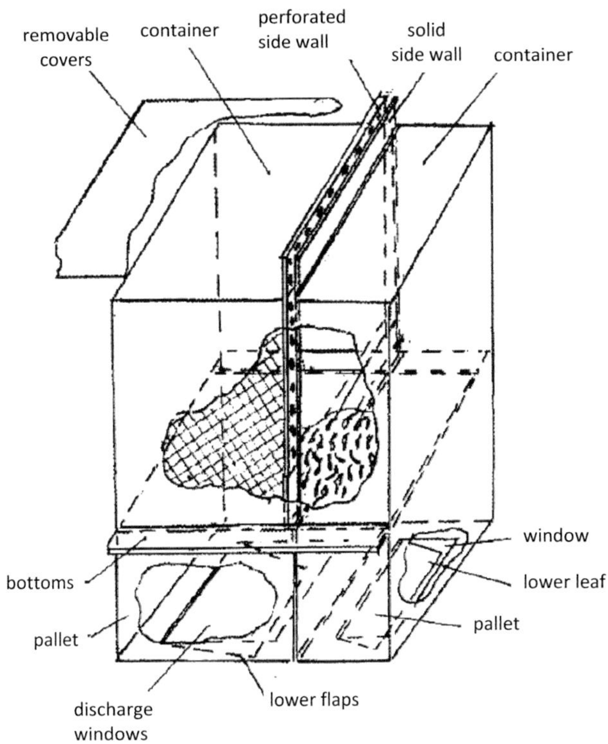

FIGURE 5.25 An apparatus for organic waste vermicomposting.

5.3.24 Drawer Type Vermireactor (Gitt, 2006)

This system (Figure 5.26) has composting drawers with apertures at their bottom. Organic material is mixed and loaded in the drawer along with earthworms. The system is left undisturbed until the organic material gets converted into vermicast. Harvesting is done by the agitator placed below each drawer. Trays are stacked one upon another in order to maximize space use.

5.3.25 Method and Device for Processing Organic Waste for Vermicomposting (Couet-Lannes, 2008)

It consists of an assembly that includes three or more containers providing a loop-type horizontal processing line. Each container has a lid with ventilation holes, and two perforated vertical walls allow the worms to move from one container to the next. There is a provision on the inner side of each perforated wall for inserting a plate to close off the wall. Additionally, there is a perforated bottom for liquid flow and a liquid collecting tray.

The first container is filled with waste to which earthworms are added. The following containers are filled in turn. After vermicomposting, the worms migrate

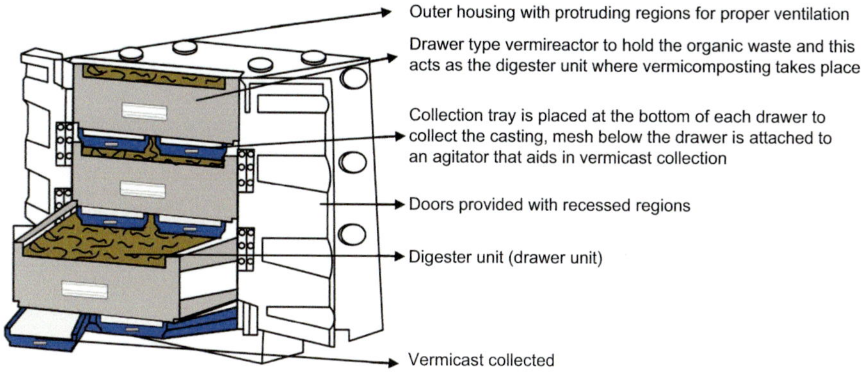

FIGURE 5.26 The drawer type vermireactor.

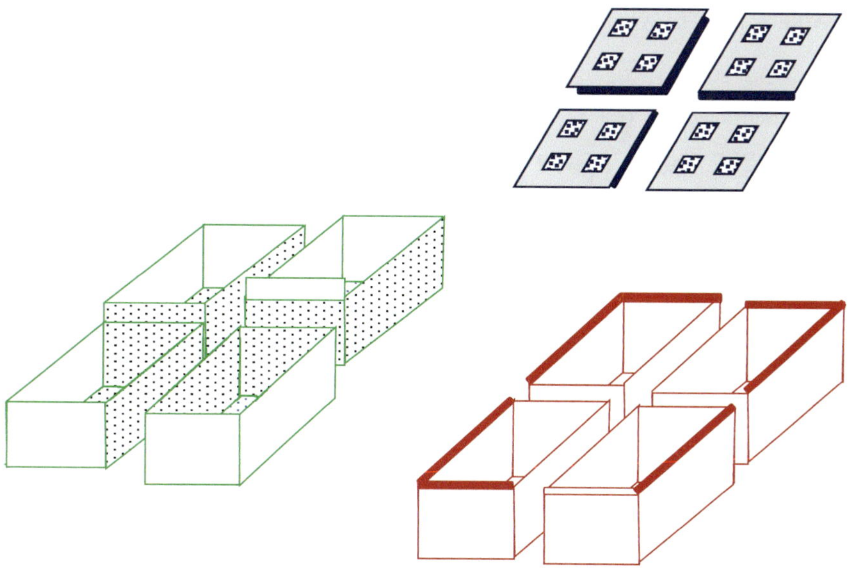

FIGURE 5.27 Method and device for processing organic waste for vermicomposting.

through the vertical perforated walls to the next container. The previous container, now full of vermicast, is emptied and repositioned in the process line. The containers are so arranged that the following container becomes at a given moment the initial container, which is again filled with waste. The authors recommend *Dendrodrilus rubidus* as the species most suitable to their design (Figure 5.27).

5.3.26 Worm Bin (Oregon, 2010)

A Worm Bin (Figure 5.28) is a rectangular or square box of wood or plastic. It has holes to drain excess water. Organic material is mixed and loaded in the bin along

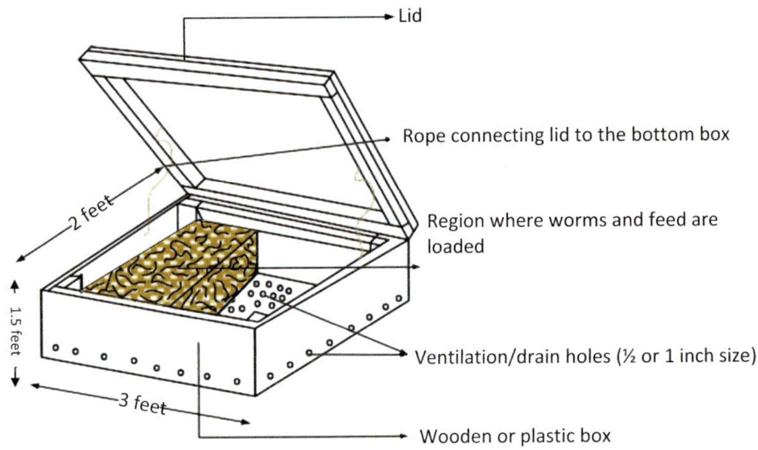

FIGURE 5.28 The Worm Bin.

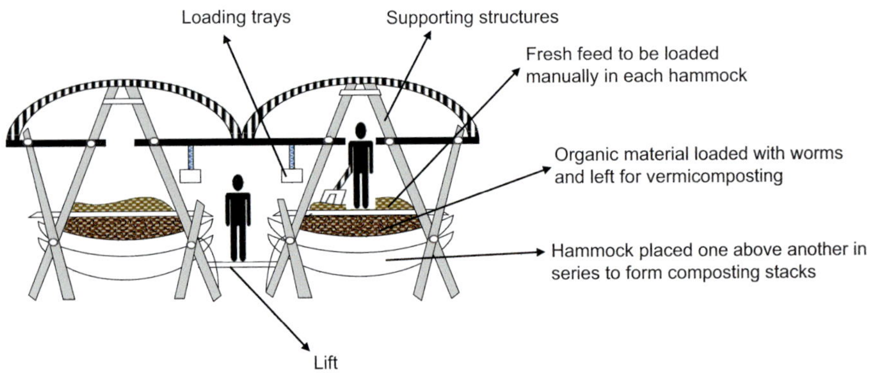

FIGURE 5.29 The Apparatus Used for Processing Organic Waste.

with earthworms. The bin is then left undisturbed until the organic material is converted to vermicast. Harvesting is done manually or by using hand sieves. The lid protects the bin from insects and other pests. The bins can be kept indoors or outdoors, as convenient and feasible.

5.3.27 Apparatus Used for Processing Organic Waste (Jaques et al., 2010)

This apparatus (Figure 5.29) consists of a supporting structure with a series of receptacles called hammocks that can hold organic material and worms together. These hammocks are slung in series, attached to the supporting frame. Organic material is loaded manually on the hammock along with the earthworms. The system is left undisturbed until the organic material is converted to vermicast. Each hammock is then unloaded using gravity chains/lift. Vermicast is harvested manually.

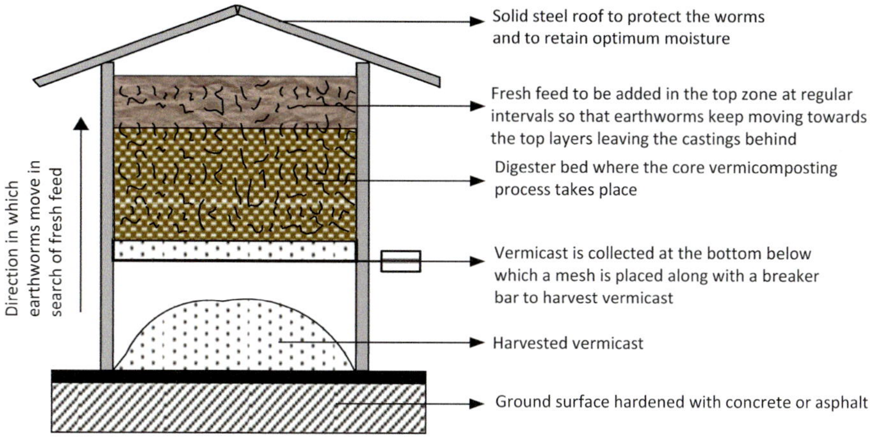

FIGURE 5.30 The Vermitech 200.

5.3.28 Vermitech 200 (Edwards, 2011)

In this system (Figure 5.30), the digester bed (<3 m width), which is mostly rectangular in shape, is placed well above the ground level. A steel roof protects the worms and retains moisture. The surface is laid with concrete or asphalt to prevent the leachate from entering into the ground. Organic material is mixed and loaded in the raised beds along with the earthworms. The beds are left undisturbed until the organic material is converted to vermicast. Earthworms move upwards towards fresh feed as the latter is added, leaving the vermicast at the bottom of the bed. Harvesting is done by using the hydraulic breaker bar placed beneath the beds.

5.3.29 Multiple-Tray Vermicomposter (Rhoads and Anderson, 2011)

This is a multiple-tray system (Figure 5.31) with a base and a lid and can be of various shapes – square, circular, or polygonal. The typical reactor is square-shaped and has stacking trays with dimensions $0.362 \times 0.108 \times 0.175\ m^3$. An air gap of 6.35 mm is provided along the full length of the tray. A thermal siphon air flow is attached along with the light-tight air-permeable connections to assist in vermicast production. The organic material is mixed and loaded in each tray along with earthworms. The trays are left undisturbed until the organic material is converted to vermicast. A mesh is provided to enable the earthworms to move from substrate-lean to substrate-rich trays. Harvesting is done by simply removing the tray containing digested substrate, which consists mostly of vermicast and some undigested material.

5.3.30 Multiple-Tray Vermicomposter with Thermal Siphon Airflow (Rhoads & Anderson, 2011)

This is a multiple-tray system consisting of a base and a lid. In between, trays with freshly loaded food and bedding are stacked on top of older trays containing worms

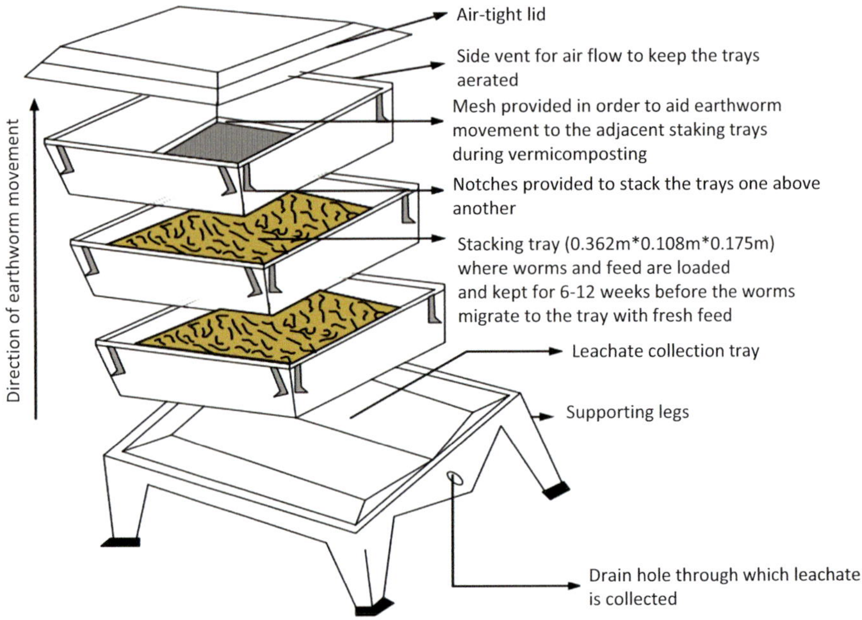

FIGURE 5.31 The Multiple-Tray Vermicomposter.

and their castings. The objective is to induce a continuous upward migration of the worms from substrate-lean to substrate-rich trays, thereby prompting continuous compost production. For this, compost trays are provided with permeable bottoms to allow worms to pass through. A thermal siphon airflow is provided to aerate the worms without the requirement of a power source. Light-tight but air-permeable connections are provided between the trays, lid, and base, to accelerate compost production by keeping the worms completely in the dark (Figure 5.32).

5.3.31 Digester Beds (Herlihy, 2012)

In this system (Figure 5.33), organic material is added at the top of the digester bed. It is then processed aerobically to remove excess moisture in a thermophilic regime lasting for 3 days. A second conditioning cycle, ranging from 7 to 14 days, is also done to the feedstock. Earthworms are then added to the feedstock, and the bed is left undisturbed for 40–45 days until the organic material is converted to vermicast. The temperature of the worm bed is kept in the mesophilic range during the vermicomposting process by providing adequate humidity. There are sprinklers to maintain adequate moisture in the bed.

5.3.32 Worm Culture Systems (Hughes, 2012)

This system consists of devices for the culture of worms and insect larvae (Figure 5.34) that are modular, light-weight, stackable, and transportable. They are 30–120 cm long, 30–120 cm wide, and 30–240 cm high, with an internal volume of 3.6 m^3. Organic

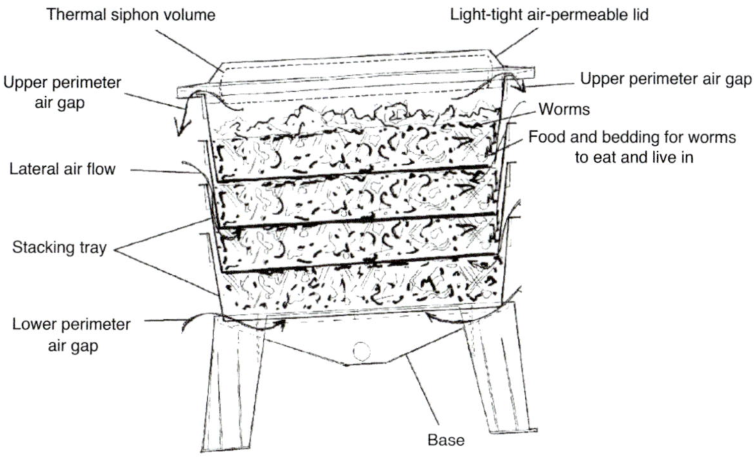

FIGURE 5.32 Multiple-tray vermicomposter with thermal siphon airflow.

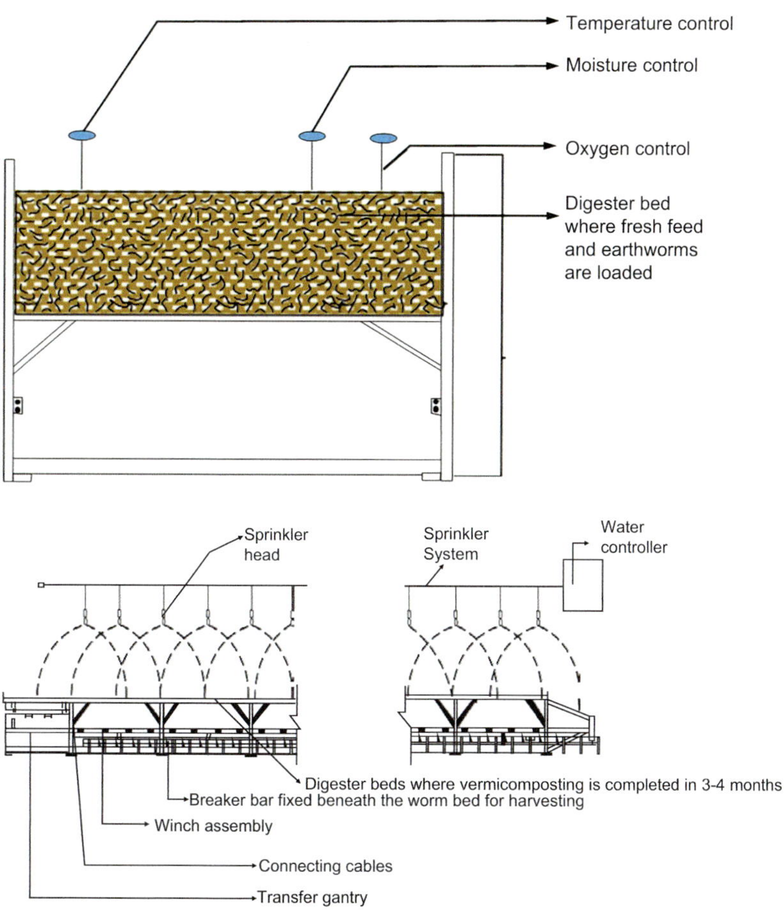

FIGURE 5.33 The digested bed system.

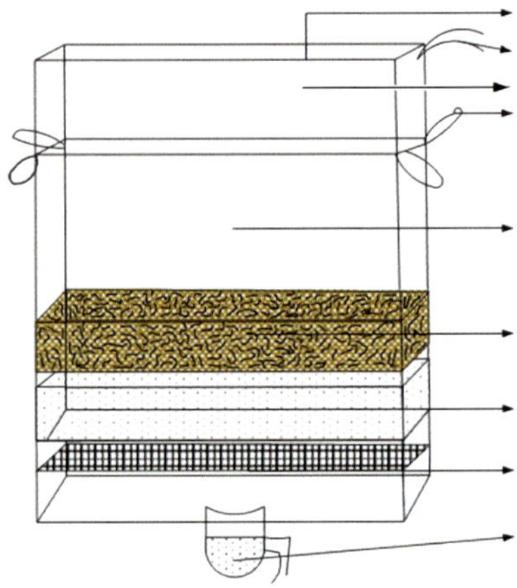

FIGURE 5.34 The Worm Culture System.

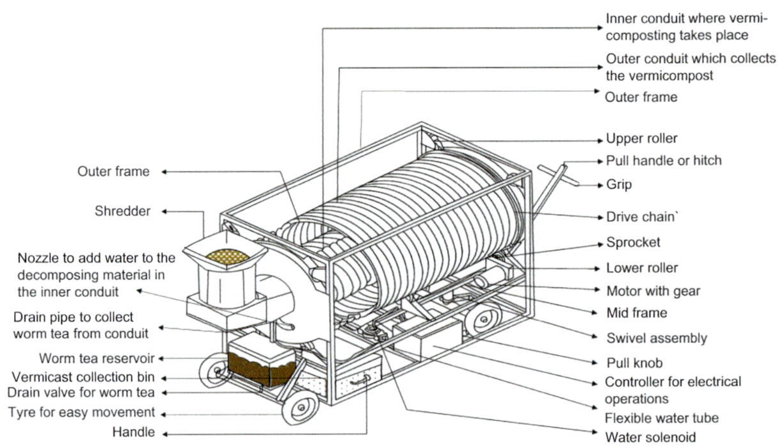

FIGURE 5.35 The Composting Apparatus and Method.

material is mixed and loaded onto the tray along with the earthworms. The entire setup is covered by a lid. The system is left undisturbed until the organic material is converted to vermicast. The product is removed by placing a mesh below the tray. It also allows the release of any liquid that is leached from the vermicast.

5.3.33 Composting Apparatus and Method (Charles et al., 2013)

This system (Figure 5.35) has inner and outer conduits of circular shape mounted on a common axis. Organic material is passed through a shredder and is loaded onto the inner conduit. Water and earthworms are then added. The vessels are rotated at a predetermined frequency, enabling worm castings and worm tea to be collected continuously. The system is automated and can be deemed a continuously fed vermireactor. The system is 150 cm long, 120 cm tall, and 750 cm wide. It can handle about 12 kg of waste per week.

5.3.34 Dual-Mode Vermicomposter (Pliva, 2013)

The dual-mode vermicomposter (Figures 5.36 and 5.37) has two modules, which are separated by a movable and a perforated wall and function in self-contained settings. There is a main control unit, which supplies water and regulates the temperature and humidity of the two modules. Organic material is mixed and loaded in module 1,

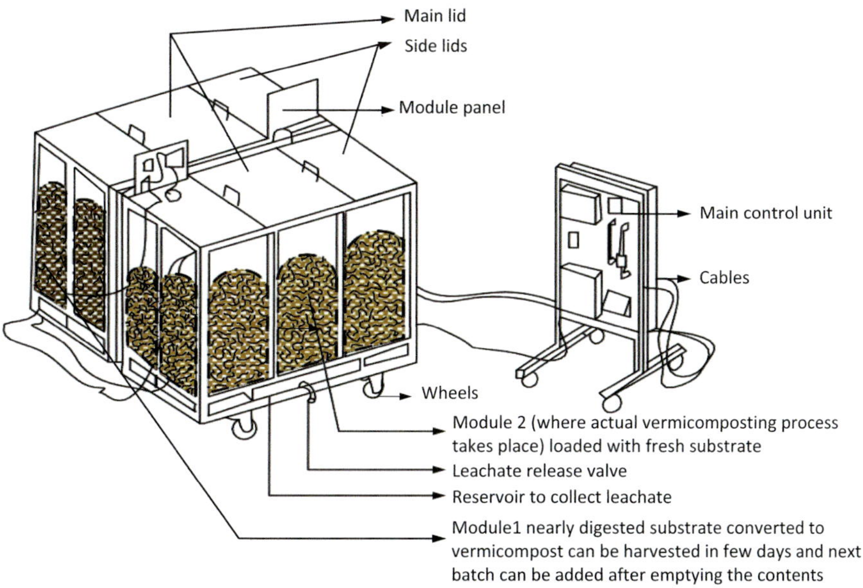

FIGURE 5.36 The Dual-mode Vermicomposter.

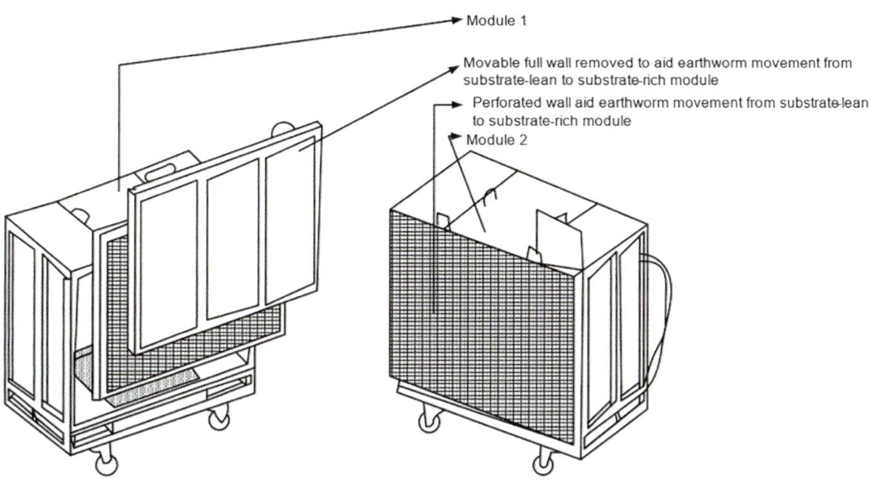

FIGURE 5.37 Components of the Dual-mode Vermicomposter.

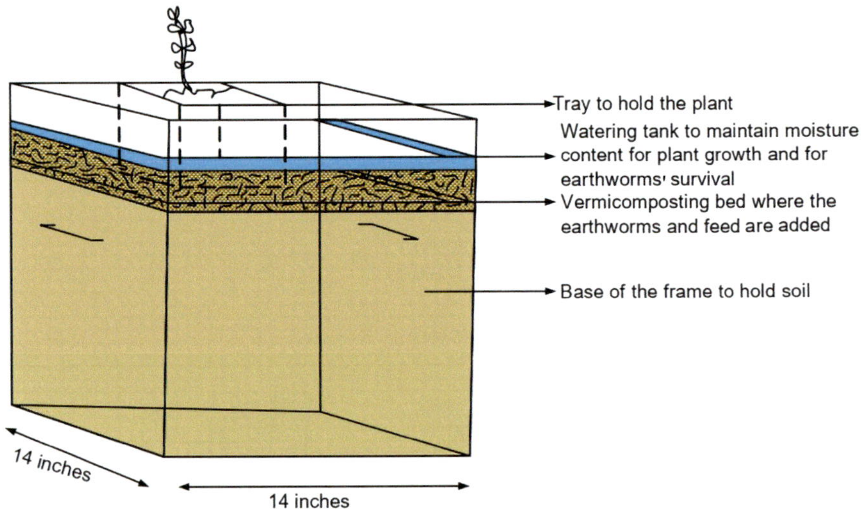

FIGURE 5.38 The Vermicomposting Growing System.

along with earthworms. Once the undigested substrate is converted to vermicast, the perforated and movable walls are removed, allowing the earthworms to migrate to module 4. After the earthworms have moved from the substrate-lean module 1, the walls are placed again, and module 1 is emptied. The vermicast is harvested and packed. Module 1 is reloaded to continue the process. The system is essentially batch-fed.

5.3.35 Vermicomposting Growing System (Almeida, 2014)

It is essentially a contraption that enables plants to be grown, which are directly fertilized by the earthworms incorporated within the contraption (Figure 5.38).

The system (Figure 5.30) has a planting unit of a typical area of 35–10 cm and is similar to a pot used for growing plants. The base of the unit is filled with soil and organic matter, such as vegetable and fruit waste, which are mixed and loaded onto the feeding tray placed above the base. Earthworms are released into this substrate. There is a water tank placed above the feeding tray to keep the system adequately moist. Since the vermicast that is generated directly provides nutrients to the plants, it is not harvested.

5.3.36 Vermicomposter Apparatus and Method (Berkson, 2014)

This contraption (Figure 5.39) consists of a chamber that holds the organic material and the earthworms. The system is left undisturbed until the organic material is converted to vermicast. Harvesting is done using an agitator, which is attached to the grids placed below the chamber. The vermicast falls onto the collection tray.

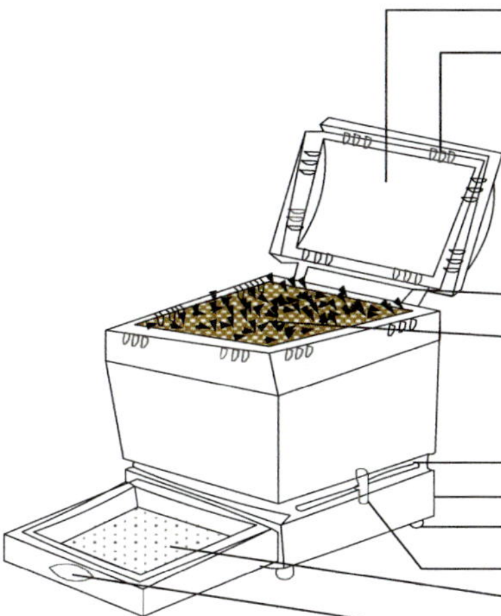

FIGURE 5.39 The Vermicomposter Apparatus and Method.

5.3.37 Vermicomposting Device (Branham, 2015)

In this device (Figure 5.40), organic material is filled in a system of hollow rings (Figure 5.32). Earthworms are introduced in the base ring. After they convert the waste into vermicast in the base ring, they migrate to the second ring. After converting the substrate in the second ring, they migrate to the third ring. In this manner, the substrate is vermicomposted all the way up to the top ring. The rings are fitted with a grid at their base, while the top is opened to maintain proper ventilation. A removable base is provided to collect the liquid byproduct leaching out of the rings.

5.3.38 Vermicomposting Apparatus and System (O'Donnell, 2015)

Vermicomposting Apparatus and System is specially designed to decompose kitchen waste with the use of earthworms inside a container (Figure 5.41). The apparatus is loaded with kitchen waste from the top, below which there is a layer of partially decomposed waste mixed with earthworms. The entire setup is placed upon a mesh, which removes excess moisture and also facilitates harvesting. The system is slightly sloped and contains vent pipes through which the excess water is drained into honeycomb filters and gravel beds. This enables filtering out large particles of decomposed waste. A spigot, which is attached to the bottom of the container, helps in collecting the effluent.

5.3.39 Vermicomposting Method and Apparatus (Toet, 2015)

Appropriate for regions of cold climate, this machine (Figure 5.42) has the provision of transferring heat from its decomposing manure to its worm growth zone.

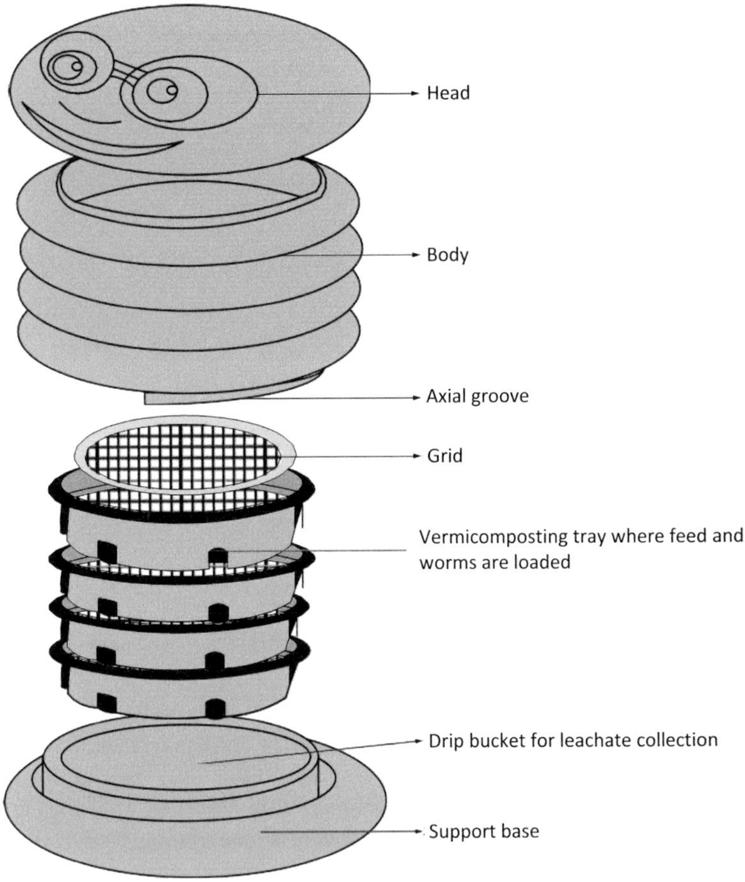

FIGURE 5.40 The Vermicomposting Device.

The worm growth zone is covered with organic material, especially animal manure, along with earthworms. The heat generated from the manure bed is diverted to the worm bed using radiant heat transfer or by air convection. The manure is harvested from the decomposing bed and is transferred to the worm bed. Proper controls of temperature and humidity are needed for efficient operation. The heat generated during composting digests the organic material to a certain extent, facilitating earthworms in generating vermicast.

5.3.40 Worm Composting Machine (Chambers et al., 2015)

In this machine (Figure 5.43), organic material is mixed and loaded in containers along with earthworms. The system is left undisturbed until the organic material is converted to vermicast. Harvesting is assisted by a moving carriage fitted with blades along the length of the containers. Earthworms migrate to the top of the containers, which are substrate-rich, leaving the vermicast below. This makes harvesting easy.

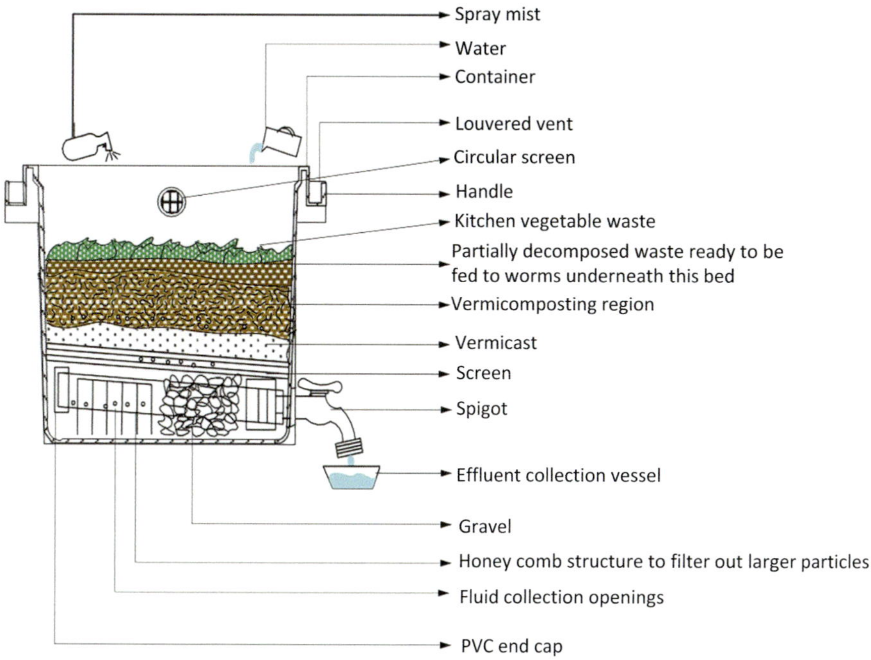

FIGURE 5.41 The Vermicomposting Apparatus.

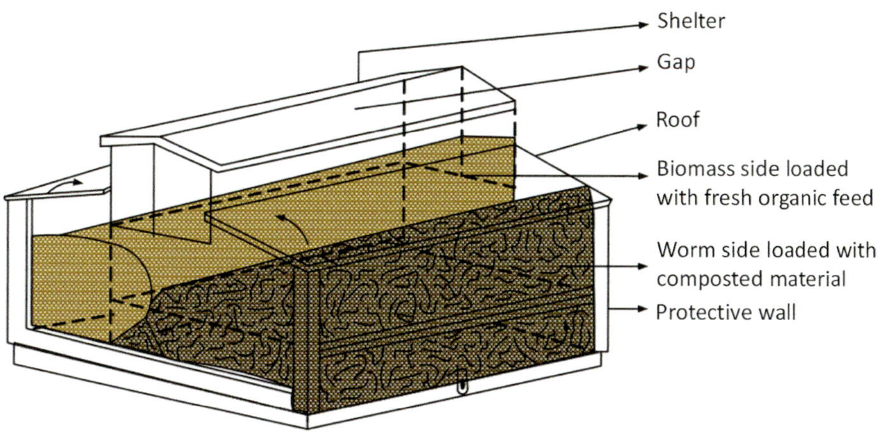

FIGURE 5.42 The Vermicomposting Method and Apparatus.

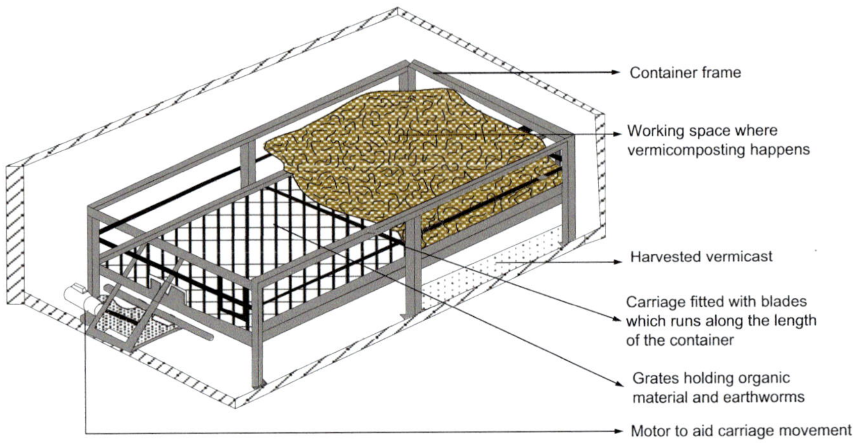

FIGURE 5.43 The Worm composting machine.

5.3.41 Method for Manufacturing Vermicompost and Device for Its Implementation (Spevak & Spevak, 2015)

This method (Figure 5.44) involves forming a ridge of the substrate and adding earthworms to it. An elaborate system of hopper, grinder, etc. is provided to separate and process the vermicompost that is generated.

5.3.42 A Housing for a Worm Farm (Mccarthy, 2016)

The housing in this system (Figures 5.45–5.47) is made of wood. Organic material is mixed and loaded into the housing along with earthworms. The material is left undisturbed till it is converted to vermicast. Harvesting is then done using a mesh/grid placed below the housing unit. The harvested vermicast falls and gets collected onto the strips and loops placed below. The strips are then opened, and the vermicast is collected in a tray.

5.3.43 Worm Harvesting Apparatus (Caesar, 2017)

This device consists of a system of perforated trays and drawers (Figures 5.48 and 5.49) to achieve vermicomposting alongside vermicast–earthworm separation. The device is simple but not efficient, needing 3 months to convert the waste into vermicast.

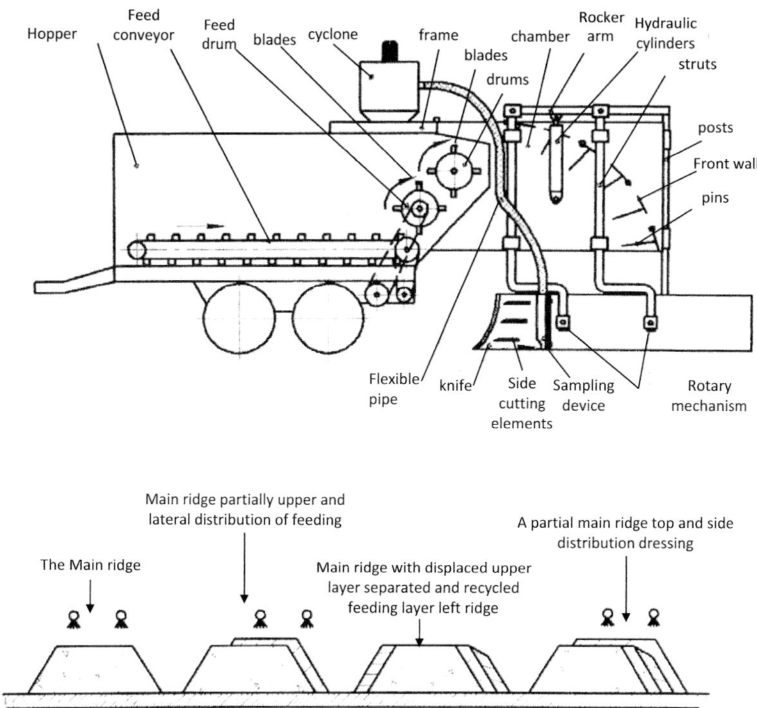

FIGURE 5.44 Method for manufacturing vermicompost and device for its implementation.

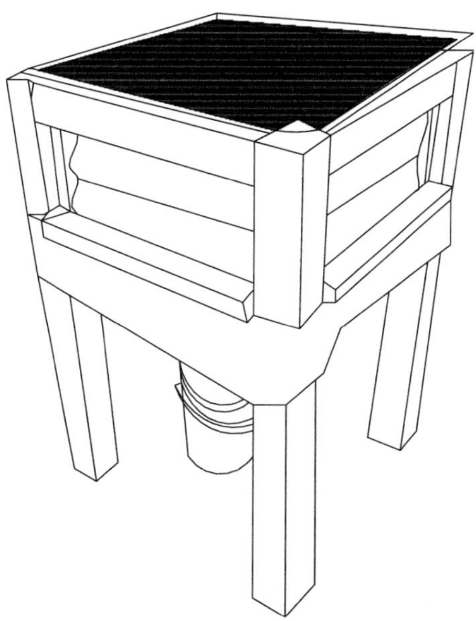

FIGURE 5.45 Front view of A Housing for a Worm Farm.

FIGURE 5.46 Side view of A Housing for a Worm Farm.

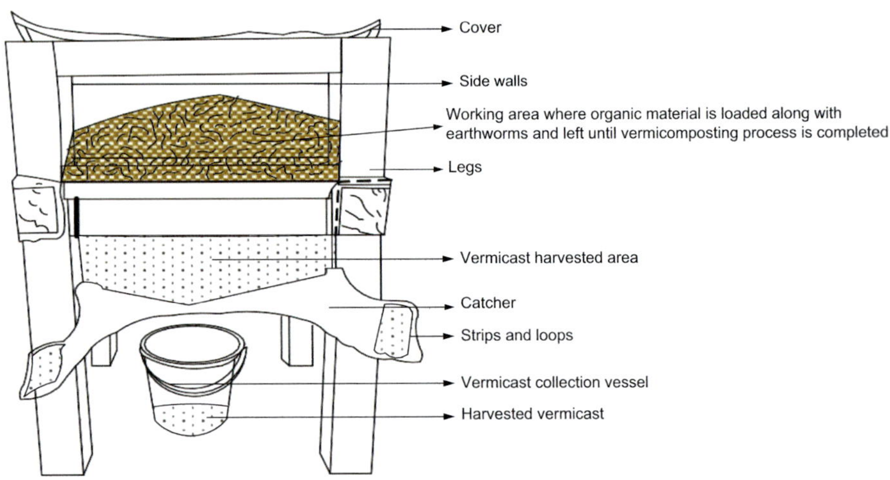

FIGURE 5.47 Side view of A Housing for a Worm Farm.

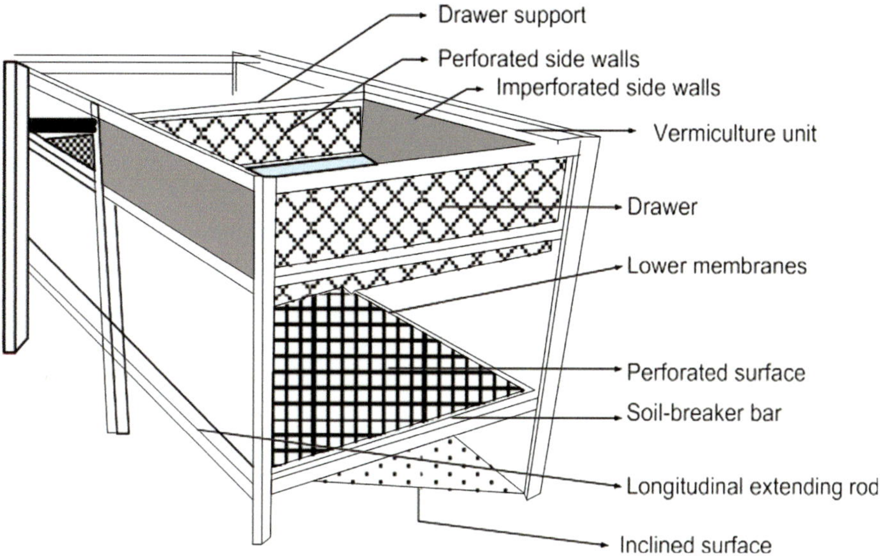

FIGURE 5.48 Individual module of the Worm Harvesting Apparatus.

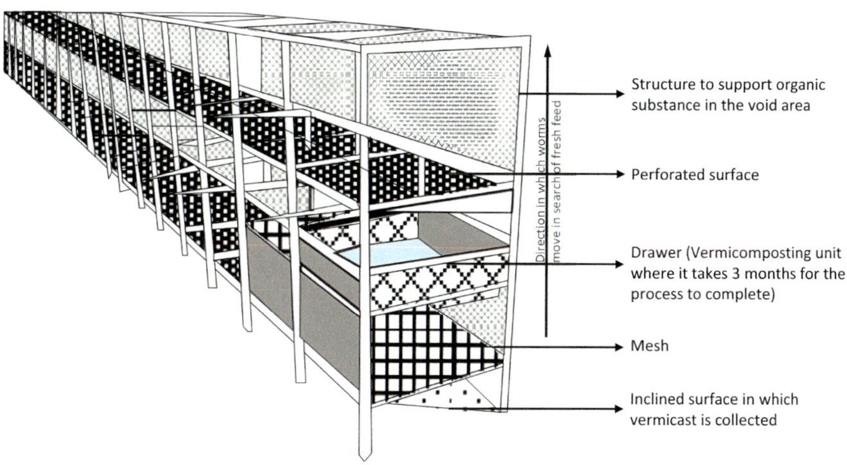

FIGURE 5.49 The multi-module Worm Harvesting Apparatus.

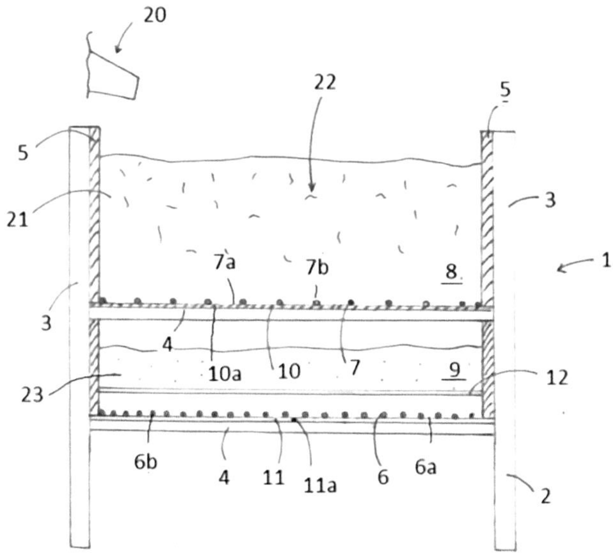

FIGURE 5.50 Consists of 1 - vermicomposting apparatus, 2 - structure, 3 - vertical supports, 4 - longitudinal supports, 5 - side walls, 6 - first grid, 7 - second grid, 8 - upper chamber, 9 - lower chamber, 10 - grid support, 11 - support, 12 - cutting means, 20 - feeding system, 21 - organic mixture, 22 - worms, 23 - humus, 10a - reinforcement, 11a - longitudinally extending reinforcement, 6a - holes, 6b - holes, 7a - recesses, 7b - holes.

5.3.44 Vermicomposting Apparatus (Romera, 2020)

The system details are as shown in Figure 5.50. The authors claim that the system is appropriate for industrial vermicomposting.

5.3.45 Vermiculture Bioreactor System and Method of Use (Price, 2020)

A portable "vermiculture bioreactor" has been reported, which is (a) climate-controlled, (b) continuous flow, and (c) computer-operated. It is claimed that conversion of organic waste to fertilizer is achieved through thermophilic composting and vermicomposting (Figure 5.51–5.54).

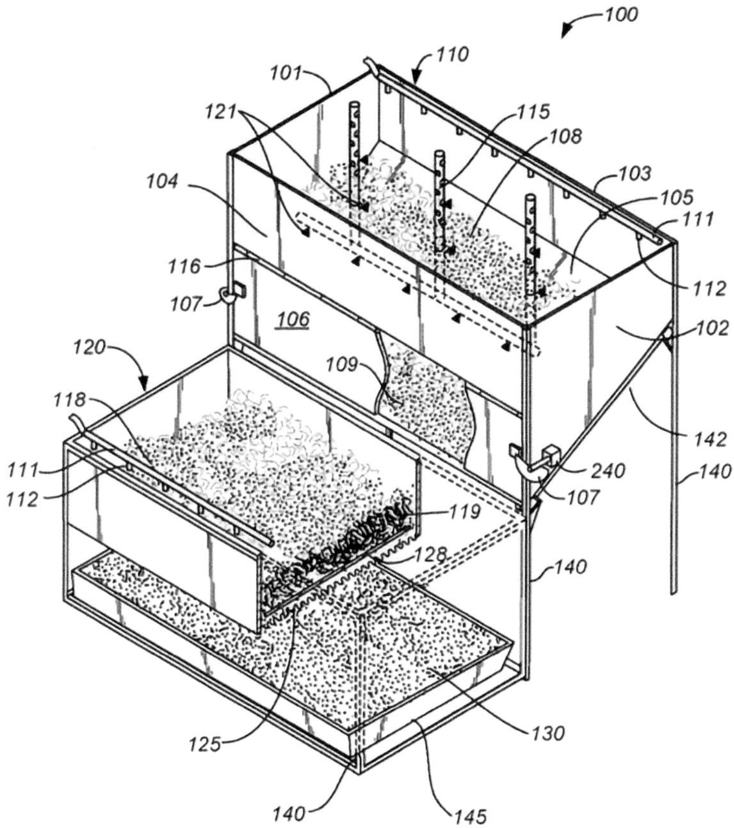

FIGURE 5.51 System details: 100 - vermiculture bioreactor, 101 - first end wall, 102 - second end wall, 103 - first side wall, 104 - second side wall, 105 - base, 106 - release door, 107 - auger-type locking mechanism, 108 - raw organic waste matter, 109 - partially decomposed organic waste matter, 110 - decomposing chamber, 111 - water piping, 112 - injectors, 115 - comprising aeration ducting, 116 - hinging mechanism, 118 - "cooled" thermophilically composted organic waste, 119 - annelids, 120 - vermicomposting chamber, 121 - plurality of sensors, 125 - porous bottom surface, 128 - cutting bar, 130 - biofertilizer, 140 - bioreactor frame legs, 142 - sloped bottom frame, 145 - removable receptacle, 240 - high-torque stepper motor.

5.3.46 Vermicompost Production Device (Anatolyevich et al., 2021)

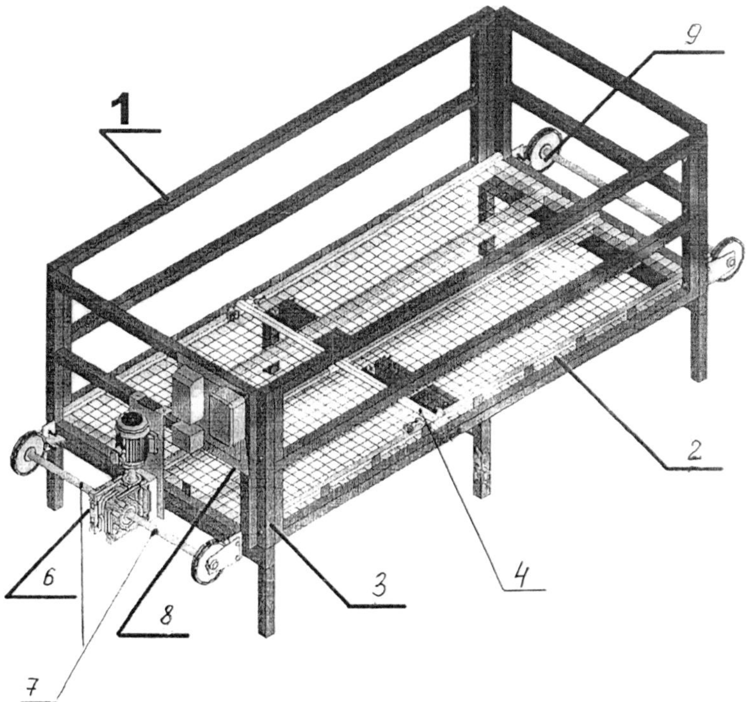

FIGURE 5.52 The main unit: 1 - section, 2 - the lower base, 3 - supports, 4 - knife made of welded blade, 6 - motor-reducer, 7 - two shafts, 8 - control device, 9 - the axle.

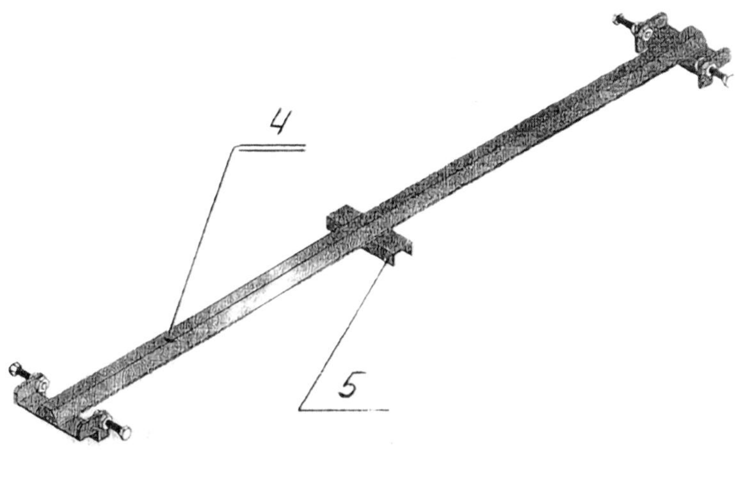

FIGURE 5.53 The knife part: 4 - knife made of welded blade, 5 - guides.

5.3.47 Installation for Complex Phytoremediation and Vermifiltration of Wastewater and Bottom Sediments (Leonidovic et al., 2022)

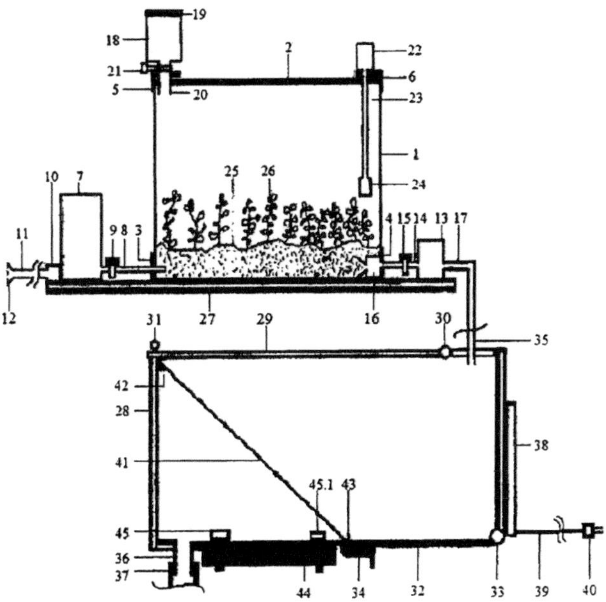

FIGURE 5.54 The phytoremediation cum vermifiltration system: 1 - aquarium; 2 - aquarium cover; 3, 4 - lower transverse corners; 5, 6 - upper transverse corners; 7 - supply electric pump; 8 - outlet pipe of the supply electric pump; 9 - outlet pipe valve of the supply electric pump; 10 - electric homogenizer of the supply electric pump; 11 - inlet pipe of the electric homogenizer of the supply electric pump; 12 - mesh filter of the inlet pipe of the electrohomogenizer; 13 - transmission electric pump; 14 - inlet pipe of the transmitting electric pump; 15 - inlet pipe valve of the transferring electric pump; 16 - electric homogenizer of the transfer electric pump; 17 - hose of the transferring electric pump; 18 - container for supplying fertilizers and suspensions of microorganisms; 19 - cover of the container for supplying fertilizers and suspensions of microorganisms; 20 - tube of the container for supplying fertilizers and suspensions of microorganisms; 21 - valve of the tank tube for supplying fertilizers and suspensions of microorganisms; 22 - aquarium aerator; 23 - aquarium aerator tube; 24 - aquarium aerator sprayer; 25 - bottom sewage sludge in the aquarium; 26 - algae planted in an aquarium; 27 - aquarium support platform; 28 - container for vermiculture; 29 - top cover of the container for vermiculture; 30 - hinged loop of the vermiculture tank cover; 31 - vermiculture container lid handle; 32 - bottom cover of the container for vermiculture for dumping vermicompost after treatment with vermiculture; 33 - hinged loop of the bottom cover of the container for vermiculture; 34 - latch of the bottom cover of the container for vermiculture; 35 - tube for supplying bottom sediments from the transferring electric pump to the vermiculture tank; 36 - tube for draining liquid from a container for vermiculture; 37 - hose for draining liquid from a container for vermiculture; 38 - electric heater tank for vermiculture; 39 - cord of the electric heater of the container for vermiculture; 40 - plug of the cord of the electric heater of the container for vermiculture; 41 - lamellar filter container for vermiculture; 42 - upper mounting of the lamellar filter; 43 - bottom mounting plate filter; 44 - support platform for vermiculture tank; 45, 45.1 - bolts of the support platform of the vermiculture tank.

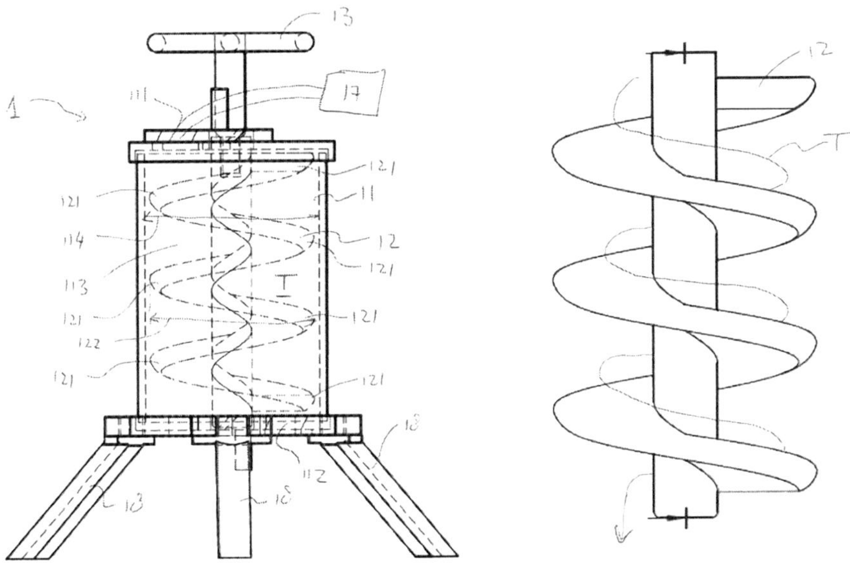

FIGURE 5.55 The system consists of 1 - a vermicomposting device, 11 - housing, 12 - screw, 13 - drive, 17 - pulverizer, 18 - legs, 111 - inlet, 112 - outlet, 113 - inner volume housing, 114 - inner diameter housing, 121 - screw blade, 122 - diameter screw, T - helical transport path.

5.3.48 Vermicomposting Device, Method for Producing Vermicompost, Compost (Marloes et al., 2022)

The specialty of this system is that it deploys a helical arrangement (Figure 5.55) for the movement of the substrate and the earthworms as the vermicomposting proceeds. It claims to solve many problems associated with tray-based systems.

5.3.49 Vermicompost Made of Aquatic Plants and Preparation Method and Use Thereof (Cui et al., 2024)

This system doesn't propose anything that has not been tried and discarded already, because even though the procedure described is technically feasible, it is totally unviable commercially. The reasons for this have been addressed in detail in the next chapter, Chapter 6.

REFERENCES

Almeida, H., 2012. Vermicomposting Growing System. U.S. Patent. Office. 13, 561876.

Anatolyevich, M.S., Vladimirovich, D.A., Valerievich, S.I. and Anatolyevich, D.R. 2021. Vermicompost production device, RU208013U1.

Appelhof, M., 1973. *Basement Worm Bins Produce Potting Soil and Reduce Garbage. Kalamazoo*, Michigan: Flowerfield Enterprises.

Berkson, R.L., 2014. Vermicomposter apparatus and method. U.S. Patent. Office. 8, 703477.

Branham, J., 2015. Vermicomposting device, US9067839B2.

Caesar, A., 2017. Worm harvesting apparatus. U.S. Patent. Office. 15, 309186.

Chambers, J.L., Craig, T.R., and Zappetini, D., 2015. Worm composting machine. U.S. Patent. Office. 9, 102574.

Chareyre, A.R., Naturba, I., 1996. Process for treatment by lumbricid-composting and device for its implementation. U.S. Patent. Office. 5, 527373.

Charles, E.M., Butterfield, D., Michael, B.M.M., and Erich, J., Hubenstricker, 2013. Composting apparatus and method. U.S. Patent. Office. 2, 263786.

Couet-Lannes, G., 2008. Method and device for processing organic waste by vermicomposting, WO2008129158A2.

Cui, J., Li, J., Yao, D., Cui, J., Wang, W., Chang, Y., and Liu, X., 2024. Vermicompost made of aquatic plants and preparation method and use thereof, US12049412B2.

Edwards, C.A., and Neuhauser, E.F., 1988. *Earthworms in Waste and Environmental Management*. The Hague, The Netherlands: SPB Academic Publishing, p.126.

Edwards, C.A., Norman, Q.A., and Sherman, R., 2011. *Vermiculture Technology, Earthworms Organic Waste and Environmental Management*. CRC Press, Boca Raton, USA, pp.220–231.

Facchi, S., and Ferdinando, B., 1998. Earthworm breeding. E.P. Patent. Office. 8, 853878.

Gaddie, R.E., 1976. Habitat for earthworm cultivation, US3961603A.

Gajalakshmi, S. and Abbasi, S.A., 2008. Solid waste management by composting: state of the art. *Critical Reviews in Environmental Science and Technology*, 38, pp.311–400.

Gitt, B., 2006. Composting apparatus and method, US7018831B2.

Herlihy, T.E., 2012. Organic waste treatment system utilizing vermicomposting. U.S. Patent. Office. 8,304227.

Holocombe, S., 1995. Holocombe earthworm reactor. https://www.worm-composting.ca/product/worm-wigwam/worm-wigwam-2/. accessed 17 February 1995.

Hughes, K.D. 2012. Worm culture systems. US20120214223A1.

Hwan, B.Y., 2005. Multi-storeyed vermicomposting system. KR100474138B1.

Jaques, S., Philippe, P.A., and Roger, F., 2010. Apparatus and method for processing organic waste. W.O. Intell. Prop. Office. 8, 131202.

Kingmoor, P., 2001. BiolaneTM. https://www.biosystemsbio.com/item/products/biolanetm.html. accessed 13 July 2001.

Leonidovich, K.Y., Mikhailovna, G.M., Viktorovich, K.A., Borisovich, K.A., Yurievich, T.M., and Iosifovich, S.D., 2022. Installation for complex phytoremediation and vermifiltration of wastewater and bottom sediments, RU2774704C1.

Lepeltier, L., Pierre, R., 2004. Apparatus and method for upgrading organic waste into natural fertilizer (or earthworm compost). W.O. Intell. Prop. Office. 6, 043879.

Marloes, P., Pieter, J.V., and Alexandre, C.R.M., 2022. Vermicomposting device, method for producing vermicompost, compost., NL2027676B1.

Mccarthy, T., 2016. A housing for a worm farm. W.O. Intell. Prop. Office. 5, 179662.

Nattrass, N. 1994. A method and apparatus for the treatment of organic matter. Butterfly's Pty Limited, WO1994019296A1.

Neklyudov, A.D., BaerAndrey, N.A., IvankinOleg, N., Mikhailov, I., and Leonov, A.Y., 2006. Device for vermicomposition of organic waste, RU51615U1.

O'donnell, L.P., 2015. Vermicomposting apparatus and system. U.S. Patent Office. 8, 951787.

Okada, T., 1978. Method of, and apparatus for producing fertilizer by utilizing earthworms, US4108625A.

Oregon, S., 2010, Wormbin. https://www.attra.ncat.org/events/wormbin.html. accessed 24 January 2010.

Patnaik, P., Ghazala, D., Abbasi, S.A., Tabassum-Abbasi, and Abbasi, T. 2022. Social behav ior o f earthworms in the context of their use in bioremediation. In: Siddiqui N., Tauseef S., Abbasi S., Dobhal, R., and Kansal, A. (eds) *Advances in Sustainable Development.* Springer Nature Singapore Pte Ltd. Springer, Singapore. Chapter 1, p 1–14. https://doi.org/10.1007/978-981-16-4400-9_1.

Pliva, P., Cejka, Z., and Hanc, A., 2013. Dual-mode vermicomposter. E.P. Patent. Office. 2, 604589.

Price, G., 2020. Vermiculture bioreactor system and method of use, US10577289B2.

Rhoads, R., and Anderson, G., 2011. Multiple tray vermicomposter with thermal siphon air-flow, US7998728B2. https://patents.google.com/patent/US7998728B2/en?q=(Vermicomposting+Apparatus+and+System)&oq=Vermicomposting+Apparatus+and+System.

Rodgers, M., 2000. Vermihut. https://www.vermitek.com/vermihut.html. accessed 17 Marrch 2000.

Romera, Q.P. 2020. Vermicomposting apparatus, ES 2 754 823 B2.

Spevak, N.V., and Spevak, V.Y., 2015. Method of manufacturing vermicompost and device for its implementation, RU2546893C1.

Sustainable Agricultural Technologies, Inc., 1995. Vermicomposting Bin. https://oldwormwigwam.composttea.com/international_worm_wigwam_manufacturer.php. Last accessed July 2025.

Taylor, Simon, E. and Gentle, Mark, L.E., 1998. Conversion of waste material using earthworms. E.P. Patent. Office. 3, 887328.

Thompson, J.M., 2003. Vermiculture composting device, US6576462B2.

Toet, G., 2015. Vermicomposting method and apparatus. U.S. Patent. Office. 13, 677301.

Windle, H.N., 2001. High efficiency vermiculture process and apparatus. US6223687B1.

6 The Shortcomings of the Conventional Vermicomposting Technology, which Necessitated Development of the High-Rate Vermicomposting Paradigm and the Associated Technology

6.1 INTRODUCTION: THE KEY CHALLENGES IN THE PATH TO RATIONAL AND OPTIMAL DESIGN OF VERMICOMPOSTING SYSTEMS

In preceding chapters, especially in Chapter 3, we have discussed the various features that make vermicomposting distinct from other forms of composting. When reflecting upon these features, certain questions emerged in the minds of these authors. Subsequent attempts to seek answers to the questions led us towards developing the paradigm of high-rate vermicomposting.

The biggest question was: Why do vermicomposting systems take 8–12 weeks or longer to deliver vermicompost of a substrate when the earthworms processing the substrate take under 18 hours to convert the substrate into vermicast?

The next biggest question was: Why is human-controlled vermicomposting at the commercial level largely confined to animal manure? In nature, too, earthworms process animal manure and convert it into a much better fertilizer in the form of vermicast. But their main role in nature has been the processing of leaf litter and other forms of fresh or dried plant debris (Figures 6.1 and 6.2). The quantities of phytomass

DOI: 10.1201/9780429331886-8

FIGURE 6.1 An earthworm eating through a leaf. (https://www.inaturalist.org/observations/249681008.)

earthworms processed in a certain space and time are several times more than those of the animal droppings they process. The resulting product, vermicast, not only fertilizes the soil but also makes the soil's physical properties more plant-friendly in the same manner as manure-based vermicompost has been known to achieve in nature. But till as late as the second decade of the present millennium, no commercially viable vermicomposting systems were in existence that were based on phytomass as feedstock. This appeared to be a very major knowledge gap to us, considering the fact that every year billions of tonnes of phytomass is generated across the world, which has no gainful use at present (Abbasi & Abbasi, 2010; Patnaik et al., 2024). This aspect has been discussed in detail later in this chapter and in subsequent chapters.

Other key questions were:

- What is the end product we seek in a vermicomposting system? In other words, what exactly is "vermicompost"?
- How to ensure that moisture is more or less uniform along the depth of the substrate? In conventional vermireactors, which are typically 0.5–1.5 m deep, most of the moisture percolates to the lowest strata, about 0.3 m deep, leaving the upper layers of the substrate either very low on moisture or almost dry.

FIGURE 6.2 An earthworm feeding upon grass and weeds. (https://www.inaturalist.org/observations/250700200.)

- How to ensure that the borrowing, tunnelling, and food-ward movements of earthworms keep the substrate adequately agitated and aerated, preventing the formation of anaerobic pockets? Such pockets are common in conventional vermireactors and are formed in regions in which earthworms do not move about. When anaerobic pockets are formed, there is a release of mercaptans and hydrogen sulphide (H_2S) along with methane. Even though mercaptans and H_2S are only in traces, their concentration is high enough to generate a foul smell and cause earthworm mortality.
- How to maximize earthworm-feed contact, minimize the time needed for the earthworms to search for food, and facilitate deposition of the vermicast?
- Above all, how to minimize the time taken between charging a vermireactor and harvesting the product (the vermicast) from it? In other words, how to enhance the rate of vermicomposting in terms of mass of feed processed per unit time – this being the ultimate goal because, for a set quality of the product, the faster the process and the more space efficient the reactor, the more economical the process?

6.2 PHYTOMASS, PHYTOWASTE, AND THE IMPERATIVE NEED TO UTILIZE IT

The word "biomass" is often used to denote plant biomass, but it is not a precise usage. The term biomass covers all life-derived, or biologically generated, material of both plant and animal origins. The biomass of animal origin should ideally be referred to as "animal biomass" or some other term like "zoomass", which indicates material that has been a part of the animal body. In like manner, biomass of plant origin should ideally be denoted by a word, such as phytomass, that is specific to the body of the plants.

According to a recent estimate (Bar-On et al., 2018), terrestrial plants make up 450 gigatonnes of carbon equivalent (GtC) on the Earth's land, followed by 70 GtC of bacteria, 12 GtC of fungi, 7 GtC of archaea, and 4 GtC of protists (Figure 6.3). Against this 450 Gt of phytomass and another about 93 GtC of other biomass of non-animal origin, animals contribute a mere 2 GtC of zoomass. Of this fraction, the lion's share is taken by invertebrates, led by arthropods. Humans, livestock, wild birds, and wild mammals together make up a miniscule 0.17 GtC. In other words, the quantity of the biomass covering all taxa on the Earth is about 550 GtC. Of this, as much as 80% is contributed by plants, dominated by embryophytes (land plants), and 15% by bacteria. Other groups, representing fungi, archaea, protists, animals, and viruses – in that order – account for less than 10% of the terrestrial biomass. It also follows that the Earth's land harbours 1,000 times more phytomass than zoomass. The Earth's annual phytomass production is of the order of 56.4 billion tonnes (53.8%) of carbon per year on the land and 48.5 billion tonnes of C/yr on the oceans (Field et al., 1998). Livestock zoomass is a mere 4% of all the Earth's zoomass (Figure 6.4), and the quantity of manure that the livestock generate is insignificant in comparison to plant biomass.

FIGURE 6.3 Biomass contained in different life forms, in terms of carbon fixed in them.

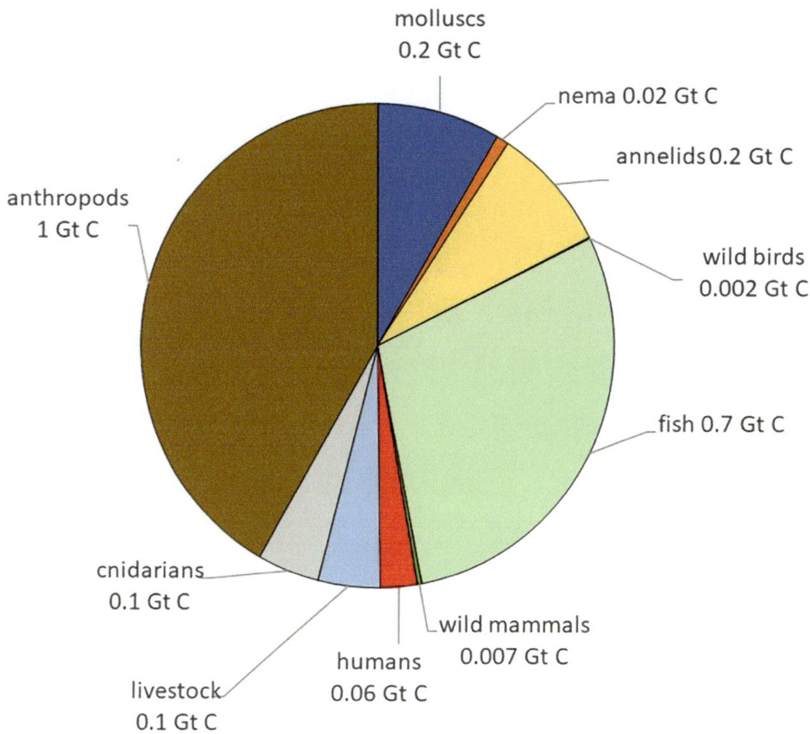

FIGURE 6.4 Contribution of different groups in the world's zoomass. Livestock make up for just 4%.

The statistics marshalled as above make it obvious that the zoomass contained in animal droppings is a minuscule – almost insignificant – fraction of the overall terrestrial biomass.

The phytomass available as possible vermireactor feedstock has two major streams: phytowaste and weeds. Phytowaste is mainly the waste leaf litter, other debris, senescenced plants, husk, stubble, and other forms of horticultural waste – which is generated in the course of agriculture. Weeds are those plants that have no recognized value but which proliferate to occupy areas meant for other purposes, often braving robust human resistance to their spread. Though not a "waste" in the true sense of the term, they are nonetheless a source of great harm.

Weeds encompass aquatic, amphibious, and terrestrial species covering diverse genera. Several weeds feature among the world's 100 most invasive plants/animals (Luque et al., 2014). Among the weeds are two of the most productive of all plants – the free-floating aquatic water hyacinth (*Eichhornia crassipes*) (Figure 6.5) and salvinia (*Salvinia molesta*) (Figure 6.6). Other highly productive species include aquatics like pistia (*Pistia stratiotes*) (Figure 6.7) and duckweed (*Lemna minor*) (Figure 6.8), amphibious species like ipomoea (*Ipomoea carnea*) (Figure 6.9), and many highly invasive and colonizing terrestrial species. A large number of the amphibious and terrestrial weeds have mammalian toxicity and/or negative allelopathy. Among the

FIGURE 6.5 Water hyacinth in bloom. (https://commons.wikimedia.org/wiki/File:Water_hyacinth,_Gyoda_Suijo_Park.JPG.)

FIGURE 6.6 Close-up of a typically dense mat of salvinia. (https://commons.wikimedia.org/wiki/File:Salvinia_natans_JdP.jpg.)

FIGURE 6.7 A close-up of pistia. (https://commons.wikimedia.org/wiki/File:Common_water_hyacinth_tree.jpg.)

FIGURE 6.8 A close-up of duckweeds. (https://commons.wikimedia.org/wiki/File:Eendekroos_dicht_bijeen.JPG.)

FIGURE 6.9 Ipomoea in bloom: Due to its pretty flowers, ipomoea was used as an ornamental before it became a serious pest. (https://commons.wikimedia.org/wiki/File:Ipomoea_aquatica_bd-2.jpg.)

most widespread of the weeds possessing these attributes are the ones we have studied extensively: ipomoea, lantana (Figure 6.10), parthenium (Figure 6.11), and prosopis (Figure 6.12). The last named also happens to be a xerophyte, which bestows upon it a resistance towards biodegradation not possessed by other types of macrophytes.

The invasiveness and colonizing ability, aided and abetted by high primary productivity, allelopathy, vegetative propagation and/or rich seed banks, hardiness, resilience, and great resistance towards chemical, biological, or mechanical control make these weeds proliferate all the time across most regions of the world (Abbasi, 1999; 2018; Abbasi et al., 2013, 2020; Tabassum-Abbasi et al., 2018; Patnaik et al., 2024). Figures 6.13–6.20 show how the weed bodies mentioned above colonize water bodies and land areas.

All through the last few centuries, especially the 20th century, very intensive efforts were made to eradicate or control the weeds by application of chemicals, biocontrol agents, and mechanical means (Tabassum-Abbasi et al., 2019). Under the integrated paste management (IPM) initiatives, combinations of the three approaches were also explored very intensively. But all the measures led to only localized success,

FIGURE 6.10 From up close, lantana can look deceptively pretty. (https://commons.wikimedia.org/wiki/File:Lantana_camara_05987.jpg.)

FIGURE 6.11 Parthenium. (https://commons.wikimedia.org/wiki/File:Parthenium_hysterophorus_2.jpg.)

FIGURE 6.12 A grown-up prosopis tree.

that too temporarily. Soon, either the same species staged a comeback, or another species of similar intrusiveness took over. The worldwide net loss of forests and associated eco-degradation that has occurred with increasing severity all through the 20th century – and is continuing at present as well – has made it progressively easier for the weeds to intrude and proliferate. The degraded landscapes do not have the biodiversity and the resilience that is possessed by natural forests to withstand the onslaught of invasive plants. The increasing eutrophication of rivers and lakes across the world – especially the developing world – has contributed to very widespread infestation of water-courses by aquatic weeds, especially *E. crassipes*, *S. molesta*, and *P. stratiotes*.

FIGURE 6.13 A lake covered by water hyacinth so comprehensively that water has been put out of sight. (https://commons.wikimedia.org/wiki/File:Pontederia_crassipes_310157893.jpg.)

FIGURE 6.14 Typical salvinia infestation, which covers the entire surface of the water body it invades. (https://commons.wikimedia.org/wiki/File:Salvinia_molesta_covering_lake.jpg.)

FIGURE 6.15 A pistia-infested pond covering all nooks and corners. (https://commons.wikimedia.org/wiki/File:Railsjf.JPG.)

FIGURE 6.16 A pond overwhelmed by duckweed invasion. (https://commons.wikimedia.org/wiki/File:Lemna_minor_15-p.bot-lemna.minor-4.jpg.)

FIGURE 6.17 An ipomoea thicket advancing into a pond. Due to its amphibious nature, ipomoea harms water bodies as much as it harms land areas. (https://www.inaturalist.org/observations/29112183.)

FIGURE 6.18 Heavy infestation of forest lands by lantana. (https://commons.wikimedia.org/wiki/File:Lantana_camara_IMG20240715131040.jpg.)

FIGURE 6.19 Parthenium infestation taking over the free land and weakening the hold of trees. (https://commons.wikimedia.org/wiki/File:Parthenium_infestation_in_Kanha_Tiger_Reserve,_India.jpg.)

In this manner, truly massive quantities of phytomass are being generated every year, which, at present, have almost nil commercial value. It is technically feasible to convert phytomass into solid fuel like woody chips, liquid fuels like ethanol and methanol, or gaseous fuels like methane and hydrogen. It is also possible, at the laboratory scale, to extract medicinals, cosmetics, and other chemicals from different weeds, or to use the weeds in the synthesis of nanoparticles, in pulp-making, and in generating household items like baskets, carry-bags, and decorative artefacts. But most options are not economically viable. The few sources of some useful chemicals, which are viable, have very low market demand. As a result, only a miniscule function of the total phytomass that is generated can be used via those options (Tabassum-Abbasi et al., 2019, 2023).

Coming to horticultural waste, also, there are massive streams. For instance, cultivation of dates (Figure 6.21) leads to an estimated 105 million tonnes of phytowaste per year. The stubble, which remains in the farmer's fields after grains like rice and wheat have been harvested, is another very major solid waste stream emanating from horticulture.

FIGURE 6.20 A typical prosopis thicket, with no other vegetation allowed in its midst.

For these wastes also, very intense efforts have been made to find such utilization options with which each of the waste streams can be utilized as it is generated. But no such option has emerged.

But the generation of phytowaste is but one side of the problem. An equally severe – perhaps bigger – problem is the adverse impact of this phytowaste. We will try to illustrate this with the example of salvinia, parthenium, and prosopis.

As earlier presented by us in Hussain et al. (2018), to quote,

> Giant salvinia (*Salvinia molesta*) is an aquatic fern native to southeastern Brazil and northern Argentina (Abbasi & Nipaney, 1986, 1993). In its native habitat, it has been existing in competition with other species of plants and performing its share of ecosystem services in the manner normal of other species. During the latter half of the previous century, it was taken out of its native habitat to various other regions as an aquarium plant. There, partly due to the absence of natural enemies and partly due to the attributes of ease of reproduction, high rate of growth, and hardiness, *S. molesta* became an invasive species. The rapid anthropogenic deterioration of the wetlands that began occurring from 1950s onwards, and which made it difficult for sensitive and delicate species of aquatic plants to survive, gave an opportunity for hardy species like *S. molesta* to aggressively invade new habitats and colonize them. *S. molesta* has proved to be a particularly dominant plant and has colonized wetlands of several regions to the exclusion of all but a very few other plant species.
>
> *Salvinia molesta* (D. S. Mitchell) is an exceedingly invasive and dominant aquatic weed, capable of multiplying and growing faster than most other known botanical species (Mitchell & Tur, 1975; Abbasi & Nipaney, 1986; Room et al., 1981). It reproduces

FIGURE 6.21 Date palm cultivation generates over 105 million tonnes of phytowaste per year. With countries like India, which have no history of date palm cultivation, now joining the fray due to innovations in date horticulture, the quantities of this waste are set to rise further.

vegetatively; a tiny bit of *S. molesta* leaf can lead to daughter plants which then multiply so rapidly that a bank-to-bank coverage of a water body by *S. molesta* can occur in a matter of a few weeks (Mitchell & Tur, 1975). Harley and Mitchell (1981) have recorded that at Lake Moondarra, Australia, *S. molesta* leaves doubled within 2.2 to 2.7 days in summer. Similarly rapid biomass doubling time was recorded at Lake Kariba, Zimbabwe, by Mitchell and Tur (1975). After spreading horizontally, *S. molesta* mats thicken vertically as the internodes become shorter and leaves longer. With it, the leaves are pushed upward and can get packed into mats up to 1 meter thick (Bhat, 2016). This enables *S. molesta* to attain biomass productivity of the order of 60 dry (ash-free) tonnes per hectare per year (Abbasi & Nipaney, 1993). This level of primary production puts *S. molesta* at par with water hyacinth, known to be the most productive of all plants (Abbasi & Nipaney, 1993; Crites et al., 2006).

A typical *S. molesta* invasion of a water body soon leads to the coverage of its entire water surface, elbowing out most other free-floating aquatic plants and cutting the sunlight off from the neuston and the benthos. This restricts the photosynthetic and aerial replenishment of oxygen in the water lying beneath the *S. molesta* carpet, seriously harming the water's chemical and biological quality. In turn, this devastates the biodiversity of the invaded wetland by letting none but a few very hardy species survive. Water quality beneath the mats is quickly degraded by decreases in dissolved oxygen and pH, and increases in CO_2 and H_2S concentrations, as the weed's debris decays aerobically or anaerobically (Mitchell, 1969). This has a cascading impact: Low pH and DO help in the dissolution of metals from the underlying sediments into water (Li et al., 2013). This, in turn, toxifies sensitive organisms.

As the plants die, their debris accumulates at the bottom and threatens fisheries by creating a shallow-water environment that restricts fish breeding (Sculthorpe, 1967). The decomposition of the organic matter further diminishes dissolved oxygen levels needed to support healthy populations of fish and other biota (Hattingh, 1961; Coates, 1982; Oliver, 1993). The matters are made worse by the propensity of *S. molesta* mats to serve as breeding grounds for vectors of human diseases, notably encephalitis, dengue, and malaria (Oliver, 1993); elephantiasis (Room et al., 1989); and schistosomiasis (Bennett, 1975).

In the large number of wetlands now colonized by *S. molesta* across the tropical and sub-tropical world (Room et al., 1981; Abbasi & Nipaney, 1993; Bhat, 2016), billions of tonnes of *S. molesta* biomass is generated every year. Since no means exists to utilize any significant portion of this biomass (Ganeshkumar et al., 2014; Bhat, 2016), it remains unharvested, leading to accelerated deterioration of the wetlands colonized by *S. molesta* (Abbasi & Nipaney, 1986; Abbasi, 1997). When *S. molesta* plants die, they decay aerobically and anaerobically — mostly the latter — generating global warming gases carbon dioxide and methane (Abbasi et al., 2011a). As most of the degradation of dead plants occurs anaerobically in the anoxic epilimnion of the wetlands, it becomes a major contributor of methane emissions. Given that methane has, molecule to molecule, 34 times greater global warming potential than carbon dioxide (Shindell et al., 2009), the harm caused by *S. molesta* decay is immense.

There have been instances wherein rapidly growing spreads of *S. molesta* have been halted and repelled by the use of biological agents (Room et al., 1981), but such successes have been few and far between. Chemical and mechanical methods have seen even lesser success in controlling *S. molesta*. Attempts to find ways to utilize the weed have not been successful either, and no economically viable process has resulted from the efforts to use *S. molesta* as compost, green manure, or mulch (Raju & Gangwar, 2004; Arthur et al., 2007; Dorahy et al., 2009), as fodder (King et al., 2004; Leterme et al., 2010), as a source of biogas (Abbasi & Nipaney, 1984, 1985, 1991; Abbasi et al., 1990, 1992a,b), or of drugs (Chantiratikul et al., 2009; Choudhary et al., 2008; Li et al., 2013a).

In Hussain et al (2017), we have described the impacts of pathenium in the words, to quote

Parthenium hysterophorus, commonly called as congress grass, is among the world's seven most noxious and devastating weeds (Patel, 2011). It is an annual flowering, erect and severally branched ubiquitous herb, which grows aggressively in a wide range of habitats. Due to the absence of any effective natural enemies, its allelopathic nature, large seed bank, and fast growth rate, parthenium grows luxuriantly all through the year, infesting millions of hectares of land masses including agricultural fields, parks, orchards, railway tracks, and other open areas (Wiesner et al., 2007; Nigatu et al., 2010; Qureshi et al., 2014). This proves disastrous in terms of monopolizing of space and nutrients by parthenium at the expense of other vegetation, consequent loss of biodiversity, and associated ecological imbalances (Hussain et al., 2016a,b). Parthenium's dominance over other vegetation is fostered by the presence in parthenium of allelopathic compounds, especially parthenin, hysterin, ambrosin, and flavonoids (Maishi et al., 1998; Khan & Abbasi, 1998; Knox et al., 2011; Patel, 2011; Kaur et al., 2014). These compounds are leached when dew or rain falls on parthenium (Abbasi & Abbasi, 2011) and reach the underlying soil. There, they cause toxicity, discouraging the growth of other vegetation in the vicinity of parthneium, thereby aiding and abetting the spread of parthenium monocultures (Hussain et al., 2016a,b).

Parthenium has mammalian toxicity as well. It causes dermatitis, eczema, asthma, allergic rhinitis, hay fever, black spots, burning, and blisters around the eyes in mammals, including humans (Gunaseelan, 1987; Towers & Rao, 1992; Maishi et al., 1998; Morin et al., 2009; Akhtar et al., 2010). Exposure to parthenium also causes systemic toxicity, including loss of skin pigmentation, dermatitis, diarrhoea, and degenerative changes in the liver and kidneys in livestock who accidentally graze upon parthenium (Gunaseelan, 1987; Rajkumar et al., 1988; Lakshmi & Srinivas, 2007).

The eradication of parthenium is a major challenge, primarily because of its epidemic proliferation, strong reproductive potential, hardiness, and competitiveness, apart from its wide ecological adoptability. Efforts made across the world to find persistent methods of controlling parthenium by way of mechanical, chemical, or biological means have at best achieved only partial and temporary success (Manoj, 2014; Hussain et al., 2016a). The weed has never been eradicated from any country, and its spread in all the tropical and sub-tropical regions of the world is only increasing with time. When viewed as a resource for the generation of green manure, biogas production, biopesticides, and drugs (Abbasi et al., 1990; Kishor et al., 2010; Kumar et al., 2012; Gunaseelan, 1987; Patel, 2011; Singh & Garg, 2014; Kumar et al., 2014; Anwar et al., 2015; Hussain et al., 2016b), parthenium has proved uneconomical, expensive, and unsustainable. Thus, finding an ecologically sound and economically viable means by which parthenium can be gainfully utilized in large quantities appears to be the only recourse which can make it profitable to regularly harvest the weed, thereby keeping it under some control.

In her doctoral thesis, Patnaik (2017) has observed, to quote

Among the invasive species of plants which are spreading at an alarming rate in the tropical and sub-tropical regions of the world, prosopis (*Prosopis juliflora*, Swartz, DC; henceforth referred as prosopis) is arguably the most hardy and resilient (Hierro et al., 2005; Pasicznik et al., 2004). A xerophyte, which is adapted to survive and thrive in very harsh desert environments, it is a major boon for impoverished people subsisting in those environments as it provides them with badly needed shelter, reduces erosion, improves micrometeorology, and is a source of food, feed, fuel, medicines, and cosmetics (Tewari et al., 2013). Attempts made during the 19th and the 20th centuries to exploit these virtues of prosopis have led to its introduction in many regions which had less-harsh climates and better availability of soil and water than existed in the native habitat of prosopis (Dubow, 2011; Walter, 2011). The relatively more favorable conditions resulted in prosopis growing and spreading much faster than it could be utilized. Soon it became a serious weed in all exotic regions where it was introduced, growing so explosively that it even enters roads and homes, driving humans out of their dwellings (Osmond, 2003; Muturi et al., 2009). The resulting damage to biodiversity is enormous (Patnaik et al., 2017).

In India, too, prosopis has been spreading — and doing so alarmingly. The situation has become so grim in the state of Tamil Nadu that a concerned citizen petitioned the Madras High Court a few months ago, resulting in an order directing the government of forthwith clear prosopis from all public spaces (Saravanan, 2017). Valiant efforts then began to uproot prosopis trees and the task appeared more and more daunting before another citizen went to the Hon'ble High Court, wisely submitting that prosopis will not be eradicated in this manner as its seeds will cause a re-infestation in no time. The plea was upheld, and a stay was ordered on the uprooting of prosopis trees (Saravanan, 2017).

The woody parts of prosopis — stem and main branches — are utilized as fuel either directly or after converting it into charcoal. Indeed, the prosopis charcoal is believed to give the barbecue a special flavor (which is delicious) and is sought after by restaurants (Walter and Armstorng, 2014). But prosopis leaves have no utility value. Worse, they are rich in chemicals which repell other vegetation and toxify soil (Goel et al., 1989). Such chemicals easily leach out when rainwater trickles down a prosopis canopy or from the leaves fallen on the ground (Al-Humaid and Warrag, 1998). The problem is made still worse by the 'longevity' of the dead prosopis leaves. Unlike leaves of other species of plants, prosopis leaves are very very slow to biodegrade. Besides their robust form, a waxy coating further helps them resist biodegradation for several months and more. As a result, any and every prosopis invasion spells the doom for all other species, principally aided and abetted by its leaves.

These aspects have been discussed in much greater detail in Patnaik et al. (2017).

In summary, there are numerous ways in which weeds play havoc with biodiversity and ecosystem services, cause pollution, and contribute to global warming. This aspect, along with our failure to eradicate, or even control, any weed with chemical, biological, mechanical, or hybrid means, has left us with only one option – finding a technically and economically viable means to utilize billions of tonnes of phytowaste at the scales at which it is generated. Pre-existing vermicomposting systems that process animal manure have proved economically viable despite the fact that animal manure is a priced commodity with several competing uses. In contrast, phytowaste is not only freely available but is also a drain on the global economy due to the harm it causes. Hence, if phytowaste can be made a feedstock in human-controlled vermicomposting, and if the resulting vermicompost can be shown to be a good organic fertilizer, we can have a means to utilize as much phytowaste as we generate, or like to process. It all looks simple and straightforward, but carries a big question along: If it is so, why has no system existed till recently to economically and sustainably vermicompost phytowaste?

To seek an answer to their question, S.A. Abbasi and coworkers (Abbasi et al., 2015) carried out an extensive survey of the prior art. It revealed:

i. The first-ever report on vermicomposting of phytomass is probably the one by Frank et al. (1983). Till the time of the survey, 96 reports had been published which had involved the use of phytowaste as a feedstock.
ii. For about 10 years after the report of Frank et al. (1983), interest in this field remained low. Then, an upward trend began, leading to a surge after 2010 – over 50 reports appeared during 2010–2014 compared to half of that number during 2005–2009, and about one-fifth of that number during 2000–2004. This reflects the increasing concern towards finding ways to vermicompost phytomass.
iii. In all except 4% of the reports, phytomass could not be vermicomposted alone, and it had become necessary to add animal manure – in most cases,

cow/buffalo manure – for earthworms to survive and feed upon. In as many as a third of all studies, the authors had to even do pre-composting of phytomass. Otherwise, earthworms did not feed upon the phytomass and tended to die quickly.

iv. To quote from Abbasi et al. (2015),

> In some cases, no systematic precomposing was done, but the substrate was kept for 3 weeks or more to enable its partial decomposition (Suthar, 2009a–c). A few authors either blended the phytomass with compost or sewage sludge or laid it over compost (Frank et al., 1983; Chaudhuri et al., 2003; Jadia & Fulekar, 2008; Rajasekar & Karmegam, 2009; FernándezGómez et al., 2010, 2011) or vermicompost (Huang et al., 2014). In one of the four studies in which phytomass was directly used for vermicomposting, by Gomez-Brandon et al. (2011), the substrate happened to be grape marc, which consists of the stalk, skin, pulp, and seeds that remain after grapes are crushed to extract their juice. Grape marc is much richer in polysaccharides than most other phytomass, for example, weeds, leaf litter, grasses, vegetable waste. It is also well fragmented and softened due to wet crushing, making it easy for earthworms to ingest. It was, expectedly, acted upon swiftly by the earthworm species *Eisenia andrei* tried by the authors. They had used very high earthworm density, 500 individuals of *E. andrei* for 1 kg fresh weight of grape marc, and felt that in 15 days the substrate had been 'completely processed'. But this surmise was not supported by any verifiable argument, such as stabilized values of bioavailable nutrients. Other three authors (Chaudhuri et al., 2003; Mainoo et al., 2009; Lim et al., 2012) have studied the action of earthworms on phytomass but have not reported the extent of vermicomposting that was achieved and whether what they finally analyzed was a fully stabilized product. Considering these factors, it can be said that no quantitative study on the direct vermicomposting of archetypal phytomass exists in the literature.

The prior art survey, reported in Abbasi et al. (2015), led us to identify the following drawbacks associated with all post attempts at vermicomposting phytomass. To quote:

a. Most of the past attempts have depended on supplementing the phytomass with large proportions of animal manure, especially cow dung. This has two major disadvantages. First is that to process the very large quantities of phytomass that are available, equally large quantities of animal manure shall be needed. But it is not possible to find so much manure because of numerous competitive uses of manure already in existence (Abbasi et al., 2012; Tauseef et al., 2013c). The second major disadvantage is that, unlike waste phytomass, animal manure is not available free of cost. Hence, dependence on animal manure makes the process economically highly unfavourable.
b. Collection and transport of animal manure are among the operations which lead to massive emissions of global warming gases, methane and nitrous oxide (Abbasi et al., 2013; Tauseef et al., 2013c), besides other pollutants, like ammonia. Large-scale use of manure in vermicomposting of phytomass will add to global warming and pollution (Tabassum-Abbasi et al., 2015).
c. The reported processes, have all been very slow, taking 2 months or more to achieve substantial conversion of phytomass to vermicast (unless

pre-composting had been done). As the rate of any process is directly related to its efficiency, hence economics, this aspect further diminishes the usability of reported processes.

d. When pre-composting had to be done, it further adds time and cost to the overall process, eroding its economic viability still further.
e. There is the absence of a logical criterion to define what the product of a vermireactor is, and how to ascertain that the reactants have been fully converted to that product. Due to the absence of this criterion, there are no pointers available with which the vermireactor operation can be engineered to enhance the efficiency and the economics of the process. Nearly all past attempts have been on batch reactors with very long, and unfounded, solid retention times.

REFERENCES

Abbasi S.A. 1997. *Wetlands of India, Volume III: Kerala.* Discovery Publishing House, New Delhi, xiv + 276 pages

Abbasi, S.A., Nayeem-Shah, M. and Abbasi, T., 2015. Vermicomposting of phytomass: limitations of the past approaches and the emerging directions. *Journal of Cleaner Production, 93*, pp.103–114.

Abbasi, S.A. and Nipaney, P.C., 1984. Generation of biogas from *Salvinia molesta* (Mitchell) on a commercial biogas digester. *Environmental Technology*, *5*(1–11), pp.75–80.

Abbasi, S.A. and Nipaney, P.C., 1985. Wastewater treatment using aquatic plants. Survivability and growth of *Salvinia molesta* (Mitchell) over waters treated with zinc (II) and the subsequent utilization of the harvested weeds for energy (biogas) production. *Resources and Conservation*, *12*(1), pp.47–55.

Abbasi, S.A. and Nipaney, P.C., 1986. Infestation by aquatic weeds of the fern genus Salvinia: its status and control. *Environmental Conservation*, *13*(3), pp.235–241.

Abbasi, S.A. and Nipaney, P.C., 1991. Effect of temperature on biogas production from aquatic fern Salvinia. *Indian Journal of Technology*, *29*(6), pp.306–309.

Abbasi, S.A. and Nipaney, P.C., 1993. *World's worst weed (salvinia)-its impact and utilization.* International book distributors, Dehradun pp. x+–224.

Abbasi, S.A., Nipaney, P.C. and Ramasamy, E.V., 1992a. Studies on multi-phase anaerobic digestion of Salvinia. *Indian Journal of Technology*, *30*(10), pp.483–490.

Abbasi, S.A., Nipaney, P.C. and Ramasamy, E.V., 1992b. Use of aquatic weed Salvinia (Salvinia molesta, Mitchell) as full/partial fedd in commercial biogas digesters. *Indian Journal of Technology*, *30*(9), pp.451–457.

Abbasi, S.A., Nipaney, P.C. and Schaumberg, G.D., 1990. Bioenergy potential of eight common aquatic weeds. *Biological Wastes*, *34*(4), pp.359–366.

Abbasi, S.A., Patnaik, P., Tabassum-Abbasi, Khamrang, C. and Abbasi, T., 2020. A comparative study of the reproductive and the vermicomposting ability of three generations of epigeic and anecic earthworms when subjected to a toxic weed (Ipomoea) as the sole feed. *Compost Science & Utilization*, *28*(1), pp.28–35. doi.org/10.1080/1065657X.2019.1709918.

Abbasi, T., and Abbasi, S.A., 2010. Production of clean energy by anaerobic digestion of phytomass – new prospects for a global warming amelioration technology. *Renewable and Sustainable Energy Reviews (Elsevier), 14,* pp.1653–1659.

Abbasi, T. and Abbasi, S.A., 2011. Sources of pollution in rooftop rainwater harvesting systems and their control. *Critical Reviews in Environmental Science and Technology*, *41*(23), pp.2097–2167.

Abbasi, T., Tauseef, S.M., and Abbasi, S.A., 2012. *Biogas Energy*, Springer, New York & Londan, xiv+169 Pages; ISBN 978-1-4614-1039-3

Abbasi, T., Tauseef, S.M., and Abbasi, S.A., 2013. Energy recovery from wastewaters with high-rate anaerobic digesters. *Renewable and Sustainable Energy Reviews*, *19*, pp.704–741.

Abbasi, T., Tauseef, S.M., Abbasi, S.A. 2011a. *Global Warming and the Role of Wetlands*. Lap Lambart Academic, Germany; ISBN 978-3846556009. Xiv+264 pages

Akhtar, N., Satyam, A., Anand, V., Verma, K.K., Khatri, R. and Sharma, A., 2010. Dysregulation of TH type cytokines in the patients of Parthenium induced contact dermatitis. *Clinica Chimica Acta*, *411*(23-24), pp.2024–2028.

Al-Humaid, A.I. and Warrag, M.O.A., 1998. Allelopathic effects of mesquite (Prosopis juliflora) foliage on seed germination and seedling growth of bermudagrass (Cynodon dactylon). *Journal of Arid Environments*, *38*(2), pp.237–243.

Anwar, M.F., Yadav, D., Kapoor, S., Chander, J. and Samim, M., 2015. Comparison of antibacterial activity of Ag nanoparticles synthesized from leaf extract of Parthenium hystrophorus L in aqueous media and gentamicin sulphate: in-vitro. *Drug Development and Industrial Pharmacy*, *41*(1), pp.43–50.

Arthur, G.D., Stirk, W.A., Novák, O., Hekera, P. and van Staden, J., 2007. Occurrence of nutrients and plant hormones (cytokinins and IAA) in the water fern Salvinia molesta during growth and composting. *Environmental and Experimental Botany*, *61*(2), pp.137–144.

Bar-On, Y.M., Phillips, R. and Milo, R., 2018. The biomass distribution on Earth. *Proceedings of the National Academy of Sciences*, *115*(25), pp.6506–6511.

Bennett, F.D., 1975. Insects and plant pathogens for the control of Salvinia and Pistia. In: Brezonik, P.L. and Fox, L. (eds) *Proceedings of the Symposium on Water Quality Management through Biological Control*, Department Environmental Engneering and USEPA, University. Florida, Gainesville, FL. pp.28–35.

Bhat, M.A., 2016. *Waste Water Treatment with a Novel Clean-Green Bioreactor SHEFROL*. Pondicherry University, Pondicherry, p.116.

Chantiratikul, P., Meechai, P. and Nakbanpotec, W., 2009. Antioxidant activities and phenolic contents of extracts from Salvinia molesta and Eichornia crassipes. *Research Journal of Biological Sciences*, *4*(10), pp.1113–1117.

Chaudhuri, P.S., Pal, T.K., Bhattacharjee, G. and Dey, S.K., 2003. Rubber leaf litters (Hevea brasiliensis, var RRIM 600) as vermiculture substrate for epigeic earthworms, Perionyx excavatus, Eudrilus eugeniae and Eisenia fetida: the 7th international symposium on earthworm ecology·Cardiff Wales· 2002. *Pedobiologia*, *47*(5-6), pp.796-800.

Choudhary, M.I., Naheed, N., Abbaskhan, A., Musharraf, S.G. and Siddiqui, H., 2008. Phenolic and other constituents of fresh water fern Salvinia molesta. *Phytochemistry*, *69*(4), pp.1018–1023.

Coates, D., 1982. Salvinia-possible biological effects on fish in Papua New Guinea. *Aquatics*, *4*(3), p.2.

Crites, R.W., Middlebrooks, E.J. and Bastian, R.K. 2006. *Natural Wastewater Treatment Systems*. CRC Press.

Dorahy, C.G., Pirie, A.D., McMaster, I., Muirhead, L., Pengelly, P., Chan, K.Y., Jackson, M. and Barchia, I.M., 2009. Environmental risk assessment of compost prepared from Salvinia, Egeria densa, and Alligator Weed. *Journal of Environmental Quality*, *38*(4), pp.1483–1492.

Dubow, J., 2011. Still-life, after-life: W.G. Sebald and the demands of landscape. In: Daniels, S., DeLyser, D. Entrikin, J.N. and Richardson, D. (eds.) *Envisioning Landscapes, Making Worlds: Geography and the Humanities*, Routledge. ISBN-13, 978.

Fernández-Gómez, M.J., Nogales, R., Insam, H., Romero, E. and Goberna, M., 2010. Continuous-feeding vermicomposting as a recycling management method to revalue tomato-fruit wastes from greenhouse crops. *Waste Management*, *30*(12), pp.2461–2468.

Fernández-Gómez, M.J., Romero, E. and Nogales, R., 2011. Impact of imidacloprid residues on the development of Eisenia fetida during vermicomposting of greenhouse plant waste. *Journal of Hazardous Materials*, *192*(3), pp.1886–1889.

Field, C.B., Behrenfeld, M.J., Randerson, J.T. and Falkowski, P., 1998. Primary production of the biosphere: integrating terrestrial and oceanic components. *Science*, *281*(5374), pp.237–240.

Frank, R., Klauck, C. and Stonefield, K.I., 1983. Metal transfer in vermicomposting of sewage sludge and plant wastes, 673–679.

Ganeshkumar, T., Premalatha, M., Gajalakshmi, S. and Abbasi, S.A., 2014. A new process for the rapid and direct vermicomposting of the aquatic weed salvinia (Salvinia molesta). *Bioresources and Bioprocessing*, *1*, pp.1–5.

Goel, U., Saxena, D.B. and Kumar, B., 1989. Comparative study of allelopathy as exhibited by Prosopis juliflora swartz and Prosopis cineraria (L) Druce. *Journal of Chemical Ecology*, *15*, pp.591–600.

Gómez-Brandón, M., Lazcano, C., Lores, M. and Domínguez, J., 2011. Short-term stabilization of grape marc through earthworms. *Journal of Hazardous Materials*, *187*(1–3), pp.291–295.

Gunaseelan, V.N., 1987. Parthenium as an additive with cattle manure in biogas production. *Biological Wastes*, *21*(3), pp.195–202.

Harley, K.L.S. and Mitchell, D.S., 1981. The biology of Australian weeds. 6. Salvinia molesta DS Mitchell. *Journal of the Australian Institute of Agricultural Science (Australia)*, *47*(2), pp.67–76.

Hattingh, E.R., 1961. Problem of Salvinia auriculata Aubl. and Associated Aquatic Weeds on Kariba Lake. *Weed Research*, *1*(4), pp.303-306. https://doi.org/10.1111/j.1365–3180.1961.tb00030.x.

Hierro, J.L., Maron, J.L. and Callaway, R.M., 2005. A biogeographical approach to plant invasions: the importance of studying exotics in their introduced and native range. *Journal of Ecology*, *93*(1), pp.5–15.

Huang, K., Li, F., Wei, Y., Fu, X. and Chen, X., 2014. Effects of earthworms on physicochemical properties and microbial profiles during vermicomposting of fresh fruit and vegetable wastes. *Bioresource Technology*, *170*, pp.45–52.

Hussain, N., Abbasi, T. and Abbasi, S.A., 2016a. Vermicomposting transforms allelopathic parthenium into a benign organic fertilizer. *Journal of Environmental Management*, *180*, pp.180–189.

Hussain, N., Abbasi, T. and Abbasi, S.A., 2016b. Transformation of the pernicious and toxic weed parthenium into an organic fertilizer by vermicomposting. *International Journal of Environmental Studies*, *73*(5), pp.731–745.

Hussain, N., Abbasi, T. and Abbasi, S.A., 2017. Detoxification of parthenium (*Parthenium hysterophorus*) and its metamorphosis into an organic fertilizer and biopesticide. *Bioresources and Bioprocessing*, *4*, pp.1–9.

Hussain, N., Abbasi, T. and Abbasi, S.A., 2018. Generation of highly potent organic fertilizer from pernicious aquatic weed *Salvinia molesta*. *Environmental Science and Pollution Research*, *25*, pp.4989–5002.

Jadia, C.D. and Fulekar, M.H., 2008. Vermicomposting of vegetable waste: A biophysicochemical process based on hydro-operating bioreactor. *African Journal of Biotechnology*, *7*(20), pp.3723–3730. DOI: 10.5897/AJB08.617.

Kaur, M., Aggarwal, N.K., Kumar, V. and Dhiman, R., 2014. Effects and management of Parthenium hysterophorus: A weed of global significance. *International Scholarly Research Notices*, 2014(1), p.368647.

Khan, F.I. and Abbasi, S.A., 1998. Techniques and methodologies for risk analysis in chemical process industries. *Journal of loss Prevention in the Process Industries*, *11*(4), pp.261–277.

King, C., McIntosh, D. and Fitzsimmons, K., 2004. Giant salvinia (*Salvinia molesta*) as a partial feed for Nile tilapia (*Oreochromis niloticus*). In: R. Bolivar, G. Mair, and K. Fitzsimmons (eds) *Proceedings of the Sixth International Symposium on Tilapia in Aquaculture*, Bureau of Fisheries and Aquatic Resources, pp. 750–754.

Kishor, P., Maurya, B.R. and Ghosh, A.K., 2010. Use of uprooted Parthenium before flowering as compost: a way to reduce its hazards worldwide. *International Journal of Soil Science*, *5*(2), pp.73–81.

Knox, J., Jaggi, D. and Paul, M.S., 2011. Population dynamics of *Parthenium hysterophorus* (Asteraceae) and its biological suppression through *Cassia occidentalis* (Caesalpiniaceae). *Turkish Journal of Botany*, *35*(2), pp.111–119.

Kumar, A., Kumar, A., Kumar, V., Kumar, M., Singh, B. and Cjouhan, P., 2012. Recycling of harmful weeds through NADEP composting. *Vegetos*, *25*, pp.315–318.

Kumar, S., Pandey, S. and Pandey, A.K., 2014. In vitro antibacterial, antioxidant, and cytotoxic activities of Parthenium hysterophorus and characterization of extracts by LC-MS analysis. *BioMed Research International*, 2014(1), p.495154.

Lakshmi, C. and Srinivas, C.R., 2007. Parthenium: A wide angle view. *Indian Journal of Dermatology, Venereology and Leprology*, *73*, p.296.

Leterme, P., Londoño, A.M., Ordoñez, D.C., Rosales, A., Estrada, F., Bindelle, J. and Buldgen, A., 2010. Nutritional value and intake of aquatic ferns (*Azolla filiculoides* Lam. and *Salvinia molesta* Mitchell.) in sows. *Animal Feed Science and Technology*, *155*(1), pp.55–64.

Li, H., Shi, A., Li, M. and Zhang, X., 2013. Effect of pH, temperature, dissolved oxygen, and flow rate of overlying water on heavy metals release from storm sewer sediments. *Journal of Chemistry*, 2013(1), p.434012.

Li, S., Wang, P., Deng, G., Yuan, W. and Su, Z., 2013a. Cytotoxic compounds from invasive giant salvinia (*Salvinia molesta*) against human tumor cells. *Bioorganic & Medicinal Chemistry Letters*, *23*(24), pp.6682–6687.

Lim, S.L., Wu, T.Y., Sim, E.Y.S., Lim, P.N. and Clarke, C., 2012. Biotransformation of rice husk into organic fertilizer through vermicomposting. *Ecological Engineering*, *41*, pp.60–64.

Luque, G.M., Bellard, C., Bertelsmeier, C., Bonnaud, E., Genovesi, P., Simberloff, D. and Courchamp, F., 2014. The 100th of the world's worst invasive alien species. *Biological Invasions*, *16*, pp.981–985.

Mainoo, N.O., Barrington, S., Whalen, J.K. and Sampedro, L., 2009. Pilot-scale vermicomposting of pineapple wastes with earthworms native to Accra, Ghana. *Bioresource Technology*, *100*(23), pp.5872–5875.

Maishi, A.I., Ali, P.S., Chaghtai, S.A. and Khan, G., 1998. A proving of *Parthenium hysterophorus*, L. *British Homoeopathic Journal*, *87*(1), pp.17–21.

Manoj, E.M., 2014. Invasive plants a threat to wild life, The Hindu, https://www.thehindu.com/news/national/kerala/invasive-plants-a-threat-to-wildlife/article6432731.ece, p3

Mitchell, D.S. and Tur, N.M., 1975. The rate of growth of *Salvinia molesta* (*S. auriculata* Auct.) in laboratory and natural conditions. *Journal of Applied Ecology*, *12*(1), pp.213–225.

Mitchell, D.S., 1969. The ecology of vascular hydrophytes on Lake Kariba. *Hydrobiologia*, *34*, pp.448–464.

Morin, L., Reid, A.M., Sims-Chilton, N.M., Buckley, Y.M., Dhileepan, K., Hastwell, G.T., Nordblom, T.L. and Raghu, S., 2009. Review of approaches to evaluate the effectiveness of weed biological control agents. *Biological Control*, *51*(1), pp.1–15.

Muturi, G.M., Mohren, G.M. and Kimani, J.N., 2009. Prediction of Prosopis species invasion in Kenya using geographical information system techniques. *African Journal of Ecology*, *48*(3), pp.628–636.

Nigatu, L., Hassen, A., Sharma, J. and Adkins, S.W., 2010. Impact of *Parthenium hysterophorus* on grazing land communities in north-eastern Ethiopia. *Weed Biology and Management*, *10*(3), pp.143–152.

Oliver, JD., 1993. A review of the biology of giant Salvinia (*Salvinia molesta* Mitchell). *Journal of Aquatic Plant Management*, *12*, pp.227–231.

Osmond, R., 2003. Mesquite: Control and management options for mesquite (Prosopis spp.) in Australia. The State of Queensland (Department of Natural Resources and Mines). Cloncurry, Queensland, Australia. 29 p.

Pasiecznik, N.M., Harris, P.J. and Smith, S.J., 2004. *Identifying Tropical Prosopis Species: a field Guide*). HDRA Publishing, Coventry, p. 29.

Patel, S., 2011. Harmful and beneficial aspects of *Parthenium hysterophorus*: an update. *3 Biotech*, *1*(1), pp.1–9.

Patnaik, P., 2017. *Towards a Clean Technology for the Total Utilization of the Invasive Xerophyte Prosopis* (*prosopis juliflora*, swartz, dc.), Pondicherry University Thesis.

Patnaik, P., Abbasi, T. and Abbasi, S.A., 2017. Prosopis (*Prosopis juliflora*): blessing and bane. *Tropical Ecology*, *58*(3), pp.455–483.

Patnaik, P., Tabassum-Abbasi and Abbasi, S.A. 2024. *Prosopis juliflora: Attributes, Impact, Utilization*. Academic Press , London & USA, 430pp. eBook ISBN: 9780128198254.

Qureshi, H., Arshad, M. and Bibi, Y., 2014. Invasive flora of Pakistan: a critical analysis. *International Journal of Biosciences*, *4*(1), pp.407–424.

Rajasekar, K., and Karmegam, N., 2009. Effects of different management practices on degradability and nutrient release capacity. *Soil Biology and Biochemistry*.

Rajkumar, E.D.M., Kumar, N.V.N., Haran, N.V.H., Murthy, N.V.A. and Ram, N.V.S., 1988. Antagonistic effect of Parthenium hysterophorus on succinate dehydrogenase of sheep liver. *Journal of Environmental Biology*, *9*, pp.231–237.

Raju, R.A. and Gangwar, B., 2004. Utilization of potassium-rich green-leaf manures for rice (*Oryza sativa*) nursery and their effect on crop productivity. *Indian Journal of Agronomy*, *49*(4), pp.244–247.

Room, P.M., Harley, K.L.S., Forno, I.W. and Sands, D.P.A., 1981. Successful biological control of the floating weed salvinia. *Nature*, *294*(5836), pp.78–80.

Room, P.M., Gunatilaka, G.A., Shivanathan, P. and Fernando, I.V.S., 1989. Control of Salvinia molesta in Sri Lanka by *Cyrtobagous* salviniae. In *Proceedings of the VII International Symposium on Biological Control of Weeds*, Rome, Italy, pp.285–290.

Saravanan, L., 2017. Madras High Court orders Tamil Nadu Government to enact law on removal of seemai karuvelam trees. *The Times of India*, February 27.

Sculthorpe, C.D., 1967. *The Biology of Aquatic Vascular Plants*. Koeltz Scientific Books, Königstein.

Shindell, D.T., Faluvegi, G., Koch, D.M., Schmidt, G.A., Unger, N. and Bauer, S.E., 2009. Improved attribution of climate forcing to emissions. *Science*, *326*(5953), pp.716–718.

Singh, R.K. and Garg, A., 2014. *Parthenium hysterophorus* L.-neither noxious nor an obnoxious weed. *Journal of Non-Timber Forest Products*, *21*(4), pp.223-225. https://doi.org/10.54207/bsmps2000-2014-XNH377.

Suthar, S., 2009a. Bioremediation of agricultural wastes through vermicomposting. *Bioremediation Journal*, *13*(1), pp.21–28.

Suthar, S., 2009b. Potential of *Allolobophora parva* (Oligochaeta) in vermicomposting. *Bioresource Technology*, *100*(24), pp.6422–6427.

Suthar, S., 2009c. Vermicomposting of vegetable-market solid waste using Eisenia fetida: Impact of bulking material on earthworm growth and decomposition rate. *Ecological Engineering*, *35*(5), pp.914–920.

Tabassum-Abbasi, Abbasi, T. and Abbasi, S.A., 2015. Reducing the global environmental impact of livestock production: the minilivestock option. *Journal of Cleaner Production*, 112, pp.1754–1766.

Tauseef, S.M., Premalatha, M., Abbasi, T. and Abbasi, S.A., 2013c. Methane capture from livestock manure. *Journal of Environmental Management*, *117*, pp.187–207.

Tewari, J.C., Mathur, B.K., Tewari, P., Singh, Y., Singh, M., Ram, M. and Sharma, A., 2013. *Prosopis juliflora*: a miracle species of hot arid and semi-arid regions of India. *Popular Kheti*, *1*(2), pp.53–59.

Towers, G.H.N. and Rao, P.V.S., 1992, February. Impact of the pan-tropical weed, *Parthenium hysterophorus* L. on human affairs. In *Proceedings of the First International Weed Control Congress*, Melbourne (Vol. 2, pp. 134–138). Weed Science Society of Victoria Melbourne.

Walter, K., 2011. Prosopis, an alien among the sacred trees of South India. Doctoral dissertation (monograph), University of Helsinki, Faculty of Agriculture and Forestry, Department of Forest Sciences.

Walter, K.J. and Armstrong, K.V., 2014. Benefits, threats and potential of Prosopis in South India. *Forests, Trees and Livelihoods*, *23*(4), pp.232–247.

Wiesner, M., Tessema, T., Hoffmann, A., Wilfried, P., Büttner, C., Mewis, I. and Ulrichs, C., 2007. *Impact of the Pan-Tropical Weed Parthenium hysterophorus L. on Human Health in Ethiopia*. Institute of Horticultural Science, Urban Horticulture, Berlin, Germany.

7 The Paradigm of High-Rate Vermicomposting and Its Validation

7.1 INTRODUCTION

After identifying the key questions to be answered and the major challenges to be overcome – as detailed in Chapter 6 – the possible avenues towards re-designing and re-engineering of vermicomposting systems were explored. The effort led to the development of a new paradigm, which we have named "high-rate vermicomposting" paradigm.

7.2 SPECIFYING THE END PRODUCT, OR "AIMED PRODUCT", OF VERMICOMPOSTING

Abbasi et al. (2015) had carried out a very extensive prior art survey which had covered, among several other aspects of vermicomposting, the question: What is the desirable and controllable/achievable end product of vermicomposting? It had revealed:

i. In only 4% of all prior art, the authors have considered vermicast as the aimed product and have quantified the extent of vermicomposting on the basis of the fraction of feedstock converted to vermicast per unit time.
ii. In all other reports, no clearly identifiable and/or reproducible criteria have been used to quantify vermicomposting. In about 60% of the studies, the authors have described the progress of vermicomposting on the basis of the reduction in the carbon-to-nitrogen (C/N) ratio. But none has specified what the desirable or achievable C/N ratio is. The authors have simply deemed vermicomposting to have been completed when the C/N ratio of the vermireactor contents fell to less than 20. In another 15% of reports, vermicomposting was considered complete when the C/N ratio of the substrate had fallen to reach the 20–30 range. In other words, by 73% of the authors, completion of vermicomposting has been considered as achieved on the basis of the C/N ratio of the reactor content, irrespective of the fact that in such situations, a good part of the substrate not yet vermicomposted can also be counted.

DOI: 10.1201/9780429331886-9

As recorded by Abbasi et al. (2015), to quote,

> But composting is a process very different from vermicomposting (Abbasi et al., 2009), and the compost of any substrate has widely different characteristics than its vermicompost. For example, vermicompost is distinguished not just by the C:N ratio or the high bioavailability of the nutrients it contains, as compost is, but also contains several of the enzymes, plant growth hormones, and pest repellants which compost does not (Abbasi & Ramasamy, 2001; Edwards et al., 2011). Hence the logic that is applicable to judge completion of composting, or suitability of a compost, cannot be directly extended to vermicomposting or vermicompost. Moreover C:N ratio basically indicates the concentration of total nitrogen in relation to total carbon and even though it is indicative of the progressive stabilization of a biowaste, it does not necessarily provide a quantitative measure of change in the bioavailability of the nutrients present in the substrate. Surprisingly, none of the authors have continued their studies till C:N ratio had reached a steady state and have only assumed that vermicomposting had been completed once the C:N ratio had dropped below a premeditated level.
>
> A few of the authors have used even more subjective criterion to judge the occurrence of vermicomposting. These include change in the colour of the substrate, or appearance of casting without quantifying the casting.
>
> Several authors (for example Singh & Suthar, 2012; Suthar & Sharma, 2013) have periodically 'homogenised' the vermireactor contents in their study. This act would mix the vermicast, which is an easily distinguishable and quantifiable product of vermicomposting, with unreacted substrate. Due to this, in the subsequent vermireactor operation, the earthworms will have to perforce ingest portions of the vermicast that they had earlier produced, and will process that much of the reactants lesser. This would work against the efficiency of the process.

7.2.1 Vermicast as the Quantifiable Product of Vermireactors

When an earthworm excretes its cast, the cast represents the final product of the process (as earlier detailed in Chapters 1–3), which began with the ingestion of the substrate by the earthworm. For a given substrate, the characteristics of the resulting vermicast are fairly uniform and reproducible (Figures 7.1 and 7.2). Also, vermicast can be easily identified as it is usually very distinct in form and colour from the parent substrate or its partially biodegraded parts (Figures 7.3 and 7.4). It, therefore, appears logical that the yield of vermicast per unit substrate mass and per unit time should be made the criteria for designing, operating, monitoring, and controlling vermireactor performance.

As elaborated by Abbasi et al. (2015), to quote,

> By periodically removing the vermicast and quantifying it, two major objectives can be attained. First is to quantify the rate of vermicomposting in terms of fraction of the feed converted to vermicast in a given interval of time. Second is to free that much vermireactor space, thereby ensuring that earthworms principally feed upon the substrate rather than forced to reingest their own excrement. Thus, using vermicast production and its quantification as the basis of vermireactor operation, process monitoring as well as process control are achievable. This measure also provides a clear idea on how the process efficiency can be increased, as all those factors which will enhance vermicast production per unit time will proportionately enhance process efficiency.

FIGURE 7.1 Earthworms wallowing in their vermicast.

FIGURE 7.2 Wet vermicast (a) in contrast with its air-dried version (b).

7.3 THE HIGH-RATE VERMICOMPOSTING PARADIGM (HRVP)

The paradigm is based on the following premises. To quote from Abbasi et al. (2015):

a. Vermicomposting is a process very different from composting, whereas the latter is a quintessential batch process, the former is amenable to continuously-fed operation (Abbasi et al., 2009). The microbiology and biochemistry of the two processes also have several major differences (Gajalakshmi & Abbasi, 2008).
b. The most rational and the most easily quantifiable criterion with which vermicomposting systems can be designed, optimized, and monitored is the vermicast production. This is because vermicast is the finished product of vermicomposting, and fresh vermicast is believed to be more soil-friendly and plant-friendly than aged vermicast (Edwards et al., 2011; Karthikeyan et al., 2014a,b).

FIGURE 7.3 Vermicast with some unconverted part of a biowaste before it is sieved.

c. Based on the earthworm species and the nature of substrate, it takes only 18h for vermicomposting to occur because this is the time that is taken between the commencement of ingestion of a substrate by an earthworm and its exit as the vermicast (Abbasi & Ramasamy, 2001). Hence, the upper limit of the speed of a vermireactor, defined as the time taken to convert a feed into vermicast, is only 12 ± 6 hours. But conventional vermireactors take up to 12 weeks for converting most of the feed to vermicast, and this indicates something basically flawed in the way conventional vermicomposting systems have been designed and operated.
d. Unless very complex instrumentation and control is done, it is not possible to have a vermicomposting system which will be able to separate from the vermireactor each grain of vermicast as it is generated. Hence, the upper limit of vermireactor speed is not achievable in practice. But it appears possible to significantly enhance vermireactor efficiency by taking it closer to its theoretical limit without compromising on simplicity (hence better economics) of vermireactor operation.

Based on these premises and extensive experimentation, modelling, and design, S.A. Abbasi and co-workers (Abbasi et al., 2009, 2011; Gajalakshmi et al., 2002, 2005; Ganesh et al., 2009; Tauseef et al., 2013, 2014) have developed the concept of high-rate vermicomposting and associated technology. It has the following attributes:

FIGURE 7.4 Air-dried and sieved vermicast as the final product of vermicomposting.

a. It relies on a reactor geometry that has been chosen to maximize earthworm–substrate contact as well as ease of cast deposition. To this end, a high surface area-to-volume ratio is set for the reactors. This also ensures mixing and aeration of the reactor content by the earthworm movements, thereby preventing anaerobic pockets from developing.
b. The vermicast harvesting is also made easier by the low aspect ratio because many species deposit their cast on top of the substrate, while some others do it at the bottom.
c. Low substrate column height makes it possible to maintain, with relative ease, uniformity in the moisture content across the reactor depth. There is little, if any, accumulation of water in the reactor bottom, saving on the need for recycling.
d. The sand-gravel "vermibed", which occupies over 25% space in conventional vermireactors, is replaced by moist jute cloth. This maximizes the use of reactor volume, proportionately reducing the system cost.
e. Earthworm density is maximized to achieve the highest sustainable population for a given feed. The high earthworm/feed ratio further helps mixing and aeration of the substrate due to earthworm movement.

7.4 VALIDATION OF HRVP

As detailed in the following chapters, the efficacy of HRVP as the enabler of direct vermicomposting of phytomass has been established on the basis of many extensive studies (Figures 7.5–7.9). HRVP is also seen to speed up manure-based vermicomposting.

It has been shown that, without any pre-treatment, sometimes even without any size reduction, and without any supplementation of animal manure or any other material, phytomass can be rapidly vermicomposted. Substrates ranging from aquatic weeds like water hyacinth (*Eichhornia crassipes*), salvinia (*Salvinia molesta*), and pistia (*Pistia stratiotes*), amphibians like ipomoea (*Ipomoea carnea*) and marsilea (*Marsilea quadrifolia*), terrestrials like lantana (*Lantana camara*), parthenium (*Parthenium hysterophorus*), and mimosa (*Mimosa pudica*), and xerophytes like prosopis (*Prosopis juliflora*) have been found easily amenable to HRVP. So are horticultural wastes like leaf litter, rice straw, and date palm waste. Even waste paper, which is mostly holocellulose and is very lean in nitrogen, can be rapidly vermicomposted on the basis of HRVP, after blending it with leaf litter (Figure 7.10) or small quantities (7%–10% by mass) of animal manure.

It has also been shown that vermicomposting transforms the phytomass of even toxic and allelopathic weeds into benign, plant-friendly, and soil-friendly fertilizer.

FIGURE 7.5 Manually operated vermireactors based on the HRVP paradigm in the author's laboratory processing different species of terrestrial and aquatic weeds.

FIGURE 7.6 HRVP-based and manually operated vermireactors processing leaf litter and weed-paper waste blends.

FIGURE 7.7 Vermicast generated from parthenium in HRVP reactors.

FIGURE 7.8 Vermicast produced from leaf litter employing the HRVP concept.

FIGURE 7.9 Near complete conversion of lantana into vermicast in HRVP reactors within 25 days.

FIGURE 7.10 Blends of waste paper and leaf litter ready for vermicomposting in HRVP reactors.

REFERENCES

Abbasi, S.A., Nayeem-Shah, M. and Abbasi, T., 2015. Vermicomposting of phytomass: limitations of the past approaches and the emerging directions. *Journal of Cleaner Production*, *93*, pp.103–114.

Abbasi, S.A., and Ramasamy, E.V., 2001. *Solid Waste Management with Vermitechnology*. Discovery Publishing House, New Delhi xiv þ 178 pages.

Abbasi, T., Gajalakshmi, S., Abbasi, S.A., 2009. Towards modeling and design of vermicomposting systems: mechanisms of composting/vermicomposting and their implications. *Indian Journal of Biotechnology, 8*, 177–182.

Abbasi, T., Tauseef, S.M. and Abbasi, S.A., 2011. The inclined parallel stack continuously operable vermireactor. *Official Journal of the Patent Office*, *22*(9571), p.201.

Edwards, C.A., Norman, Q.A., Sherman, R., 2011. *Vermiculture Technology, Earthworms, Organic Waste and Environmental Management*. CRC Press, Boca Raton, pp.17–19.

Gajalakshmi, S. and Abbasi, S.A., 2008. Solid waste management by composting: state of the art. *Critical Reviews in Environmental Science and Technology*, *38*(5), pp.311–400.

Gajalakshmi, S., Ramasamy, E.V. and Abbasi, S.A., 2002. High-rate composting–vermicomposting of water hyacinth (*Eichhornia crassipes, Mart. Solms*). *Bioresource Technology*, *83*(3), pp.235–239.

Gajalakshmi, S., Ramasamy, E.V. and Abbasi, S.A., 2005. Composting–vermicomposting of leaf litter ensuing from the trees of mango (*Mangifera indica*). *Bioresource Technology*, *96*(9), pp.1057–1061.

Ganesh, P.S., Gajalakshmi, S. and Abbasi, S.A., 2009. Vermicomposting of the leaf litter of acacia (*Acacia auriculiformis*): Possible roles of reactor geometry, polyphenols, and lignin. *Bioresource Technology*, *100*(5), pp.1819–1827.

Karthikeyan, M., Gajalakshmi, S. and Abbasi, S.A., 2014a. Effect of storage on the properties of vermicompost generated from paper waste: with focus on pre-drying and extent of sealing. *International Journal of Energy and Environmental Engineering*, *5*, pp.291–301.

Karthikeyan, M., Hussain, N., Gajalakshmi, S. and Abbasi, S.A., 2014b. Effect of vermicast generated from an allelopathic weed lantana (*Lantana camara*) on seed germination, plant growth, and yield of cluster bean (*Cyamopsis tetragonoloba*). *Environmental Science and Pollution Research*, *21*, pp.12539–12548.

Singh, D. and Suthar, S., 2012. Vermicomposting of herbal pharmaceutical industry solid wastes. *Ecological Engineering*, *39*, pp.1–6.

Suthar, S. and Sharma, P., 2013. Vermicomposting of toxic weed—Lantana camara biomass: Chemical and microbial properties changes and assessment of toxicity of end product using seed bioassay. *Ecotoxicology and Environmental Safety*, *95*, pp.179–187.

Tauseef, S.M., Abbasi, T., Banupriya, D., Vaishnavi, V. and Abbasi, S.A., 2013. HEVSPAR: a novel vermireactor system for treating paper waste. *Official Journal of the Patent Office*, *24*, p.12726.

Tauseef, S.M., Abbasi, T., Banupriya, G., Banupriya, D., Abbasi, S.A., 2014. A new machine for clean and rapid separation of vermicast, earthworms, and undigested substrate in vermicomposting systems. *International Journal of Energy and Environmental Engineering*, *56*, pp.495–498.

8 Vermireactors Designed on the Basis of the High-Rate Vermicomposting Paradigm

8.1 INTRODUCTION

The presently offered vermicomposting devices, described in the previous chapter, are based on conventional reactors characterized by high depth-to-surface-area ratios and high solid retention times. They suffer from one or more of the following drawbacks, which the inventions described in this chapter, based on the high-rate vermicomposting paradigm (HRVP), have intended to overcome:

1. No capability for processing multiple substrates and usage of multiple earthworm species.
2. Non-efficient usage of floor space.
3. Non-efficient surface-to-volume ratio.
4. Need for an additional aeration system.
5. Formation of anaerobic patches in the vermicomposting system.
6. Non-efficient maintenance of moisture content.
7. High adaptability period for earthworms.
8. Requirement of skilled workers for maintenance.
9. High operation cost.
10. Non-scalability and flexibility of systems.
11. Low system efficiency.
12. Not useful for low-scale processing.

All these systems have been patented (Abbasi et al., 2011; 2018; Tauseef et al., 2013a,b; 2020) and opened to commercialization.

DOI: 10.1201/9780429331886-10

8.2 A FLAPPABLE UNITS VERMIREACTORS TRAIN SYSTEM (FLUVTS)

8.2.1 The Need for FLUVTS

This system, described in detail in Abbasi et al. (2018), is suitable for vermicomposting all forms of horticultural and other forms of phytomass, and also waste paper. It had become necessary to design and develop FLUVTS in view of the fact that, to quote from Abbasi et al. (2018),

> Waste paper contains predominantly cellulose and hemicellulose (85%–90%), with such small quantities of major, medium, and trace elements that earthworms cannot survive on a paper-alone diet for long (Edwards, 1997; Van Wyk et al., 1999; Hummel et al., 2016). Due to this reason, it is necessary to blend paper waste with some other substrate which can supply nutrients adequate for the earthworm's needs. In all the studies reported so far, the substrate has been cow manure (cow dung), but none of the studies have shown the potential of being commercially viable due to the following reasons:
>
> 1. They all need 50% or more of cow dung with or without additives like rock phosphate, in the cow dung–waste paper blends (Gupta & Garg, 2009; Mupambwa et al., 2016; Unuofin & Mnkeni, 2014; Unuofin et al., 2016). But cow dung is a prized commodity with a myriad competing uses and high demand (Tauseef et al., 2013b). This, on one hand, adds to the cost of vermicomposting paper waste and, on the other hand, limits the quantities of paper waste that can be vermicomposted by the quantities of cow dung that could be available.
> 2. In addition to the use of 50% or more cow dung, some authors have found it necessary to include a pre-composting step (Nath & Deb, 2008; Mupondi et al., 2011; Ravindran & Mnkeni, 2016; Sharma & Garg, 2018). It was apparently done to convert paper sheets into a powder well-mixed with stabilized animal manure so that earthworms can ingest it easily. But the composting step not only increases the cost per unit mass of paper waste utilized but also adds 30–45 days of processing time (Abbasi et al., 2009). In contrast, FLUVTS does not need any such pre-processing of the feed. This is because the high aspect ratio of its modules and other advantages associated with the high-rate vermicomposting protocol, as detailed in Abbasi et al. (2015), make the pre-processing unnecessary.
> 3. Even without the pre-composting step, the reported processes have required 45 days or more to accomplish the vermicomposting. Such long processing times further jeopardize the commercial viability of these processes because the larger the solids' retention time, the greater the requirement of reactor space per unit mass processed, and the higher the process costs".

8.2.2 Distinguishing Features of FLUVTS

The FLUVTS design is based on a train of modular reactors that can be moved along a railing and can be overturned using a flip mechanism (Figure 8.1). If a large set of modules is used, the substrate to be vermicomposted can be applied to the system in short pulses to have a near-continuous operation. The flip mechanism is equipped

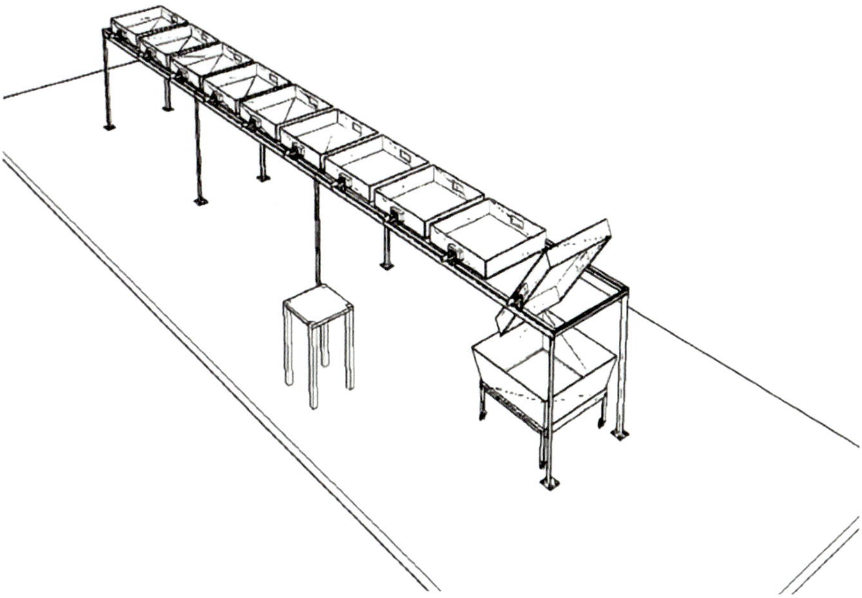

FIGURE 8.1 A typical example of modular reactors used in FLUVTS tracks.

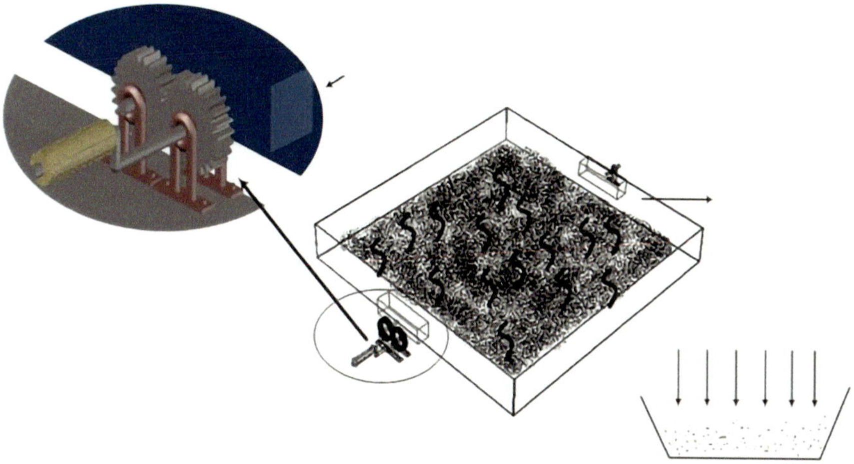

FIGURE 8.2 The gear system provided in FLUVTS to enable smooth harvesting.

with gears (Figure 8.2) for the smooth harvesting of the reactor content. The system flow chart is shown in Figure 8.3. Due to resource constraints, we have experimented only with a manually operated version of FLUVTS, but the system can be easily mechanized. As is true with automation in general, the larger the scale of operation is, the more cost-effective the mechanized system will be. The distinguishing features of the FLUVTS design are as follows:

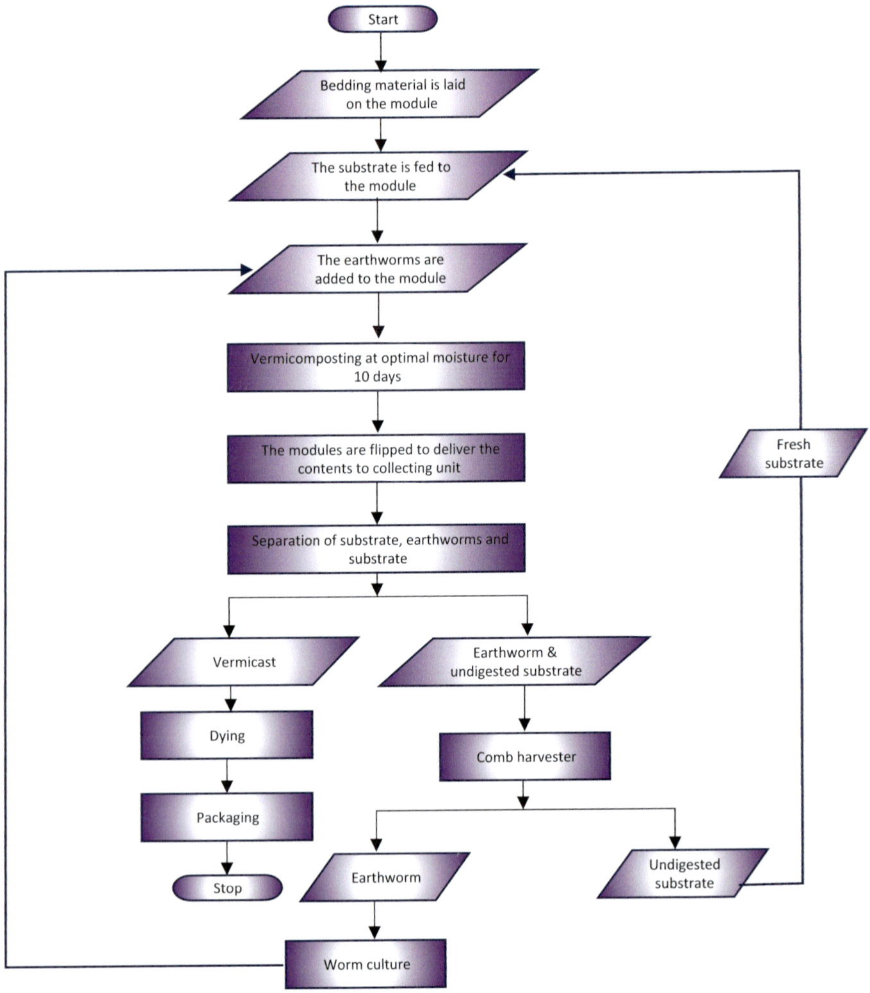

FIGURE 8.3 The FLUVTS flow chart.

1. It is modular, which uses a discrete set of vermireactors in an easily manoeuvrable fashion.
2. It has a simple and easy-to-use harvesting system, which comprises a mechanized harvesting unit including a sieve and a collecting vessel below (Figure 8.2).
3. It uses a novel flip mechanism to harvest the contents of the vermicomposting modules after the processing of organic waste. The modules can be turned upside down to deliver their contents to the harvesting unit, which is slightly bigger in dimension than the module. This is to avoid the spillover of the contents of the module.

The modules are made square-shaped so as to optimize the material cost per unit reactor volume and achieve a maximum surface area-to-volume ratio; it also facilitates easy handling.

In a typical commercial unit of FLUVTS, each module can hold 20 kg of the substrate, as each module provides adequate volume for that amount of substrate and is also not too heavy to hinder the smooth movement of the modules along the tracks. Further, in a typical commercial unit of FLUVTS, each module would have a similar square-shaped form, as it provides maximum base area constrained to optimal earthworm feeding, optimal surface to substrate volume ratio, and a substrate thickness not more than 8 cm. These steps ensure reactor operation without causing any handling problems or endangering the mechanical stability of the designed module.

The modules are typically made up of mild steel with fibre reinforced plastic (FRP) lamination, which offers several advantages, including high strength, fatigue and corrosion resistance, tailorability, and low maintenance requirements. The use of modules is designed to keep the waste more contained and therefore more controlled in terms of environmental conditions. This makes FLUVTS easy to maintain.

8.2.3 Comparison with Conventional Designs

In Chapter 5, we have described a large variety of pre-existing vermicomposting systems. These systems, to quote from Abbasi et al. (2018),

> Typically use thick beds in which large quantities of organic material are worked by worms in a relatively stationary mode. Thick beds typically become stratified with regions of active worms and regions of compacted material. This leads to suboptimal use of reactor space. In contrast, FLUVTS operates at low substrate thickness, achieved by high aspect ratio modules. It prevents anaerobic pockets from forming, facilitates mixing, prevents sharp stratification of moisture, and facilitates harvesting. All these factors, except the last one, ensure that gases like hydrogen sulfide and volatiles like mercaptans, which cause a foul smell around conventional vermireactors, are not formed in FLUVTS modules.
>
> In pre-existing systems, the trays are stacked in closed structures, which consume more materials for construction; in contrast, FLUVTS uses open tracks. In conventional systems using stacked bins, the bins typically have a depth of 30–45 cm, which leads to stratification of substrate, resulting in suboptimal processing of waste, formation of anaerobic patches, and considerably long processing time of up to six months. In contrast, FLUVTS has none of these drawbacks and enables vermicomposting at a high rate of 30 days or less per batch. This manner of processing the substrate at 3–4 times faster rates than conventional systems means that proportionally smaller reactors can achieve the same throughput as larger but proportionally slower reactors. Moreover, multi-tier tracks can be utilized to achieve vermicomposting of several times larger masses of the substrate per unit floor area.
>
> In all pre-existing larger-scale vermicomposting systems, it is necessary to keep a provision for ventilation. This increases capital and operational costs. In comparison, FLUVTS does not require a ventilation system for aeration, as the substrate depth is only 8 cm and the earthworm movement is adequate to keep the substrate agitated and aerated. On the other hand, presently marketed systems such as the "High-Efficiency Vermiculture Apparatus" and "Worm Gin" need conveyor belts, which are very costly to install and maintain.

Further, unlike some of the vermicomposting systems, such as windrows and wedges, which are only suited to rural areas, FLUVTS is as appropriate for rural areas as it is for urban regions. This is due to the low manufacturing cost, high flexibility of scale, ease of operation, and compactness of the FLUVTS design.

8.3 THE HIGH-EFFICIENCY VERMICOMPOSTING SYSTEM FOR PAPER WASTE AND OTHER REFUSE (HEVSPAR)

The HEVSPAR consists of two principal units – the automated paper soaking unit (APSU) and the high-efficiency vertical stacked vermireactor (VSV) unit (Tauseef et al., 2013). The VSV unit comprises a set of modular reactors arranged in series; sets of such reactors are also arranged in multiple rows over a fixed frame. The VSV unit is provided with a sprinkler system, a loading–unloading arrangement, and a harvesting arrangement.

8.3.1 Novelty of the APSU

The APSU has the following novel features:

1. The machine is the first of its kind, which enables automatic soaking of paper waste in cow-dung slurry to accelerate the subsequent vermicomposting process.
2. The APSU is unique as it is easy to load/unload and maintain. The overall system also enables economy of space and materials.
3. The APSU has been designed to minimize manual intervention. It achieves an automated transport of paper waste, in a thin layer, to the soaking module where it is soaked effectively and quickly in cow-dung slurry using a sprinkler system.
4. The system is capable of scaling up to meet any capacity requirement.
5. The automation of the separation and soaking of paper waste in cow-dung slurry makes the unit efficient and easily amenable to utilization at a large scale.

8.3.2 Novelty of the VSV Unit

The VSV unit is distinguished by the following attributes:

1. The design makes for a continuous, automated operation, which minimizes the need for regular manual intervention. At the same time, the design is utilizable even without the provision of automation by low-income groups who would like to work with low capital inputs.
2. The design includes modular units placed in series over the tracks. The tracks are stacked one over the other, supported on a strong frame. This ensures very efficient space utilization.
3. This design incorporates a simple and easy-to-use harvesting system, which comprises a conveyor belt and a collecting vessel.

4. The system is capable of processing other organic waste such as manure alongside paper waste. It is also capable of utilizing several different earthworm species simultaneously.
5. The paper waste soaked in cow dung is applied in thin layers in the modular reactors, optimized for efficient use of reactor volume. This helps in improving the process efficiency, the uniformity of vermicomposting, and the quality of the final product.
6. The system runs on the high-rate vermicomposting concept developed by us. It uses a combination of reactor geometry and high stocking density of earthworms to maximize the conversion of the paper waste into vermicast at high rates.
7. The overall process takes 15 days or less to convert paper waste into vermicast. Thus, the process is 4–6 times faster than the pre-existing systems processing organic waste.
8. The vermicomposting modules have been so designed as to ensure adequate aeration and optimal temperature maintenance, thereby providing a good environment for the earthworms to thrive and swiftly process the paper waste.
9. The use of a conveyor-based harvesting system ensures easy collection and transport of the vermicast for final processing.
10. The use of modular design makes the process robust. Infection or contamination in one part of the system does not affect the rest of the system, and the affected module can be easily replaced without disrupting the ongoing process.
11. The automated system makes the loading and unloading of the modules easy, and a single person can efficiently handle the whole system.
12. The system requires minimal maintenance of about 10 min/day.
13. The pre-existing, large-scale systems typically use thick beds, which are charged with large quantities of organic material to be worked by worms in a relatively stationary mode. Such thick beds typically become stratified with regions of active worms and regions of compacted material, which leads to suboptimal use of reactor space. At times, unintended composting sets in, lifting the substrate temperature and killing the earthworm.
14. Generation of earthworm zoomass, which is a valuable poultry feed, is a bonus.

8.3.3 Advantages Over Other Known Alternatives, National and Global

- There is no pre-existing process for vermicomposting that specializes in paper waste. Hence, this is the first of its kind.
- Compared to other ways of disposing of or utilizing paper waste, this option is gainful, is eco-friendly, and does not lead to any other waste. It has the added virtue that it assists in global warming control.
- The process is versatile in terms of ease, effectiveness, and flexibility.

8.3.4 Detailed Description of the Invention

8.3.4.1 The APSU

The purpose of this unit is to rapidly soak a large number of waste paper sheets one by one with dilute cow-dung slurry and feed them to the VSV unit (Figure 8.4). The system consists of a hopper that spreads the paper waste evenly in thin layers over a conveyor belt, which transports it to a surface inclined at an angle of 15°. The inclined surface is fitted with a set of overhead sprinklers and an array of taps to spread a thin layer of cow-dung slurry over the inclined surface. This not only ensures proper soaking, reinforced with the help of a set of overhead sprinklers, but also facilitates transportation and deposition of the paper waste into the collection module placed at the bottom of the 2 m long inclined surface. The paper soaked in cow-dung slurry is collected and transferred to the vermicomposting unit. The excess slurry is collected and recycled to the overhead tank, used to store and supply cow-dung slurry to the sprinkler system.

Among the distinguishing features of the design is the feed arrangement, which consists of a hopper fitted with a vibrator and a mechanism to move it to-and-fro. This to-and-fro movement, along with the vibration, helps in evenly spreading the paper waste over the conveyor belt. The overhead sprinklers, along with the array of taps fitted on the inclined surface, ensure proper soaking and transport of paper waste into the collection vessel. The excess cow-dung slurry is collected

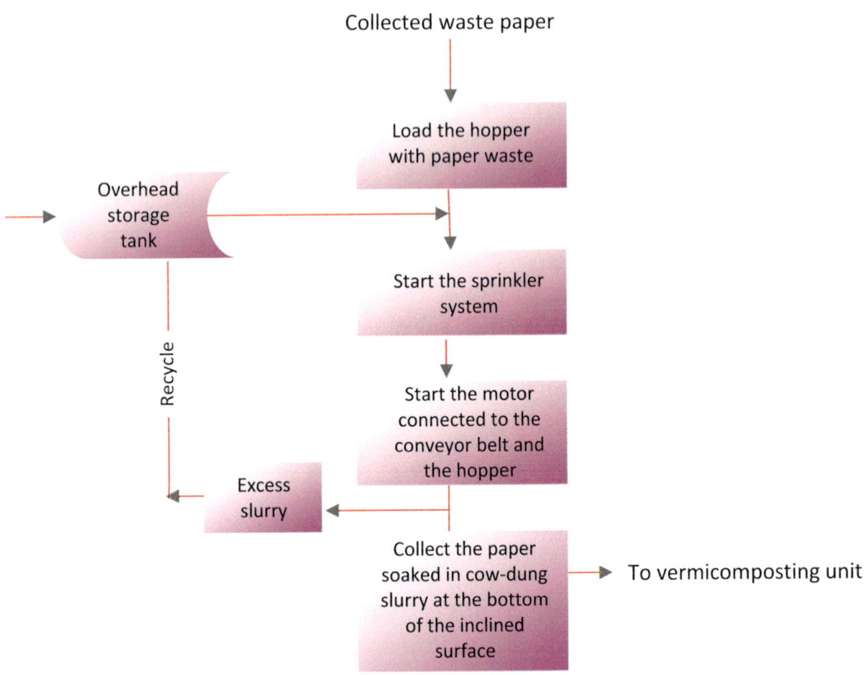

FIGURE 8.4 The APSU.

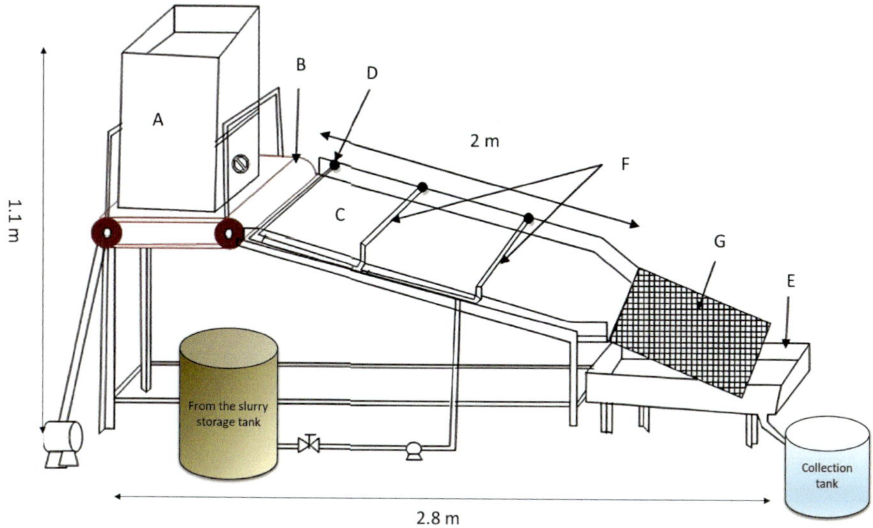

FIGURE 8.5 Flow chart of the paper soaking machine APSU.

underneath and is recycled to the overhead storage tank. The flow chart of the system is depicted in Figure 8.5.

Each APSU, shown in Figures 8.4 and 8.6, typically includes a hopper **A** mounted over a conveyor belt **B** of 0.6 m in length, and a 2 m long metal sheet **C** placed at the end of the conveyor belt. The metal sheet **C** is placed such that it has an inclination of 15°. An array of taps **D** is positioned on the top of the inclined surface **C**. These taps apply a thin layer of cow-dung slurry over the inclined surface. As the cow-dung slurry flows down the inclined surface **C**, under the force of gravity, it carries the paper waste, transported from the conveyor belt **B**, to the collection module **E**. The proper soaking of the paper waste is ensured by a set of overhead sprinklers **F**. The collection module is fitted with a mesh **G** under which a collection tank is placed to collect the excess slurry. The collected slurry is then recycled to the overhead tank, which is used to store and supply the cow-dung slurry to the APSU. The height of a typical paper soaking unit is 1.1 m.

The hopper (**A**; 0.4 m* 0.4 m* 0.4 m) is mounted on a set of rods; it can slide freely over these rods. A set of levers is used to couple the movement of the conveyor belt and the hopper. Thus, only a single motor is required to drive both the hopper and the conveyor belt. In order to regulate the flow of paper waste, the hopper is provided with a set of shutters that are adjusted to provide a clearance of 5 mm between the conveyor belt and the hopper. The thin layer of paper waste is then conveyed onto the inclined metal sheet fitted with the sprinkler system. The sprinkler system is designed to minimize the energy required to run the unit. The gravity-driven flow of cow-dung slurry over the inclined sheet provides a thin film of fluid over which the paper waste is easily transported down the slope. The overhead sprinklers ensure that the paper waste is properly soaked before it floats out of the unit and into the collection module. A collection module is provided at the end of the inclined surface to facilitate the collection and recycling of excess cow-dung slurry to the overhead tank.

FIGURE 8.6 A prototype of the APSU unit.

The functions of the main components are defined as follows:

a. *Side-wise movable hopper*: The hopper is designed to spread thin layers of paper waste onto the inclined surface. It moves sideways to aid in the equal distribution of paper waste on the conveyor belt placed below. Hopper can hold up to 5–7 kg of paper waste at a time.
b. *Conveyor belt*: Below the hopper is the conveyor belt, which is attached to the mechanical handle to rotate the belt. When the conveyor belt is rotated, the paper waste falls on the inclined surface.
c. *Inclined surface*: This is specially designed to hold thin sheets of paper waste. The paper waste slides downwards due to gravity as it gets soaked in the cow-dung slurry.
d. *Sprinkler system and the cow-dung slurry pump*: Cow-dung slurry is loaded into a slurry tank, and as the pump runs, the slurry is sprayed on the

inclined surface, which has paper waste, through the sprinklers attached on the inclined surface.

e. *Mesh and collection vessel*: As the soaked paper slides down, the slurry drains into the collection vessel placed below. The collection vessel has a drain, which is attached to the slurry tank through which the excess slurry is sent to the slurry tank and reused.

The mesh filters out the soaked paper waste, which can be sent to the vermicomposting unit for further processing. Figure 8.6 presents the APSU after fabrication.

8.3.4.2 Distinguishing Features of the APSU

- The proposed unit is designed using a set of optimized parameters to ensure proper soaking of paper waste in cow-dung slurry, thus accelerating the vermicomposting process.
- The simple design can be easily scaled up or down to suit the desired plant capacity.
- The paper can be easily fed into the hopper manually or with the help of an automated loading system.
- A system of levers is used to couple the movement of the conveyor and the hopper to cut down the energy requirement.
- The unit exploits the gravity-driven slurry flow over an inclined surface to cut the energy required to run the unit.
- The excess slurry is collected and recycled to the overhead tank.
- The sprinkler system is convenient to maintain and operate.
- A typical unit can easily handle 25 kg of paper waste per hour.
- Being only 1.1 m in height, the unit enables easy loading and operation.
- The automated process thus saves the labour required to separate waste paper sheets prior to soaking, thus making the overall process more efficient and economical.

8.3.4.3 The High-Efficiency VSV

The paper waste soaked in cow dung from the APSU unit is vermicomposted in the VSV unit, which consists of a set of modular reactors arranged in series as well as parallel. The unit is provided with loading, unloading, and harvesting arrangements (Figure 8.7).

The unit comprises a fixed frame **B** to hold modules **A** loaded in series as well as in parallel. The modules **A** move over the fixed frame **B** with the help of wheels **C** provided on either side. The wheels move within the brackets to prevent the module **A** from moving in the vertical direction at the time of harvesting. The unit consists of a sprinkler system **D** used to maintain the moisture content in modules **A**; the sprinkler system has nozzles **E** positioned above the modules to serve the purpose.

A rod **F** is placed below the modules **A,** which can be rotated 180° using the gear mechanism **G.** It helps in emptying the contents of the modules onto the conveyor belt **H** placed below each track. The guiding mechanism at one end of the conveyor belt **H** serves in collecting the content of the modules without spillage into the collection tray. The loading **J** and unloading **K** systems help in loading and unloading of the modules **A** onto the fixed frame **B** using rack and pinion arrangement **R** and **T**.

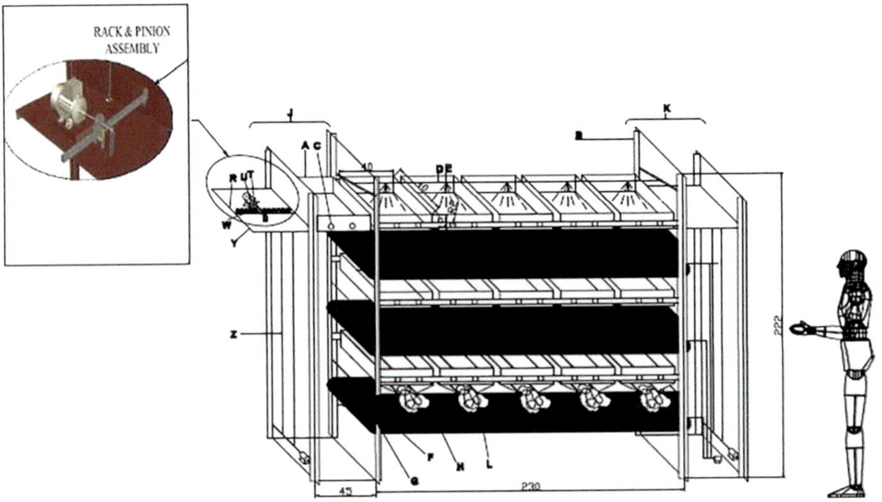

FIGURE 8.7 A representation of the VSV system.

Module **A** is placed over the fixed frame **B** with the help of wheels **C** fixed on either side of the module **A**; the base of the module **A** opens to empty its content onto the conveyor belt **H** placed below.

The harvesting system of the VSV unit consists of a conveyor belt **H** and a guiding mechanism at one end of the conveyor belt **H.** The guiding mechanism aids in collecting the content of the conveyor belt **H** without spillage into the collection tray**.** A roller attached to a motor helps in rotating the conveyor belt **H** at the time of harvesting.

Module **A** is filled with substrate and placed on the loading end **J.** The motor aids in the lifting of the module with the help of a rope **Z.** The rack and pinion **R** and **T** arrangement, driven by motor **U,** helps in placing the module **A** onto the fixed frames **B**. Motor **U** for the rack and pinion arrangement is supported on a frame, and the rod **W** is attached to a hinge **X**. The whole setup is placed on a frame **Y**.

8.3.4.4 Working of the VSV Unit

1. The substrate along with the specified earthworm species is loaded onto the modules **A** (0.4 m × 0.4 m × 0.12 m), which is filled up to 8 cm height and placed in the loading end **J** where the module is carried over to the fixed frame **B** by the rope **Z** with the help of a motor **U** and the rack and pinion **R** & **T** arrangement.
2. The modules **A** with wheels **C** on either side move within the bracket, which restricts the vertical movement of the modules.
3. Each module **A** is split into two equal halves at its bottom to aid in the easy emptying of the modules during harvesting.
4. Each reactor is left for vermicomposting for 10–15 days. The substrate is almost fully utilized at this time by the worms, making it easy to harvest the product.
5. When the module is ready for harvesting, the rod **F** below the modules **A** (dissected at the base) is rotated to 180° downwards using the gear mechanism **G**. The supporting rod moves, and the base of module **A** opens to empty the contents onto a conveyor belt **H**.

6. The conveyor **H** is then rotated with the help of a motor. The vermicast, along with worms and the remaining undigested matter, gets collected in the collection vessel.
7. The rod **F** is rotated back to its original position. This brings the base of the modules **A** together.
8. The modules are unloaded at the other end using the unloading mechanism **K**.
9. Moisture content is maintained using sprinkler systems **D** attached to modules. The temperature is monitored regularly using a digital thermometer.

8.3.5 Distinguishing Features of the VSV Unit

This design is based on a set of modular reactors placed in series as well as in parallel, which maximizes the use of space.

The loading of the modules is made easy with the help of a rack and pinion arrangement and by the motor used to lift the modules to their respective tracks.

The loading of organic material into the system and harvesting of the vermicast can be done at desired intervals.

Discrete modules containing the waste material to be processed are supported on a frame structure, and the modules are fitted with wheels for easy movement/loading on the fixed tracks.

The modules used as the reactors are square-shaped with dimensions 0.4 m × 0.4 m × 0.12 m. They are made of 0.5 mm thick aluminium sheets to ensure adequate high strength at low material density and flexibility.

The modules are split into two equal halves at their base, making it easy to empty the contents onto a conveyor belt placed below.

All the modules are closed at the bottom with the help of a single rod. When the rod placed below the modules is rotated to 180° using the gear mechanism, all the modules empty their contents onto the conveyor belt.

The conveyor belt is rotated with the help of a motor, and a collection vessel is placed at the loading end to collect the contents of all the reactors placed in series. This makes it very easy and convenient to operate the unit.

8.3.6 Fabrication of a HEVSPAR Prototype

Prototypes of the APSU (Figure 8.6) and the VSV units (Figures 8.8 and 8.9) of HEVSPAR were fabricated and utilized by us in the studies presented in the subsequent chapters. Visuals of some key parts are given in Figures 8.10–8.12.

8.4 THE INCLINED PARALLEL STACK CONTINUOUSLY OPERABLE (IPSCO) VERMIREACTOR SYSTEM

8.4.1 The IPSCO Vermireactor System

This design (Abbasi et al., 2011) is based on a set of modular reactors arranged in multiple sloping stacks. It includes loading and unloading mechanisms, several vermicomposting modules, a sprinkler system, and a harvesting unit.

FIGURE 8.8 A prototype of the VSV unit.

The distinguishing features of the design include discrete modules that may be fed with different substrates and can be loaded onto the system using a loading mechanism that delivers the modules to the respective stacks. The frequency of loading and unloading of modules can be adjusted to suit the rate of vermicomposting of the given substrate–species combination. The product (vermicompost) is harvested using a harvesting unit for sun drying, packing, and storing. The flow chart of the system is depicted in Figure 8.13.

Each IPSCO vermireactor unit typically includes five stacks, each with a modest slope but parallel to each other. Each stack is 11 m long and 1.2 m wide. The length is chosen to facilitate ten 1 m long reactor modules to move through and reach the end of the stack. The stacks have tracks made up of mild steel. Each stack can typically hold 10 modules. A distance of 10 cm with a support structure is provided to prevent each module from clashing with the next module. The stacks are inclined at an angle of about 5° to enable the modules to move at the desired speed by gravity. The height of a typical IPSCO unit is 4.1 m (Figures 8.14–8.17).

FIGURE 8.9 A HEVSPAR unit system in operation.

The modules are square-shaped (1 m × 1 m × 0.12 m), made up of fibre-reinforced plastic to ensure high strength at low material density, adequate elasticity, and flexibility. The square shape is chosen for ease of handling, too. The feed is laid up to a height of 8 cm with a buffer space of 4 cm; the total depth of the module has been set as 12 cm. This, in turn, is based on the "high-rate vermicomposting" concept earlier developed by us. A support structure has been provided at the end of each module, which controls the movement of the modules and prevents the modules from clashing with each other. Each module has four wheels attached to enable it to move through the inclined stacks.

The loading mechanism carries the modules and drops them in the respective stack. It includes a base tray that carries the modules, a lifting rope, a motor assembly, and a rack and pinion assembly.

The unloading mechanism, which works on a principle similar to the loading mechanism, is equipped with a base plate, a lifting rope, and a lock. A motor has been provided to help in bringing down the modules to be sent to the harvester.

The plant has a sprinkler system which includes a motorized pump, and horizontal and vertical pipe lines with sprinklers. Each module is provided with a set of four sprinklers to release the water required in each module.

FIGURE 8.10 A typical module showing the arrangement at the base with which the content can be emptied. The guiding wheel is seen on the side.

FIGURE 8.11 The conveyor belt with a slant at its unloading end.

FIGURE 8.12 A view of the sprinkler system placed above each module.

A drainage system runs parallel to the stacks. The modules have a small hole at the bottom wherein the excess moisture gets collected. It is then led through a trough to a drain pipe, which collects excess moisture from all the stacks and sends it into a collecting vessel.

8.4.2 Distinguishing Features of the IPSCO Vermireactor Design

The design is modular, with a multiple parallel stack system that can be easily scaled up or down to suit the desired plant capacity.

It maximizes the use of floor area by optimal use of vertical space.

The parallel stack design has separate loading and unloading points for each stack. The modules are loaded at the loading point to travel the whole stack and reach the unloading point after a pre-determined number of days (usually 10–15). This duration can be easily altered to suit the rate of the process by alterations in the frequency of loading and unloading.

It has an efficient loading mechanism that employs easy-to-fabricate steel channels, bearing wheels, a gearbox, a rope drum, and a motor to lift the modules to the respective stacks.

The shape of the modules is such that it optimizes the material cost and maximizes the surface area-to-volume ratio. It also facilitates handling.

The modules are 1 m × 1 m in length and breadth, a size which is adequately large yet convenient to handle. It is not too large to endanger the mechanical stability of the overall structure.

Wheels are fixed to each module for easy movement along the tracks.

An unloading mechanism is provided to easily bring down the modules from their respective stacks.

The sprinkler system is convenient to operate and control.

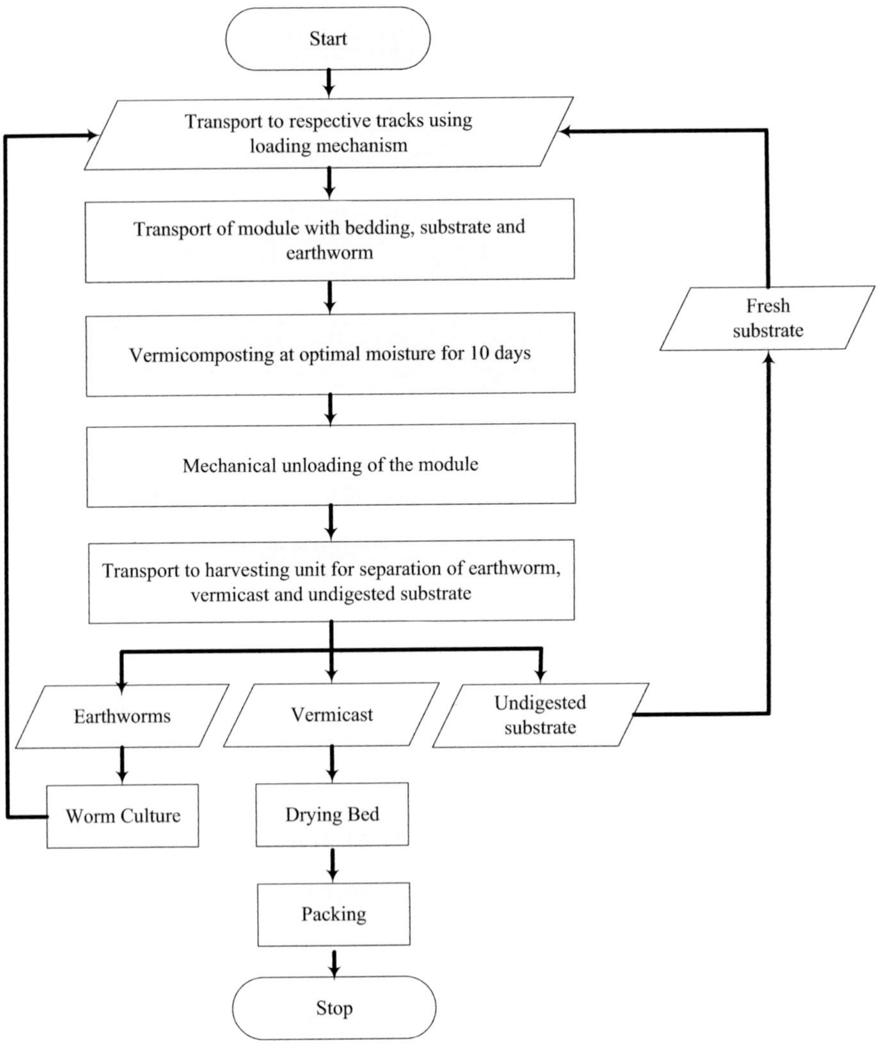

FIGURE 8.13 Flow chart of the IPSCO system.

8.4.3 Novelty of the IPSCO Vermireactor Design

1. The inclined parallel stack system is unique as it makes loading/unloading automatic and easy. The overall system also enables economy of space and materials.
2. The reactor dimensions enable vermicomposting at a much faster rate than the 4–6 months time taken in existing systems.
3. An IPSCO vermicomposting system is scalable both stack-wise and module-wise; its ability to extend vertically enables it to process a greater quantity of waste in a smaller floor area.

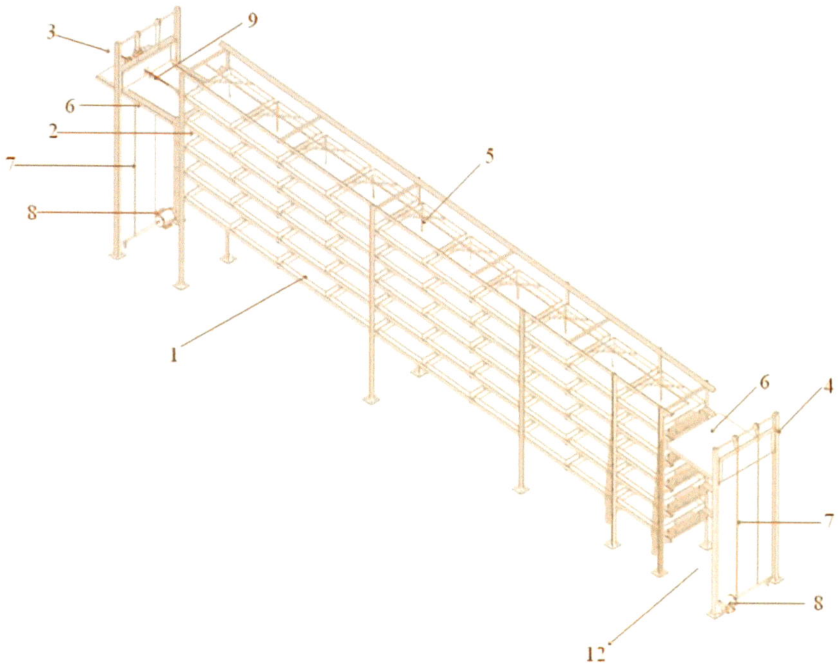

FIGURE 8.14 Isometric representation of IPSCO.

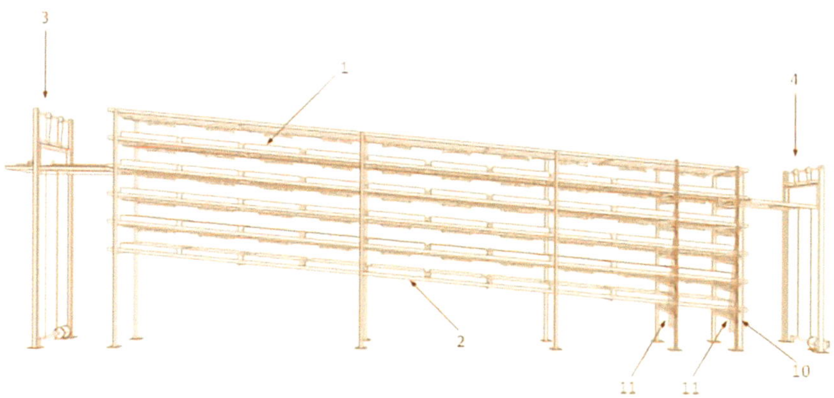

FIGURE 8.15 Side view of the IPSCO.

4. As against systems based on large bins, the low-height modular reactor system of IPSCO vermireactors protects any contamination or pest attack occurring in a small fraction of the feed from spreading into the rest of the feed.
5. The IPSCO vermireactor system is also unique in that it utilizes moist jute cloth of certain porosity and thickness as the earthworm bedding instead

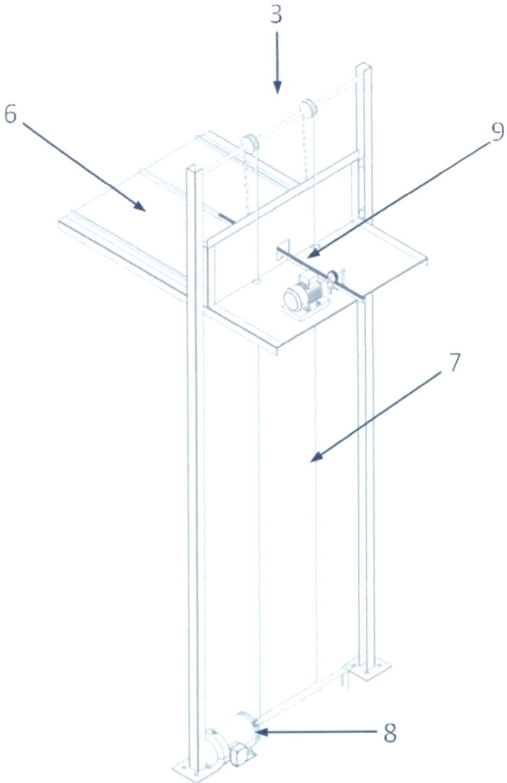

FIGURE 8.16 Loading unit in IPSCO.

of the elaborate gravel-sand-soil bedding used conventionally. This enormously enhances the effective reactor space and reduces bedding costs.

6. The modular system enables different substrates to be vermicomposted simultaneously and also different species to be utilized simultaneously without the danger of interfering with each other.
7. In the pre-existing institutional and industrial size flow-through vermireactors, ventilation systems have to be provided for proper aeration, which increases the capital costs. But the IPSCO system does not require any extra ventilation system for aeration, as the substrate depth is only 8 cm and the module is open. Hence, it can't get heated up as deeply and widely as vermicomposting bins often do.
8. In pre-existing designs such as Worm Wigwam, there is the probability of the earthworms getting injured or killed as a crank is used for harvesting. In case of a high-efficiency vermiculture process and apparatus, a strong light source or a fan blowing air is required to make the worms move away from the unloading end. These contraptions are not needed with IPSCO.
9. Some of the existing processes depend on the movement of the conveyor belt, which is very costly compared to the IPSCO vermireactor system.

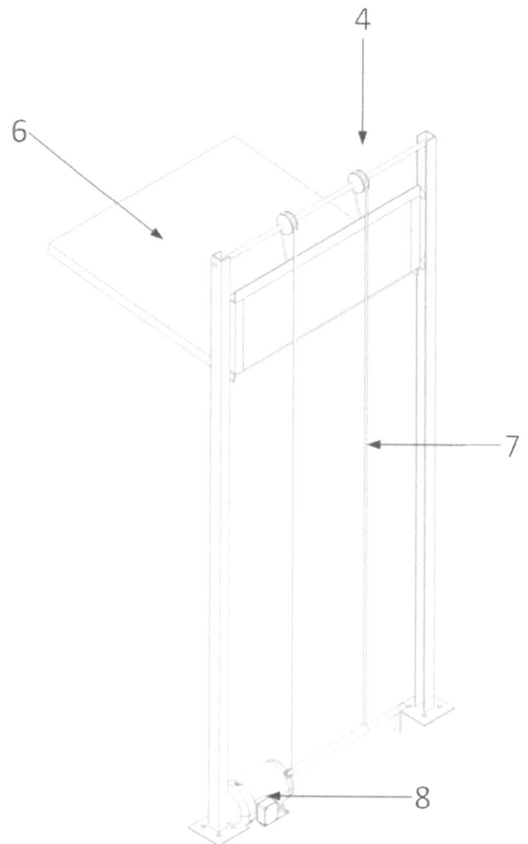

FIGURE 8.17 Unloading unit of IPSCO.

10. The IPSCO vermireactor system is by far the most versatile and flexible *vis-à-vis* the variety of substrates it can handle, species it can use, capacities it can handle, etc. It is also among the least expensive.

8.5 A MULTIPLE COMPARTMENT VERMICOMPOSTING SYSTEM (MCVS.)

8.5.1 Novelty of the Invention, in a Nutshell

The design of this MCVS is based on a set of compartments arranged in a vertical sequence with provisions for swift and even sprinkling of water, harvesting, etc (Tauseef et al., 2020). The system is appropriate for use at single households and shops, to cottage/small-scale industry levels.

Among the distinguishing features of the design is the aspect that it consists of discrete compartments that may be fed with different substrates yet processed simultaneously. The design is, on one hand, exceedingly space-efficient and, on the other, provides great flexibility in terms of the number of substrates that can be processed at a time, simultaneous use of different earthworm species, and scale of operation. The other major distinguishing feature is the system efficiency: it can achieve vermicast production at rates 4–6 times faster than pre-existing systems.

8.5.2 Description of the Invention

Each MCVS unit (Figures 8.18 and 8.19) typically includes four or more compartments arranged in a vertical sequence on a frame made of mild steel (Figure 8.18). The compartments are made of fibre-reinforced plastic to ensure high strength at low material density, adequate elasticity, and flexibility. The square shape is chosen for ease of handling, as well as economy of space. They can be sized to any length or breadth but must be 0.15 m high, have a modest slope of about 20, and be parallel to each other. The slope is to ensure easy collection and disposal of excess moisture in the modules. The height of a typical MCVS unit is 0.8 m.

The modules **(A),** which are loaded onto the stacks, are square-shaped (1 m × 0.6 m × 0.12 m) and made up of fibre-reinforced plastic to ensure high strength at low material density, adequate elasticity, and flexibility. The square shape is chosen for ease of handling, as well as economy of space. The feed is laid up to a height of 8 cm with a buffer space of 4 cm; the total depth of the module has been set as 12 cm. Each module has ball bearings arranged on the sides to enable it to slide to-and-fro over the inclined stacks.

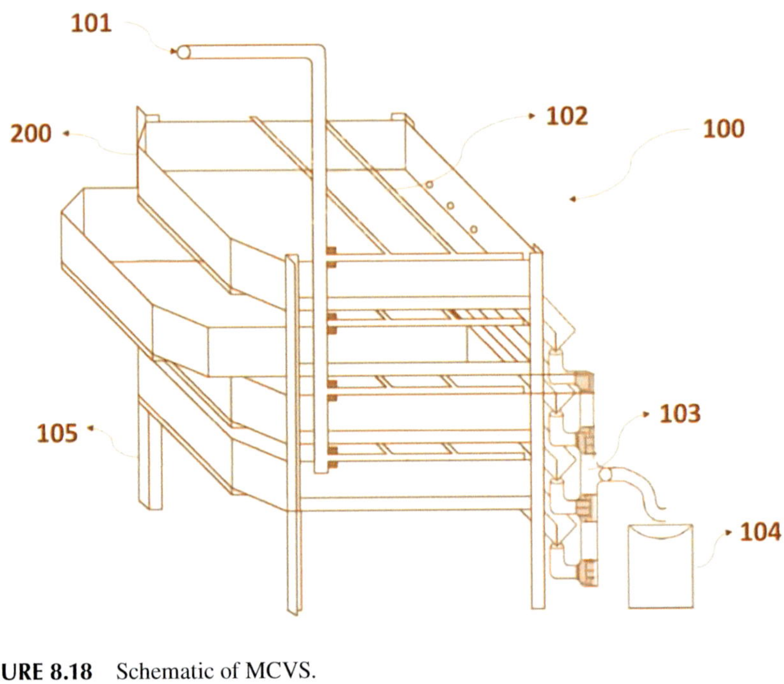

FIGURE 8.18 Schematic of MCVS.

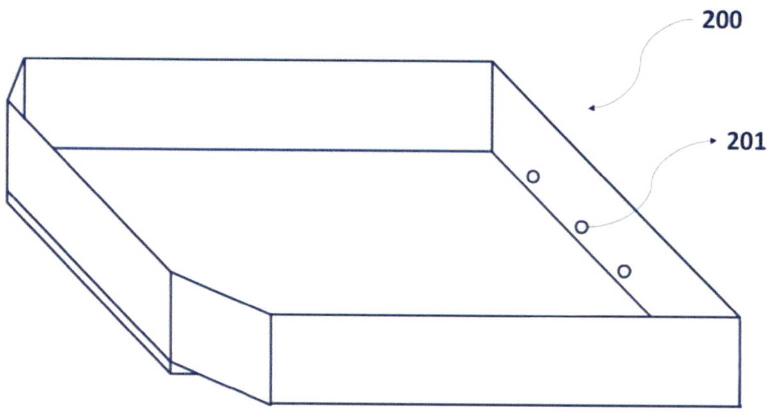

FIGURE 8.19 A typical MCVS drawer.

The unit has a sprinkler system **(B)**, which includes a motorized pump, and horizontal and vertical pipe lines with sprinklers. Each module is provided with a set of four sprinklers to release the water required in each module.

A drainage hole **(C)** is provided at the bottom of each module to facilitate the collection and disposal of excess moisture using a network of connected pipes **(D)**. The drain pipes collect excess moisture from all the stacks and send it into a collecting vessel **(E)**.

8.5.3 Distinguishing Features of the Invention

- The design is modular, with a multiple parallel stack system that can be easily scaled up or down to suit the desired plant capacity.
- It maximizes the use of floor area by optimal use of vertical space.
- Different modules can be loaded onto the unit simultaneously or at different times, depending on the availability of the substrate, and are ready for harvesting after a pre-determined number of days (usually 10–15).
- The shape of the modules is such that it optimizes the material cost and maximizes the surface area-to-volume ratio. It also facilitates handling.
- The modules are 1 m × 0.6 m in length and breadth, a size which is adequately large yet convenient to handle. It is not too large to endanger the mechanical stability of the overall structure, yet large enough to yield as much as 8´5 kg of vermicast per week.
- Modules are supported on an arrangement of ball bearings for easy movement along the tracks.
- Being only 0.8 m in height, the unit ensures easy loading and unloading of the modules.
- The sprinkler system is convenient to operate and control.
- The MCVS design is unique as it is easy to load/unload and maintain. The overall system also enables economy of space and materials.

- The reactor dimensions have been decided based on extensive experimentation. These dimensions are such that, on one hand, they prevent undesirable anaerobic conditions from developing and, on the other hand, facilitate very rapid vermicomposting. It enables vermicomposting at rates 4–8 times faster than the 4–6 months taken in existing systems.
- MCVS is scalable horizontally as well as vertically; its ability to extend vertically enables it to process a greater quantity of waste in a smaller floor area. The modular stacks can be replicated horizontally to process as much total waste as desired.
- As against systems based on large bins, the modular design of MCVS protects any contamination or pest attack occurring in a small fraction of the feed from spreading into the rest of the feed.
- MCVS is also unique in that it utilizes moist jute cloth of certain porosity and thickness as the earthworm bedding instead of the elaborate sawdust–sand–soil bedding used conventionally. This enormously enhances the effective reactor space and reduces system costs.
- MCVS enables different substrates to be vermicomposted simultaneously and also different species to be utilized simultaneously without the danger of interfering with each other.
- In the pre-existing domestic vermireactors, ventilation systems have to be provided for proper aeration, which increases the capital as well as operational costs. But MCVS does not require any extra ventilation system for aeration, as the substrate depth is only 8 cm and the module is open. Hence, it doesn't become oxygen deficient or get heated up (as deeply and widely as vermicomposting bins often do).
- In pre-existing designs such as Worm Wigwam, there is the probability of the earthworms getting injured or killed as a crank is used for harvesting. In case of a high-efficiency vermiculture process and apparatus, a strong light source or a fan blowing air is required to make the worms move away from the unloading end. These contraptions, which are stressful and potentially injurious, are not needed with MCVS.

9) The MCVS vermireactor system is by far the most versatile and flexible *vis-à-vis* the variety of substrates it can handle, species it can use, capacities it can handle, etc. It is also among the least expensive. The MCVS flow chart is presented in Figure 8.20.

8.6 DISTINCTIVE FEATURES OF THE FOUR DESIGNS

FLUVTS, HEVSPAR, IPSCO, and MCVS are all based on the HRVP, just as a very large number of previously reported reactors, described in Chapter 5, are based on conventional vermireactors. But in terms of substrate preference, market niche, range of viable scales, and associated operational details, each is clearly distinct from the others. This is reflected in the comparison between IPSCO and MCVS presented in Table 8.1.

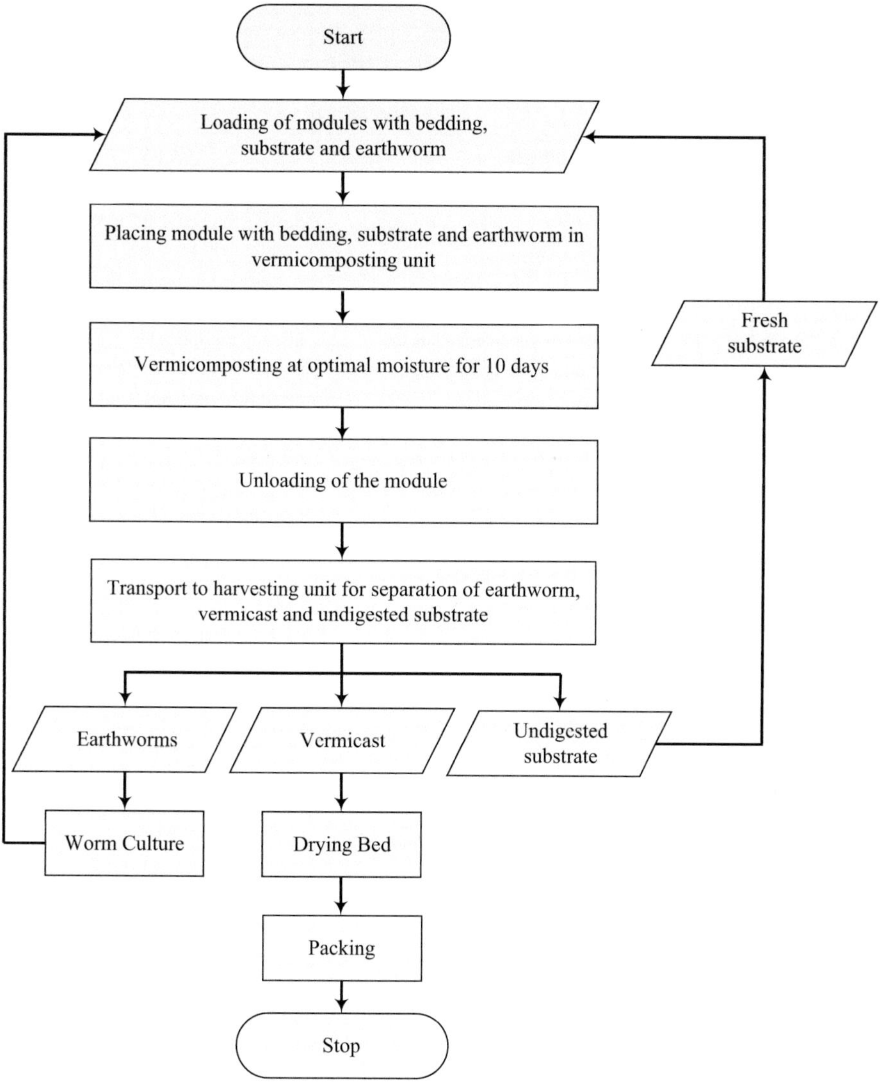

FIGURE 8.20 Flow chart of the MCVS vermireactor system.

TABLE 8.1
A Comparison of the Distinctive Features of IPSCO and MCVS

	Specialty of	
Aspect	***IPSCO***	***MCVS***
Scale of operation appropriate for the system	Medium to large industrial level	Single household and shop to cottage/industry/small industry level
Reactor configuration and operation	The system has been designed such that it can be increasingly mechanized as the scale of operation goes up and can be made almost fully mechanized.	The system remains labour-intensive, with only the sprinkler system having been mechanized, and other units not amenable to mechanization.
	The core design consists of a vertical frame supporting several rows of rails. Each rail can hold as also transport multiple rectangular, wheeled, vermireactors – called modules. It is in these modules that the substrate is loaded.	As the name suggests, it is a "stacked multiple drawer" system, wherein pull-out drawers are provided in a rectangular matrix. Each drawer serves as a stand-alone vermireactor.
	When the system is in operation, the wheeled modules loaded over the rails travel down the slightly inclined (5o) plane over a certain fixed duration (which corresponds to the solid retention time of the reactor). The movement occurs under the force of gravity, from the loading point to the unloading point, due to the mild gradient that has been provided.	There is no such arrangement here; each vermireactor is individually pulled out from its holding in the MCVS system, to be fed or emptied.
	The 5° slope of the rails is for the movement of the modules, as above.	The 2° slope of the individual drawers is to facilitate drainage if, accidentally, more water gets sprinkled than necessary.
	A typical IPSCO unit is over 5 m tall and needs a specially designed automated system for loading and unloading the vermireactors.	An MCVS is under 1.5 m tall, and its reactors are drawn out and placed back in manually.
	IPSCO is a pseudo-continuous operation; essentially, a pulse-fed semi-continuous process operated in a sequential batch manner to achieve the ambiance of a continuous process.	MCVS is operable only in fully batch mode or semi-continuous mode.

(*Continued*)

TABLE 8.1 (*Continued*)
A Comparison of the Distinctive Features of IPSCO and MCVS

	Specialty of	
Aspect	***IPSCO***	***MCVS***
Nature of scaling up	Can be scaled up horizontally as well as vertically.	Individual units can only have a limited range of sizes (as is for fridges and almirahs) because they are essentially meant, at the lowest end of the scale, to be operated by a single lady or gent of a normal height. Moreover, the nature of the system's operation is essentially manual. Hence, scaling up of production is basically in terms of using a larger number of stand-alone systems.
System economics	Relatively much higher capital costs but much greater profitability due to more favourable economies of scale.	Relatively much lesser capital costs. Units at the household level can be put up with a few hundred rupees.

REFERENCES

Abbasi, S.A., Hussain, N., Tauseef, S.M. and Abbasi, T. 2018. A novel FLippable Units Vermireactor Train System – FLUVTS – for rapidly vermicomposting paper waste to an organic fertilizer, *Journal of Cleaner Production, 198*, pp.917–930. doi.org/10.1016/j.jclepro.2018.07.040

Abbasi, S.A., Nayeem-Shah, M., Abbasi, T., 2015. Vermicomposting of phytomass: limitations of the past approaches and the emerging directions. *Journal of Cleaner Production*, *93*, pp.103–114.

Abbasi, T., Gajalakshmi, S., and Abbasi, S.A., 2009. Towards modeling and design of vermicomposting systems: mechanisms of composting/vermicomposting and their implications. *Indian Journal of Biotechnology, 8*, pp.177–182.

Abbasi, T., Tauseef, S.M. and Abbasi, S.A., 2011. The inclined parallel stack continuously operable vermireactor, *Official Journal of the Patent Office*, 22, p.*9571*. Patent No. 302643.

Edwards, J.H., 1997. Composition and Uses of Uncomposted Wastepaper and Other Organics. In: Rechcigl, J.E. and MacKinnon, H.C. (eds) *Agricultural Uses of By-Products and Wastes*. ACS Publications, Washington, USA, pp.163–184.

Gupta, R. and Garg, V.K., 2009. Vermiremediation and nutrient recovery of nonrecyclable paper waste employing *Eisenia fetida*. *Journal of Hazardous Materials*, *162*, pp.430–439.

Hummel, Ma, Y., Maattanen, M.M.M.S.A., Sarkilahti, M., Harlin, A. and Sixta, H., 2016. Upcycling of waste paper and cardboard to textiles. *Green Chemistry*, *18*, pp.858–866.

Mupambwa, H.A., Ravindran, B. and Mnkeni, P.N.S., 2016. Potential of effective microorganisms and Eisenia fetida in enhancing vermi-degradation and nutrient release of fly ash incorporated into cow dungepaper waste mixture. *Waste Management, 48*, pp.165–173.

Mupondi, L.T., Mnkeni, P.N. and Muchaonyerwa, P., 2011. Effects of a precomposting step on the vermicomposting of dairy manure-waste paper mixtures. *Waste Management and Research*, *29*, pp.219–228.

Nath, G. and Deb, U.K., 2008. Vermicomposting-an effective technique for paper mill solid waste utilisation and value addition. *IPPTA: Quarterly Journal of Indian Pulp and Paper Technical Association A–, 20*, pp.127–135.

Ravindran, B. and Mnkeni, P.N.S., 2016. Bio-optimization of the carbon-to-nitrogen ratio for efficient vermicomposting of chicken manure and waste paper using *Eisenia fetida. Environment and Science Pollution Control Series, 23*, pp.16965–16976.

Sharma, K. and Garg, V.K., 2018. Comparative analysis of vermicompost quality produced from rice straw and paper waste employing earthworm *Eisenia fetida* (Sav.). *Bioresource Technology, 250*, pp.708–715.

Tauseef, S.M., Abbasi, T., Banupriya, D., Vaishnavi, G., and Abbasi, S.A., 2013a. HEVSPAR: A novel vermireactor system for treating paper waste. *Official Journal of the Patent Office, 24*, pp.*12726*, Patent No. 312587.

Tauseef, S.M., Abbasi, T., Tabassum-Abbasi, Vasigui, T., and Abbasi, S.A., 2020. A novel multiple compartment vermicomposting system (MCVS) for simultaneous and rapid processing of different substances. *Official Journal of the Patent Office 39*, p.1–23. Appl. No. 201911010438 K&S (Nikita Anand) Registered.

Tauseef, S.M., Premalatha, M., Abbasi, T. and Abbasi, S.A., 2013b. Methane capture from livestock manure. *Journal of Environmental Management, 117*, pp.187–207.

Unuofin, F.O., and Mnkeni, P.N.S., 2014. Optimization of Eisenia fetida stocking density for the bioconversion of rock phosphate enriched cow dungewaste paper mixtures. *Waste Management. 34*, pp.2000–2006.

Unuofin, F.O., Siswana, M. and Cishe, E.N., 2016. *Enhancing Rock Phosphate Integration Rate for Fast Bio-Transformation of Cow-Dung Waste-Paper Mixtures to Organic Fertilizer*. Springer Plus 5, 1986.

Van Wyk, J.P.H., Mogale, M.A. and Moroka, K.S., 1999. Bioconversion of waste paper materials to sugars: an application illustrating the environmental benefit of enzymes. *Biochemistry and Molecular Biology Education, 27*, pp.227–228.

9 Design and Performance Evaluation of Machines for Separating Earthworm-Vermicast-Substrate Mixtures

9.1 INTRODUCTION

In Chapter 4, we have presented a large variety of conventional vermicomposting systems in which, upon conversion of most of the present batch of feed into vermicast, earthworms are separated by making them migrate towards a new batch of feed. This is accomplished either by providing ways by which earthworms migrate from the present batch to the new one, or a conveyor belt is used, which slowly but continuously keeps bringing new feed to the earthworms even as they finish feeding upon most of the previous feed.

But this manner of separation is not only very slow but is unable to separate undigested substrate from the vermicast.

As earlier elaborated in Chapters 6 and 7, the high-rate vermicomposting paradigm (HRVP) – which has enabled direct and rapid vermicomposting of phytomass, phytowaste, and other forms of cellulosic waste such as discarded paper – has brought billions of tonnes of feedstock within the preview of vermicomposting at a commercial scale, which had been largely confused to animal manure. Given that at any time there is 1,000 times more plant biomass (phytomass) in the world than zoomass, the total quantity of manure that can be processed by vermicomposting is miniscule in comparison to the enormous quantities of phytomass which is not only available for vermicomposting into organic fertilizer but is in crying need for being so utilized.

These developments have made it imperative that not only vermicomposting is made simpler and faster – hence more remunerative – with the HRVP, but better ways should also be found to carry out rapid and clean-cut separation of vermicompost, earthworms, and undigested substrate (if any).

9.2 THE SEPARATOR OF EARTHWORMS, VERMICOMPOST, AND SUBSTRATE (SEVS) MACHINES

Figures 9.1 and 9.2 show two embodiments, or models, of the SEVS machines. The machines consist of (a) a light source, (b) a shaker, (c) a set of sieves of appropriate pore sizes, and (d) supporting forms, a cam, a pulley, and a matter.

DOI: 10.1201/9780429331886-11

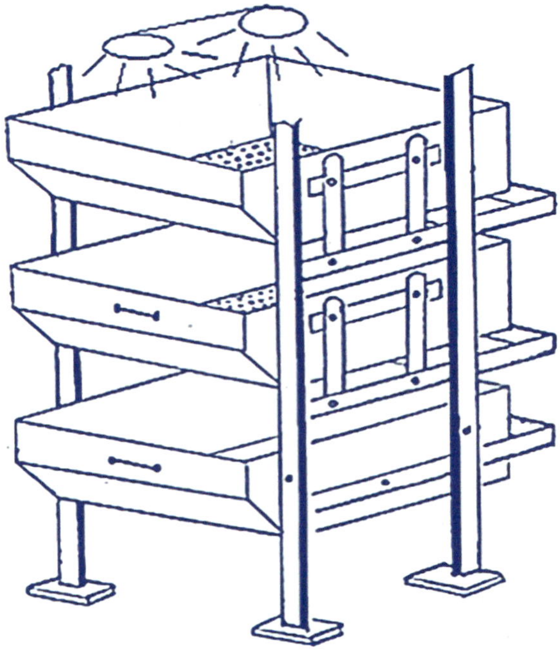

FIGURE 9.1 An embodiment of a SEVS system.

A typical SEVS consists of three modules, which are placed one over the other vertically with a gap of 0.06m. A typical SEVS has dimensions 1.1×0.6×1.1m, incorporating the first module of dimensions 0.75×0.45×0.3m, while the other two modules cover 0.75×0.45×0.2m each.

The separation of earthworms is speeded up in SEVS by a source of light, which hastens the migration of the earthworms from the undigested substrate in the upper module through a 6mm sieve to the middle module with a 2mm sieve. The separation of the vermicast occurs in the middle module. To facilitate and expedite the sieving, a motor is provided, which runs at 200rpm, causing the sieves to vibrate. Due to the cam mechanism (Figure 9.3), a shaking action takes place only in the first and second modules, while the third module serves as a vermicast collection unit. After the vermicast, substrate, and earthworm mixture is added to the first module, the entire setup is kept under light for 5min. The intensity of light and consequent increase in the air temperature cause the earthworms to migrate towards the middle module, and undigested substrate remains in the first sieve. The harvester is made to run for 5 minutes so that the earthworms and the vermicast fall below to the next sieve of size 2mm. The vermicast particles, which are smaller than the pore size of the sieve, fall into the collection unit.

After the SEVS removal of the undigested substrate from the earthworms and the vermicast, the corresponding modules are taken out by disassembling the respective containers.

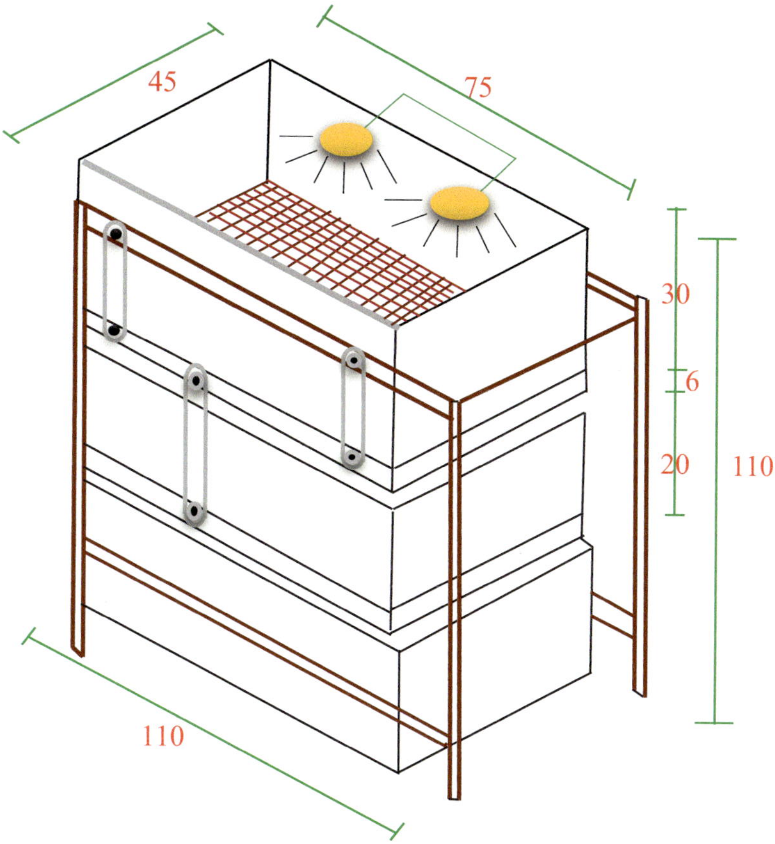

FIGURE 9.2 Another embodiment of a SEVS machine.

9.3 TESTING OF THE SEVS

The efficacy of the SEVS machine was tested on different types of earthworm–vermicast–substrate (EVS) combinations, resulting from the vermicomposting of paper waste, cow dung, and different species of weeds. As fresh vermicast tends to be sticky, the EVS mixtures were air dried for a few minutes before transferring them into the upper module of the SEVS. Their moisture content was also measured. The SEVS was then exposed to light for 5 minutes, which induced the earthworms to move from the upper module to the middle one.

The SEVS motor was then run for 5 min. It caused the separation of the contents in their respective modules, which were emptied and weighed.

To get an idea of the time needed for the adequate drying of the vermicast, 2, 4, 6, 10, 15, and 20 kg of the cow dung vermicast–earthworm–substrate mixtures were air dried for different durations: 0, 1, 4, 6, and 24 hours. They were then subjected to SEVS separation.

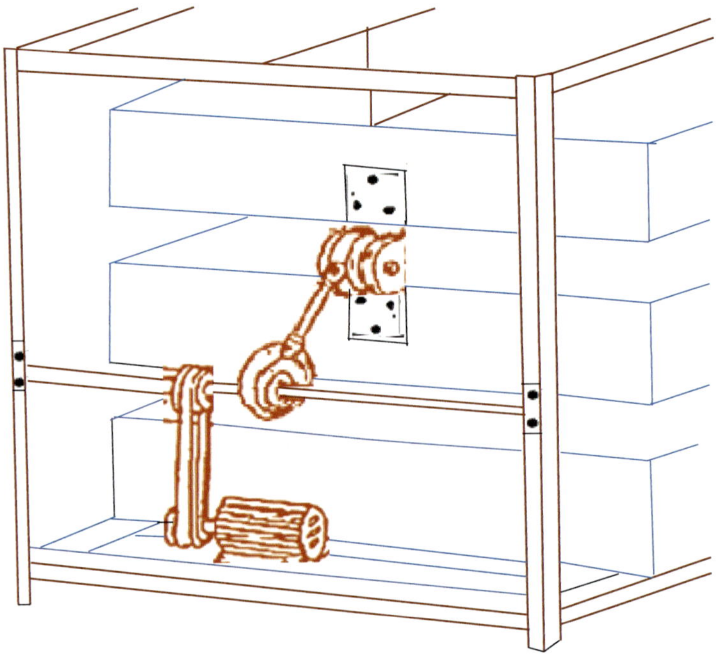

FIGURE 9.3 The cam mechanism used in the SEVS machine.

9.3.1 Earthworm–Vermicast–Paper Waste Mixtures

The results of duplicate runs, each with 5 kg of EVS mixture, are presented in Table 9.1. The substrate and the earthworms were concentrated in Module 2, while the bulk of vermicast was in Module 3. The separation of most of the substrate and the earthworms in Module 2 is particularly useful because the mixture can be recycled in the vermireactor along with fresh feed to continue the vermicomposting.

9.3.2 Earthworm–Vermicast–Salvinia Mixtures

As seen from Table 9.2, most (66.0%, 73.0%, 73.9%, and 79.0%) of the vermicast was separated in the third module, with negligible undigested substrate (<2.9%) and earthworms (<4%) coming there. As a significant fraction (17.1%–32.5%) of the vermicast was found in the second module, that mixture was run on SEVS again for 5 minutes. It led to near-total (≥93.5) separation of earthworms.

9.3.3 Earthworms–Vermicast–Ipomoea Mixtures

Quadruplicate runs (Table 9.3) show very clean separation of earthworm (≥92.2%) in the second module. The bulk of the substrate (97.3–99%) was also separated into the first or second module. As appreciable quantities (17.1%–32.5%) of vermicast had

TABLE 9.1
Separation of Paper–Earthworm–Vermicast Mixtures by the SEVS Machine

		Proportion (%) of		
PEV Mix (kg)	*Module*	*Vermicast Mass (%)*	*Substrate Mass (%)*	*Earthworm Numbers*
Run 1:5	1	6.6	16.2	8.0
	2	7.5	83.6	90.1
	3	85.9	0.3	2.0
Run 2:5	1	2.6	16.3	4.5
	2	2.9	82.9	91.9
	3	94.6	0.8	3.5

TABLE 9.2
Separation of salvinia–earthworm–vermicast by the SEVS machine

		Proportion (%) of		
Salvinia–Earthworm–Vermicast Mix (kg)	*Module*	*Vermicast Mass (%)*	*Substrate Mass (%)*	*Earthworm Numbers (%)*
Run 1 2	1	1.53	1.57	1.7
	2	2.5	65.49	94.9
	3	66.0	2.91	3.4
Run 2 1.5	1	2.8	39.3	2.6
	2	23.3	58.7	93.5
	3	3.9	2.0	3.9
Run 3 1.5	1	3.6	45.5	2.7
	2	23.4	52.1	93.6
	3	3.0	2.4	3.6
Run 4 1.5	1	3.9	54.0	2.4
	2	17.1	43.7	93.7
	3	9.0	2.4	4.0

remained in the second module, one more run of the SEVS machine was carried out, which led to near complete separation.

9.3.4 Earthworm–Vermicast–Lantana Mixtures

The findings of quadruplicate runs were similar to those obtained with salvinia and ipomoea. Once again, clean separation (≥93.3%) of earthworms occurred in the middle module, and the third module had most of the vermicast with negligible undigested substrate (≥2.2%) or earthworm (≤3.9%). Once again, the middle module had significant (22.5%–53.5%) vermicast. Hence, another round of SEVS operation was carried out to achieve near-total separation.

TABLE 9.3
Separation of Ipomoea–Earthworm–Vermicast by the SEVS Machine

		Proportion (%) of		
Ipomoea–Earthworm–Vermicast Mix (g)	*Module*	*Vermicast Mass*	*Substrate Mass*	*Earthworm Numbers*
Run 1 2,076	1	1.4%	8.8%	1.1%
	2	45.02%	88.5%	96.0%
	3	53.6%	2.7%	2.82%
Run 2 1,640.4	1	0.8%	20.2%	3.1
	2	49.2%	78.9%	92.4%
	3	50.0%	0.8%	4.6%
Run 3 1,781.9	1	1.1%	25.6%	3.5%
	2	58.9%	73.4%	92.2%
	3	288.9 g, 40.0%	14.3 g, 1.1%	5, 4.3%
Run 4 1,809.4	1	7.7 g, 0.9%	459.3 g, 31.2%	4, 2.9%
	2	329.6 g, 40.2%	995.98 g, 67.7%	128, 92.8%
	3	482.1 g, 58.8%	16.97 g, 1.2%	6, 4.3%

9.4 EFFECT OF MOISTURE ON THE EFFICIENCY OF VERMICAST RECOVERY

The findings are summarized in Figure 9.4. Reducing the moisture contents of the weeds led to improved vermicast recovery of the order of 14% in the case of salvinia, 19% in the case of ipomoea, and 30% in the case of lantana were reduced. It follows that drier vermicast gets sieved more easily and completely in the SEVS machine than does wetter vermicast.

9.5 EFFECT OF THE BATCH SIZE

Earthworm–vermicompost–cow dung mixtures of batch sizes ranging from 2 to 20 kg were processed by SEVS with air drying for periods ranging from 0 to 24 hours. As seen in Table 9.4, the extent of separation improved with the duration of the drying. But for the SEVS machine of the size we were using, only batches up to 15 kg achieved separation efficiencies reaching a maximum of 99%. However, in the 20 kg batches, only up to 21% vermicast could be separated. The earthworm recovery (Table 9.5) had a similar general trend. But, in contrast to the erratic pattern of vermicast harvest (Table 9.4), the earthworm recovery displayed a uniform pattern showing better retrieval with increasing duration of air drying (Table 9.5). Even the 20 kg batch achieved 96% recovery when air dried for 24 hours.

To achieve still better separation, the contents of all the second modules covered in Table 9.4 were re-processed with SEVS for 3 minutes each. It led to near quantitative separation in most batches (Table 9.6). This indicates that, in general, an

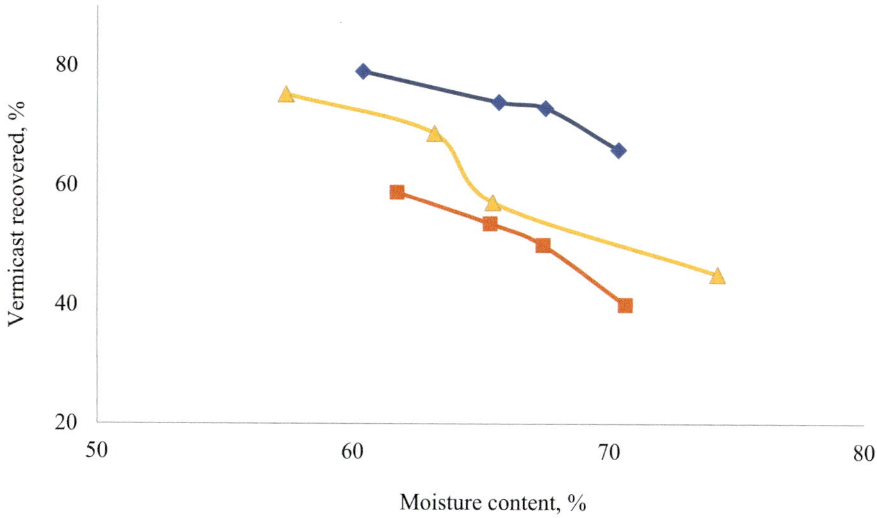

FIGURE 9.4 Vermicast recovery as influenced by the moisture content. —■— ipomoea, —▲— lantana, —◆— salvinia).

TABLE 9.4
Effect of Batch Size and Its Drying on Vermicast Recovery

	Vermicompost Recovered, (%), After Drying for				
Batch Size, kg	***0 hour***	***1 hour***	***4 hours***	***6 hours***	***24 hours***
2	72	77	88	95	98
4	75	78	67	62	99
6	79	54	63	62	98
10	77	64	79	66	65
15	43	46	48	85	99
20	4	9	21	20	21

TABLE 9.5
Effect of Batch Size and Its Drying on Earthworm Recovery

	Earthworm, (%), Recovered After Drying the Batch for				
Batch Size	***0 hour***	***1 hour***	***4 hours***	***6 hours***	***24 hours***
2	73	91	90	85	91
4	85	92	92	93	95
6	92	93	94	94	94
10	95	93	96	95	96
15	91	93	95	98	99
20	23	36	54	61	96

TABLE 9.6
Improvement in the Vermicast Recovery upon a 3-Minute SEVS Re-Processing of the Second Module Contents

Size of the Batch to Which the Second Module Contents Belonged	*Vermicompost Recovery, %, Corresponding to the Second Module Contents of the Batches which Had Been Air Dried for*				
	0 hours	*1 hours*	*4 hours*	*6 hours*	*24 hours*
2	95	99	99	99	100
4	95	99	98	93	100
6	96	99	88	90	100
10	96	92	98	99	98
15	–	67	84	99	–
20	–	–	–	–	81

initial 5-minute operation of SEVS, followed by 3 minutes of SEVS operation of the contents of the second module as obtained in the initial separation, can ensure near complete separation of vermicast on one hand, and substrate-cum-earthworms on the other hand.

10 Effect of Vermicomposts Derived from Phytomass and Waste Paper on Different Species of Commonly Used Plants[1]

10.1 INTRODUCTION: THE PIVOTAL ROLE OF MANURE AS A FEEDSTOCK IN VERMICOMPOSTING

Traditionally, both vermiculture and vermicomposting have revolved around the use of animal manure (Ramos et al., 2022; Anandyawati et al., 2023; Kinigopoulou et al., 2023). Vermiculture, which aims to generate earthworms for use as baits, as poultry feed, as parts of folk medicine, etc, is generally done over animal manure. This is because epigeics like *Eisenia fetida*, *E. andrei*, and *Eudrilus eugeniae* – which have come to be called "manure worms" – tend to get naturally attracted towards animal manure and thrive in it, reproducing well and growing vigorously. As for vermicomposting, animal manure has traditionally been the first choice as a substrate due to these reasons:

a. The manures of cows, buffaloes, pigs, horses, and other livestock are decent fertilizers by themselves. They have been used in agriculture for centuries. They provide nutrients to the soil, improve its porosity and water retention ability, act as pest repellents and as anti-pathogens, and attract such invertebrates whose feeding and burrowing activities improve soil fertility (Rostaei et al., 2024). All these advantages increase multi-fold if the manure is first composted or vermicomposted before it is applied. Vermicomposting is more effective than composting because it leads to greater bioavailability of nutrients, better impact on soil texture, greater ability to suppress pathogens, and greater potential to repel pests. Additionally, it provides enzymes, hormones, and microflora, which are beneficial to the growth and yield of the fertilized plants (Wong et al., 2020). The worm mucus contained in the vermicompost (VC) prevents the washing away of nutrients when it rains. Processing by earthworms also renders heavy metals less bioavailable. All these factors, together, make the VC of manure a more prized commodity than the manure's compost.

[1] This chapter has been co-authored with Naseer Hussain, School of Life Sciences, B.S. Abdur Rahman Crescent Institute of Science and Technology, Chennai 600048, India.

DOI: 10.1201/9780429331886-12

b. In contrast to plant debris, many streams of horticultural waste (such as stubble, husk, straw), waste paper, and animal manures are easier to ingest for earthworms. Most have a lower C/N ratio than phytomass generally has, and usually have a higher concentration of nutrients. For these reasons, earthworms tend to begin voraciously feeding upon animal manure, especially of cows/buffaloes, from day one (Raza et al., 2022). On the other hand, several days of incubation are needed before epigeics take to substrates other than manure.
c. When manure is the source of earthworm nutrition, there is almost never a sudden earthworm die-off. In contrast, die-offs can occur with other feeds if one or other factor – such as stratified moisture, some leachate, or sudden change of habitat – causes a sudden shock to the earthworms.

Due to the special place manures have in the choice of earthworm feed, it is not uncommon to begin vermicomposting with one or the other animal manure, even when the end objective is to vermicompost food, vegetable, or some other type of waste. After beginning with animal manure as the sole feed, the proportion of other waste is slowly increased, but it is also not uncommon to keep 50%–75% of the feed as animal manure, with the rest made up of non-manure components.

10.2 BUT WHAT ABOUT VC OF OTHER SUBSTRATES?

We had carried out (Hussain & Abbasi, 2018) an extensive assessment of prior art in which studies have been described on the effect on germination, growth, and fruit yield of various species of plants as influenced by VCs. The survey shows that, barring 16 studies of these authors, very few other reports exist in which VC of a substrate other than animal manure has been described. The vast majority of the past reports are either on VC of animal manure or on substrates which come either pre-composted and/or blended with animal manure before being subjected to the action of earthworms. Moreover, as has earlier been dwelt at length in Chapters 6 and 7, the majority of past reports do not represent VC as a finalized and reproducible product but as something that has been worked upon by earthworms for an arbitrarily decided duration. Worse, this duration has varied from author to author (Abbasi et al., 2015). Some authors have deemed vermicomposting as accomplished when the C/N ratio of the vermireactor contents has fallen below 20. Some others have even relied on visual observation of colour change of the substrate–earthworm–vermicast mixture as a sign of completion of vermicomposting. A similar manner of ad-hoc and inconsistent approach is witnessed in most studies on the effect of non-manure VCs on different species of plants. For instance, Kadam and Pathade (2014) inoculated 1 kg of shredded and pre-composted leaves of tendu (*Diospyros melanoxylon*) with just 20 individuals of *Eudrilus eugeniae* and left the mixture for 10 days. They then sieved the mixture, calling it "vermicomposted material", and used it to assess its effect on the germination/growth of French bean (*Phaseolus vulgaris*). There is no mention of the pore size of the sieve and no detail on what fraction of the starting substrate had passed through the sieve. This has left the methodology difficult to understand and use by other authors.

In another paper published in the same year as the above-mentioned paper, by Mendoza-Hernandez et al. (2014), a substrate seeded with "approximately 25,000 individuals of *Eisenia fetida* per m^3" was kept for as long as 198 days before the authors assumed that vermicomposting had occurred. No details have been given on why the duration chosen was 198 days and not shorter or longer. For another VC, the same authors provided 261 days of substrate–earthworm interaction, without giving any rationale as to why the duration this time was 63 days longer than that of the other vermicomposting.

In like manner, one finds that in most other past studies, no rational or reproducible criterion and/or method has been used to produce that which the authors have termed "vermicompost". Due to this, it is very difficult to evaluate from the past studies the efficacy of VCs of non-manure substrates as fertilizers. It is well-known (Silva & Uehida, 2000; Jones, 2012; Hussain and Abbasi, 2018) that fertilizers benefit plants and soil when applied up to certain concentrations. Beyond those levels, fertilizers can cause adverse impacts. Unless the dividing line that separates beneficial concentrations from excessive ones is known with certainty for a given fertilizer, based on its well-defined and consistent characteristics, it can never be safe to apply that fertilizer. This makes it imperative that the performance of VCs is assessed with a rationally developed and consistently applied methodology. And such a methodology has to be necessarily based on the use of a clearly defined and reproducible material of fairly consistent characteristics, which we designate as the fertilizer.

VCs have been demonstrated to offer a range of benefits that surpass those of traditional composts and manures in agricultural applications. These advantages, supported by various studies, make VCs a valuable resource for improving soil fertility and crop productivity:

1. *Enhanced nutrient content:* VCs exhibit superior nutritional profiles compared to conventional composts. This is attributed to more efficient mineralization processes and a higher degree of humification, which is facilitated by earthworm activity (Patra et al., 2022).
2. *Increased availability of essential nutrients:* VCs are rich in readily available plant nutrients, including nitrates, phosphates, exchangeable calcium, soluble potassium, and trace elements. When incorporated into soil, VCs increase the availability of these micro- and macronutrients, thereby promoting healthier plant growth and improving yields (Singh et al., 2020; Ahmad et al., 2022).
3. *Humus enrichment for nutrient retention:* VCs contribute substantial amounts of humus, which plays a critical role in nutrient retention and release. The negatively charged components in humus effectively bind cations, allowing VCs to function as slow-release fertilizers, ensuring a sustained nutrient supply for plants (Thakur et al., 2021; Villarreal et al., 2021).
4. *Improvement of soil physical properties:* Rich in organic matter, VCs enhance soil structure by increasing porosity, improving aeration, and enhancing water retention. These changes lead to reduced soil bulk density and particle density, contributing to better root development and overall plant performance (Chatterjee et al., 2020; Oyege & Bhaskar, 2023).

5. *Stimulation of soil microbial activity:* VCs introduce a diverse array of beneficial microorganisms that produce plant growth regulators, enzymes, and hormones such as cytokinins and auxins, which enhance plant growth and health (Munoz-Ucros et al., 2020; Iqbal et al., 2024).

In contrast, VCs can exert a negative influence if applied in concentrations that are higher than necessary. The possible reasons for it, to quote from Hussain and Abbasi (2018), are:

1. *Nutrient overload:* When VCs are applied beyond the necessary concentrations, the excess nutrients can adversely affect plant growth, similar to how over-fertilization inhibits germination. Excessive nutrient availability can disrupt the nutrient balance in the soil, impairing plant growth (Silva & Uchida, 2000; Jones, 2012; Hussain et al., 2015).
2. *Salinity stress:* High levels of VC application can increase soil salinity, which inhibits seed germination and plant growth. Elevated salinity reduces water uptake by seeds and roots, leading to impaired root elongation and stunted growth in plants (Vibhuti et al., 2015; Uçarlı, 2020).
3. *Phenolic and humic compounds:* Excessive levels of active phenolic and humic substances present in VCs can inhibit plant growth. These compounds, involved in plant allelopathy, can interfere with multiple physiological processes:
   a. *Membrane permeability*: Phenolic compounds increase cell membrane permeability, leading to leakage of cell contents and subsequent tissue death. This impairs nutrient absorption and slows overall plant development.
   b. *Root and cell growth*: High concentrations of phenolics disrupt root elongation and plant cell division, and alter cellular ultrastructure, hampering normal plant growth (Ievinsh, 2011; Manivannan & Selvamani, 2014).
   c. *Photosynthesis and respiration*: Phenolic compounds weaken oxygen absorption and reduce chlorophyll content, impairing photosynthesis and slowing plant respiration. In soybean (*Glycine max*), compounds such as caffeic acid and ferulic acid have been shown to reduce chlorophyll content and inhibit growth at concentrations of 10–30 μmol/L (Patterson, 1981).
   d. *Enzyme activity*: Certain phenolic compounds inhibit enzymatic activities, such as phosphorylase and ATPase, disrupting metabolic processes. Chlorogenic acid and cinnamic acid derivatives are known to impede these vital functions, further affecting plant development (Sun et al., 2020; Pratyusha, 2022).
   e. *Hormonal imbalance*: Phenolic compounds interfere with plant hormones like indole acetic acid and gibberellin, inhibiting their decomposition and altering normal hormonal regulation. This can significantly hinder plant physiological activities (Leslie & Romani, 1998; He & Lin, 2001; Zeng et al., 2001; Tanase et al., 2019).

4. *Heavy metal toxicity*: Excessive VC application can also lead to elevated levels of heavy metals, which, in conjunction with increased salinity and nutrient concentrations, can suppress plant growth. Heavy metals are known to inhibit plant physiological and biochemical activities, disrupt cellular structures, and limit overall growth when present above threshold concentrations (Atiyeh et al., 2000; Arif et al., 2016).

Given this reality, no study on the effect of a fertilizer on the germination/growth/yield of any plant can be meaningful unless it covers a wide enough spectrum of VC concentrations. Only such studies can reveal the beneficial range of that VC's application, as well as the limit above which that VC will become harmful to a given species of plant.

10.3 STUDIES BY S.A. ABBASI AND CO-WORKERS

10.3.1 The Broad Scope

To cover the major knowledge gap mentioned above, S.A. Abbasi and co-workers had launched a programme of extensive investigations from 2013 onwards, which had the following components:

10.3.1.1 Comparative Study on the Fertilizer Value of the VC of Four Dreaded Weeds

This set of investigations involved a comparative study of the effect of VC on four of the most widely distributed weeds of the world. Each of these weeds is notorious for its very high net productivity, resilience, intransigence, and damage potential. The species included a free-floating aquatic weed salvinia (*Salvinia molesta*), an amphibious weed ipomoea (*Ipomoea carnea*), and two terrestrials lantana (*Lantana camara*) and parthenium (*Parthenium hysterophorus*). Each of these weeds is strongly allelopathic, and all except salvinia cause toxicity to soil and animals. Together, the four weeds generate billions of tonnes of phytowaste per year, which, as detailed in Chapters 6 and 7, is highly problematic.

VC of all four weeds was generated by invoking the high-rate vermicomposting paradigm (HRVP), without adding cow dung (CD) or any other supplementation. In order to make the comparison independent of the earthworm species, *Eisenia fetida* was used to separately vermicompost each weed. In all cases, the vermicast was taken as the final product of the action of earthworms on the respective weeds – i.e., vermicomposting or earthworm-based zoocomposting (EBZ).

The effect of each VC was studied on the germination and early growth of three common vegetables: ladies finger (*Abelmoschus esculentus*), cucumber (*Cucumis sativus*), and green gram (*Vigna radiata*) at VC applications ranging from 0.75% to 40% by weight of the soil.

To validate these and other bench-scale studies, field-scale investigations were carried out as mentioned in upcoming Sections 10.3.1.3–10.3.1.5, with detailed monitoring of plants and soil right from the seed germination up to the quantification of yield.

10.3.1.2 Assessment of Fertilizer Value of VCs Derived from Paper Waste

As paper waste is very lean in nutrients, earthworms cannot remain healthy if made to feed upon it for more than a few days. Past work on vermicomposting of paper waste, reviewed in Abbasi et al. (2018), has been based on overcoming this problem by fortifying paper waste with 50%–75% of animal manure by weight. This enables the earthworms to survive and grow, but makes the so-called paper waste VC (PWVC) more of a manure VC. The requirement of so much manure per tonne of paper makes the vermicomposting option for paper waste valorization unviable because manure is a prized commodity with several competing uses. We have found that if the HRVP is invoked, rapid and sustained vermicomposting of paper waste can be accomplished with the addition of just 7%–10% by mass of CD. Even the need for this much CD can be avoided by using phytowaste in place of animal manure as the source of essential nutrients.

In this component of the above-mentioned programme of studies, the fertilizer value of the PWVC, generated by us through the HRVP, has been assessed.

10.3.1.3 Field-Scale Validation of Some of the Above-Mentioned Studies Encompassing the Full Germination-Transplanting-Growth-Flowering-Yield-Harvesting Cycle

After the bench-scale experiments as above had revealed consistently beneficial impact of all the VCs on all the species of plants, and had given an indication of the range of VC concentration within which the effect of that VC is favourable, field-scale studies were then carried out in which ladies finger was cultivated from germination to fruition with lantana VC (LVC), ipomoea VC (IVC), parthenium VC (PVC), and salvinia VC (SVC) as fertilizers.

10.3.1.4 A Comparative Study of the Fertilizer Value of LVC and Equivalent Doses of Synthetic Chemical Fertilizers (SCF)

After the previous study had conclusively shown that the VCs of all four weeds stimulated germination and early growth of all three species of plants that were studied, and also exerted a positive impact on the underlying soil, this set of studies was carried out. The central question addressed was: Is the LVC as beneficial, or lesser/more beneficial, as the SCF?

10.3.1.5 A Comparative Study of the Efficacy of LVC Relative to the CD VC

Given the near-universal belief that VCs derived from animal manure are benign, plant friendly, and soil friendly fertilizers with good pest repellent properties, this study aimed to see whether LVC suffers in comparison or is as good/better.

In this manner, a systematic assessment, in adequate breadth and depth, was made on the application of clearly defined VCs of different weeds and paper waste, on different plant species, at a wide range of VC applications. This was followed by field-scale studies on all stages of cultivation of a chosen species (ladies finger, *A. esculentus*) right up to the fruit (pod) yield. In all studies, pesticidal effect of the different VCs was also studied along with the effect the VCs had on the physical characteristics of the soils to which the VCs had been applied.

10.3.2 Key Steps of the Methodology

The methodology had the following key steps:

i. Healthy, mature plants of the target weed species were selected from their natural habitats (Figures 10.1–10.4). They were carefully harvested, cleaned thoroughly with water to remove any soil or surface contaminants, and gently wiped to eliminate excess moisture. They were then directly subjected to vermicomposting using *Eisenia fetida* in high-rate, pulse-fed vermireactors, as previously described by Ganeshkumar et al. (2014) and Nayeem-shah et al. (2015). Notably, the process did not involve pre-composting or the addition of manure. Vermicast was regularly harvested in each pulse, and the product was quantified precisely by determining the dry weight after oven drying the samples to a constant weight at 105°C. For paper waste, a similar vermicomposting procedure was applied; however, it was enriched with 7%–10% fresh CD by mass to provide additional nutrients.

ii. In separate experimental setups, 60 seeds each of green gram (*Vigna radiata*), okra (*Abelmoschus esculentus*), and cucumber (*Cucumis sativus*) were randomly selected and sown into plastic trays (36 cm × 42 cm × 10 cm) filled with soil amended with varying concentrations of VC derived from the respective weeds. The VC levels applied were 0% (control), 0.75%, 1.5%, 2%, 4%, 8%, 20%, and 40% by weight. Three replicates were conducted for

FIGURE 10.1 Lantana (*Lantana camara*).

FIGURE 10.2 Ipomoea (*Ipomoea carnea*).

FIGURE 10.3 Parthenium (*Parthenium hysterophorus*).

FIGURE 10.4 Salvinia (*Salvinia molesta*).

each treatment and plant species (Figure 10.5). The trays were maintained under controlled conditions with daytime temperatures of 32°C ± 3°C and nighttime temperatures of 27°C ± 2°C, while relative humidity was kept between 55% and 70%. Germination was defined by the emergence of a radicle measuring greater than 3 mm. Similar experimental protocols were followed for other crops, including cluster beans (*Cyamopsis tetragonoloba*), chilli (*Capsicum annuum*), and tomato (*Solanum lycopersicum*).

iii. Post-germination, early growth observations were conducted daily for a period of three weeks. Throughout this period, the plants were watered regularly, and the containers were exposed to natural sunlight for 9–10 (Figure 10.6). After three weeks, the plants were harvested for analysis. Key growth parameters, including shoot length, root length, shoot fresh weight, and root fresh weight, were measured. Following these measurements, the plant material was oven-dried at 105°C until a constant weight was achieved, allowing for the calculation of dry biomass.

iv. At the end of each experiment, soil samples (with and without VC amendments) were collected from each container using a uniform cylindrical corer with a specified internal diameter and height. The volume of each sample was recorded, after which the samples were oven-dried at 105°C to a constant weight. The bulk density, particle density, and total porosity of the soil samples were subsequently determined following the standardized methods outlined by Carter and Gregorich (2007).

v. To determine the water holding capacity of the VC (vermicast), samples were placed in perforated-base cylinders. The cylinders were submerged in

FIGURE 10.5 A section of germination experiments.

water, allowing the vermicast to fully absorb moisture, and then drained. The cylinders, along with their contents, were weighed both before and after oven drying at 105°C to a constant weight. The difference in weight before and after drying was used to calculate the water retention capacity of the vermicast (Margesin & Schinner, 2005).

vi. Chlorophyll and carotenoids were extracted from the leaves using N,N-dimethyl formamide (DMF) according to the method of Moran and Porath (1980). Absorbance of the extracts was measured at 470, 647, and 664 nm using a UV-visible spectrophotometer to quantify chlorophyll and carotenoid concentrations, as described by Wellburn (1994). Tissue nitrogen content was assessed using standard analytical procedures outlined by the Association of Official Agricultural Chemists (2012). Ammonical nitrogen concentrations in the soil media were determined via the modified indophenol blue method (Bashour & Sayegh, 2007), while nitrate levels were measured using the Devarda alloy method (Jones, 2001). Microbial biomass carbon, serving as an indicator of soil microbial activity, was quantified using the fumigation-extraction method (Margesin & Schinner, 2005), as originally proposed by Jenkinson (1988) and Smith and Paul (1990). Calibration curves were constructed using the standard addition method to minimize matrix effects, and all analyses were repeated to ensure concordant results.

FIGURE 10.6 A section of early growth studies.

vii. The collected data were subjected to statistical analysis to evaluate the significance of the observed differences. A one-way analysis of variance (ANOVA) was performed, followed by the application of the least significant difference (LSD) test to identify significant pairwise differences, following established methodologies (Alan & David, 2001; Field, 2009).

viii. The field trials were conducted following the outlined methodology to assess the impact of various weed-derived VCs (LVC, IVC, PVC, SVC) on *Abelmoschus esculentus*. Seeds of *A. esculentus* were sown in soils contained within 50 L low-density polyethylene (LDPE) bags (Figures 10.7 and 10.8). Each bag was filled with soil amended at rates of 0 (control), 2.5, 3.75, and 5 tonnes per hectare (t/ha) of the respective VC. For each treatment, 35 bags were used, with five seeds sown per bag. The soil used in the experiment was collected from a previously uncultivated site to ensure that prior fertilizer applications did not influence the results. The trials were conducted from February to May, a period ideal for growing *A. esculentus* in the study area (ICAR, 2011).

FIGURE 10.7 A section of growth experiments on soil filled in 50 L plastic bags.

FIGURE 10.8 Another section of growth experiments on soil filled in 50 L plastic bags.

viii-a. Seed germination was monitored daily for up to eight days following sowing, and germination success was expressed as a percentage. Once germination data were recorded, the seedlings were thinned to one plant per bag. The growth experiment continued for 15 weeks, during which the containers were irrigated regularly with tap water. At 100 days post-sowing, five plants from each treatment group were harvested for detailed analysis. Measurements included shoot length, root length, plant biomass, number of leaves, stem diameter, and number of branches. The harvested plants were thoroughly rinsed with tap water to remove soil particles from the roots and then wiped and weighed. Finally, the plant material was oven-dried at 105°C to a constant weight to determine the dry biomass.

viii-b. Flowering success was quantified by recording the number of days until the first flower appeared, along with the total number of flowers produced per plant. Pods were harvested every other day to evaluate yield, with measurements taken for the number of pods per plant, as well as the length (cm), diameter (mm), and weight (g) of the pods.

viii-c. The protein content of the pods was determined using the Kjeldahl method, while carbohydrate content was measured using the Anthrone method. Ash content was assessed through dry ashing, and total solids were quantified by heating the pods at 105°C to a constant weight, following the procedures outlined by Nelson (2009).

viii-d. The pest and pathogen-repellent properties of the VCs were evaluated by comparing disease prevalence in VC-treated plants with control plants. During the experiments, plants showed signs of infestation by leaf miners and leaf spot disease. The leaf miner infestation, attributed to *Liriomyza* spp., was identified by feeding punctures and characteristic white speckles on the upper surface of the leaves (Ahmed, 2000). Leaf spot disease, caused by the fungus *Alternaria alternata*, was recognized by the presence of light brown spots on the leaves, which later developed into concentric dark brown rings (Cho & Moon, 1980; Werner, 1987; Tohyama et al., 1995; Arain et al., 2012). In severe cases, the leaves turned brown and eventually dropped off the plant (Canihos et al., 1999; Amenduini et al., 2003; Antonijevic et al., 2007). Additionally, pod infestation by the borer *Earias vittella* was observed (Guptha & Tayde, 2022). The extent of infestation was quantified as the percentage of infected fruit weight relative to the total fruit weight per treatment.

10.4 THE MAIN FINDINGS

10.4.1 Of the Studies Defined in Section 10.3.1.1

10.4.1.1 Effect of LVC

The first report under this programme appeared on the use of LVC on the germination and early growth of *V. radiata, A. esculentus,* and *C. sativus* (Hussain et al., 2015). It showed that the LVC enhanced germination success of the seeds of all three species, and for LVC concentrations of 0.75% (w/w), the germination was significantly higher ($p \leq 0.5$) than the control seeds for all three species (Figure 10.9). In the case of *V. radiata* and *C. sativus*, LVC concentrations of 1.5% and 2% also caused significantly enhanced germinations, while in *A. esculentus*, LVC concentrations other than 0.75% caused enhancement, but this wasn't statistically significant. In all cases, LVC at levels higher than 4% suppressed germination.

LVC at concentrations 0.75%–2% led to significantly longer shoots in all three plants (Figure 10.10), while at concentrations outside this range, there was no beneficial effect. The root length was also significantly higher in all treatments of LVC in the 0.75%–2% range. But while in *V. radiata* this benefit ceased outside this range, it persisted till 4% LVC levels in the case of *A. esculentus* and still higher levels in *C. sativus* (Figure 10.11). Similar trends were seen in respect of shoot dry weight (Figure 10.12) and root dry weight (Figure 10.13), with minor exceptions and occasional outliers. The positive impact on chlorophyll-a (Figure 10.14), chlorophyll-b (Figure 10.15), carotenoids (Figure 10.16), and tissue nitrogen concentrations (Figure 10.17) recorded significant improvement with LVC concentrations in wider bands of 0.75%–8% for *V. radiata*; 0.75%–4% for *A. esculentus*; and 0.75%–20% – except chlorophyll-b – for *C. sativus*.

In brief, LVC concentrations in the range 0.75%–2% (w/w) were beneficial to all three species of plants in terms of all nine variables that were studied. In several cases, the beneficial effect extended to 4% LVC, and in a few cases, even to 20%, but the 0.75%–2% range benefitted all without exception.

In plots of all three species, the bulk density and the particle density were seen to reduce with an increase in LVC concentrations from nil to 40% (w/w) in soil. These reductions, which occurred in a linear fashion (Figures 10.18 and 10.19), indicate a positive impact because the lower the bulk density and particle density of a soil, the better its aeration and moisture retention. This interpretation is endorsed by the findings summarized in Figures 10.20 and 10.21, which show the total porosity and water holding capacity of LVC fertilized plots linearly increasing with an increase in the LVC concentration. LVC similarly enhanced the nitrate and the ammonium levels in the soils (Figures 10.22 and 10.23). There was also a significant boost in microbial biomass caused by LVC (Figure 10.24), which, too, consistently increased with LVC levels.

There was no instance when LVC did not perform significantly better than the unfertilized control plots. Nor was there any evidence of allelopathy or other forms of negative effect on the soil, the seed germination, or the early growth of any of the plants – as long as the LVC application did not exceed certain concentrations. In terms of boosting agriculture when applied with certain concentrations, and harming it when the application became excessive, LVC behaved in the same manner as the VCs derived from animal manures, and other chemical or organic fertilizers, are known to behave (Table 10.1).

To check whether the parent lantana also has the attributes that its VC possesses, the experiments were repeated with plots fertilized by finely chopped raw lantana in the same w/w proportions with soil as were deployed for LVC. But lantana allowed significantly less germination than the control (unfertilized soil), and the few seedlings that emerged had stunted growth for a short while before they all died.

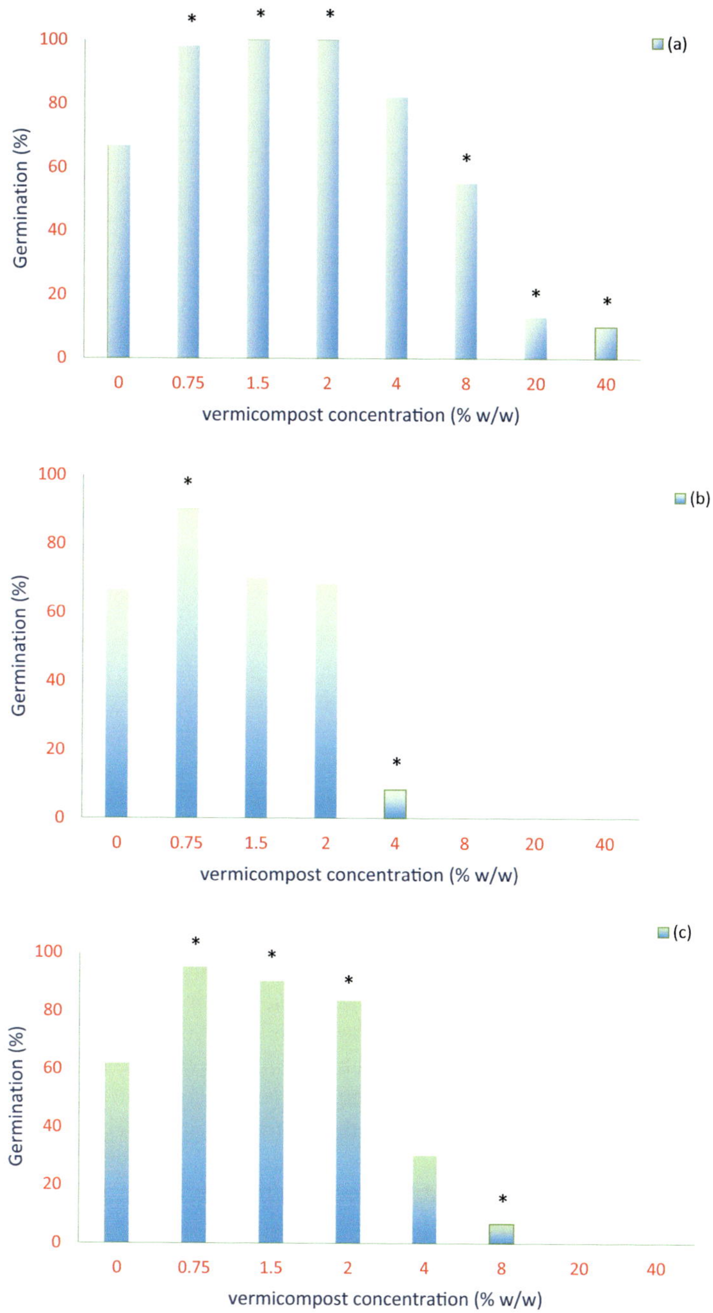

FIGURE 10.9 Germination success of seeds in the soils fertilized with different doses of LVC: (a) green gram, (b) ladies finger, and (c) cucumber. The data represented by bars accompanying an asterisk on the top are significantly different from the control (only soil) at $p \leq 0.05$.

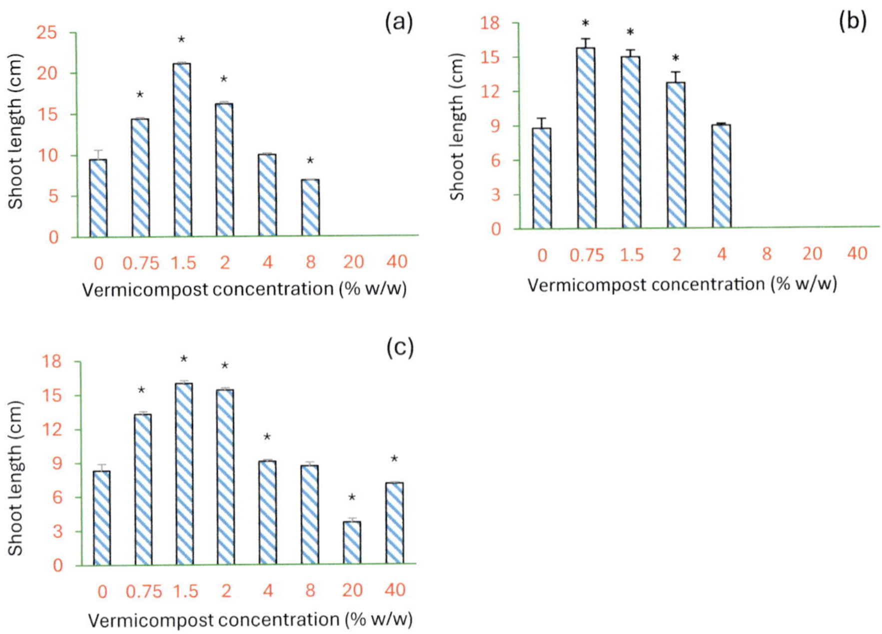

FIGURE 10.10 Effect of LVC on the shoot length of (a) green gram, (b) ladies finger, and (c) cucumber. The data on the bars with an asterisk differ significantly ($p \leq 0.05$) from the control (soil not augmented with LVC).

10.4.1.2 Effect of IVC

Despite being one of the most toxic and allelopathic weeds studied, IVC exhibited no signs of toxicity or inhibition of seed germination. In fact, IVC improved germination success compared to the control (Figure 10.25). Positive effects were observed at IVC concentrations up to 2% for *Vigna radiata*, 8% for *Abelmoschus esculentus*, and 4% for *Cucumis sativus*. Similar to the trends observed with LVC, germination declined at higher IVC concentrations, with germination completely ceasing at concentrations above 8% in *C. sativus*. The highest germination rates were 93%, 97%, and 99% for *V. radiata*, *A. esculentus*, and *C. sativus*, respectively, at IVC levels of 1.5%, 4%, and 0.75%.

Growth performance, measured by shoot length, root length, shoot dry weight, and root dry weight, was greater than the control for *V. radiata* at IVC concentrations ranging from 0.75% to 8%. The optimal growth was achieved at 2% IVC, followed closely by treatments at 1.5% and 4%. Seedlings exposed to 20% and 40% IVC did not survive beyond one week, and these treatments were excluded from further analysis.

For *A. esculentus* and *C. sativus*, growth was significantly enhanced at IVC concentrations between 0.75% and 8%, with 4% IVC generally yielding the best results. However, root growth in *C. sativus*, particularly root length and dry weight, peaked at

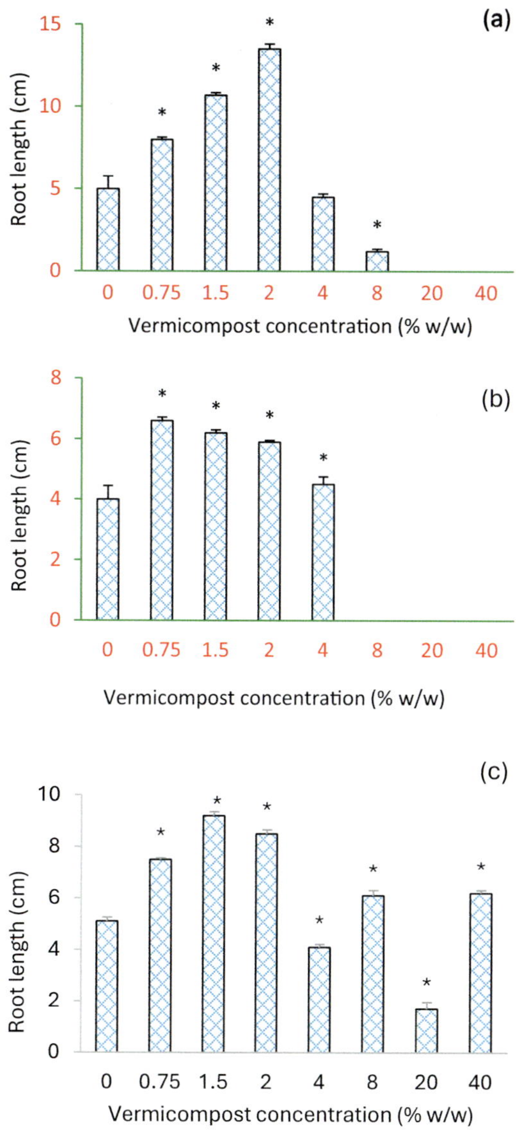

FIGURE 10.11 Effect of LVC on the root length of (a) green gram, (b) ladies finger, and (c) cucumber. The data on the bars with an asterisk differ significantly ($p \leq 0.05$) from the control (soil not augmented with LVC).

2% IVC. Growth declined sharply beyond 8% IVC for both species. Since there was no germination in the 20% and 40% IVC treatments, seedlings grown in unamended soil were transplanted into these treatments, and their growth was monitored.

In *V. radiata* and *A. esculentus*, the highest levels of chlorophyll-a were observed at 4% IVC, while in *C. sativus*, the peak occurred at 1.5% IVC. Chlorophyll-b and carotenoid content were also enhanced at IVC levels up to 4% for *V. radiata*, up to

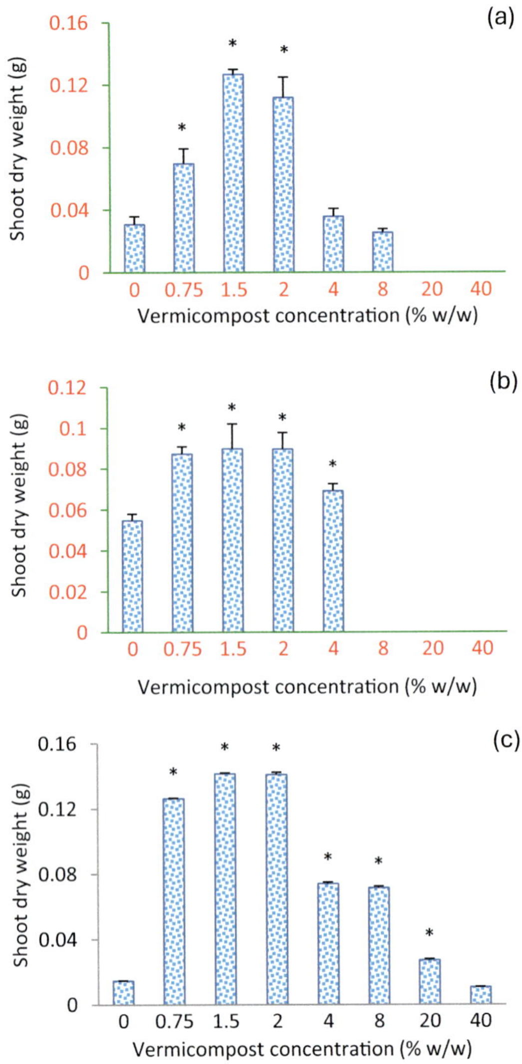

FIGURE 10.12 Effect of LVC on the shoot dry weight of (a) green gram, (b) ladies finger, and (c) cucumber. The data on the bars with an asterisk differ significantly ($p \leq 0.05$) from the control (soil not augmented with LVC).

8% for *A. esculentus*, and 2% for *C. sativus*. The tissue nitrogen content reached its highest at 4% and 8% IVC for *C. sativus*, *A. esculentus*, and *V. radiata*. Although these values declined at higher IVC levels, the decrease was less dramatic than the drop in growth metrics.

10.4.1.3 Effect of PVC

In all three species tested, PVC significantly enhanced seed germination compared to the control (Figure 10.26) when applied at concentrations between 0.75% and 2%.

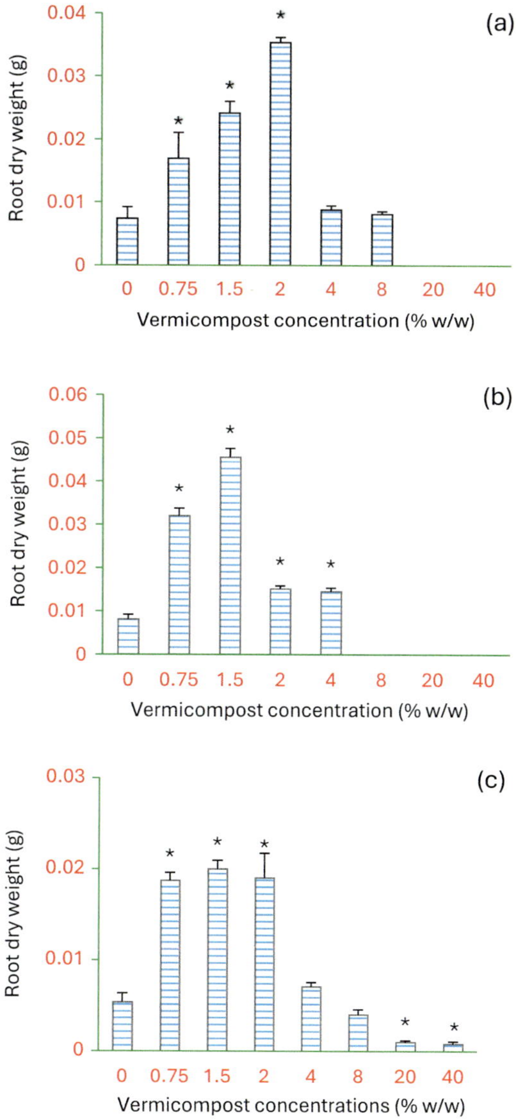

FIGURE 10.13 Effect of LVC on the root dry weight of (a) green gram, (b) ladies finger, and (c) cucumber. The data on the bars with an asterisk differ significantly ($p \leq 0.05$) from the control (soil not augmented with LVC).

The highest germination rates were recorded at 2%, 0.75%, and 1.5% for *Abelmoschus esculentus*, *Cucumis sativus*, and *Vigna radiata*, respectively, with germination rates reaching 97%, 98%, and 98%. However, when PVC concentrations exceeded 2%, germination success declined, with the most pronounced reductions seen in *A. esculentus* and *C. sativus*, where germination ceased altogether at concentrations above 8%.

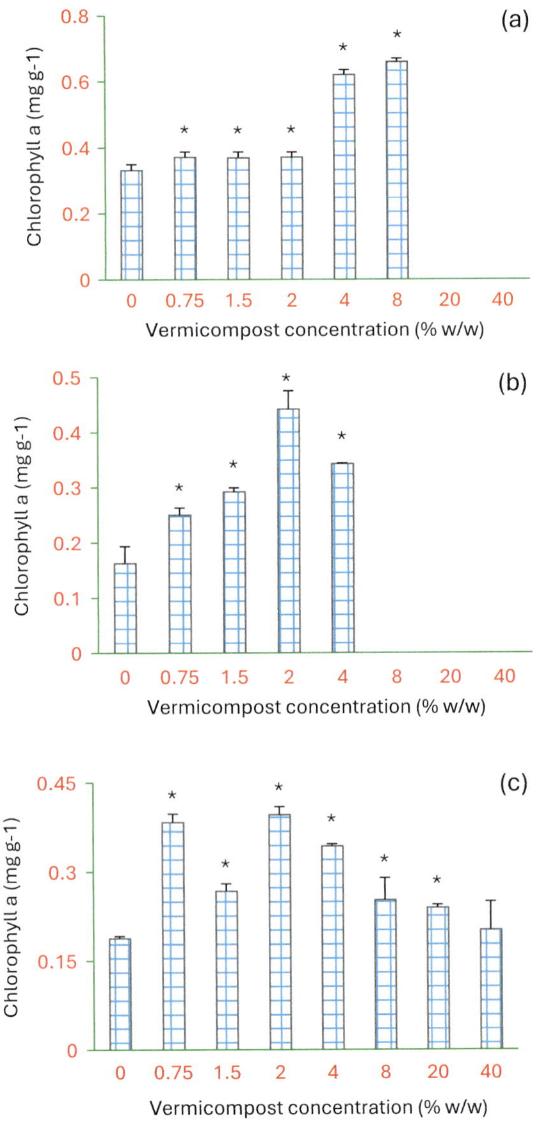

FIGURE 10.14 Effect of LVC on the chlorophyll "a" of (a) green gram, (b) ladies finger, and (c) cucumber. The data on the bars with an asterisk differ significantly ($p \leq 0.05$) from the control (soil not augmented with LVC).

In terms of growth parameters, *V. radiata* showed improved shoot length, root length, shoot dry weight, and root dry weight in PVC treatments ranging from 0.75% to 4%, after which a slight decrease in growth was observed. For *A. esculentus*, PVC application up to 8% promoted better growth compared to the control, though shoot dry weight was lower at 8% PVC, with 4% providing the highest growth in all other aspects.

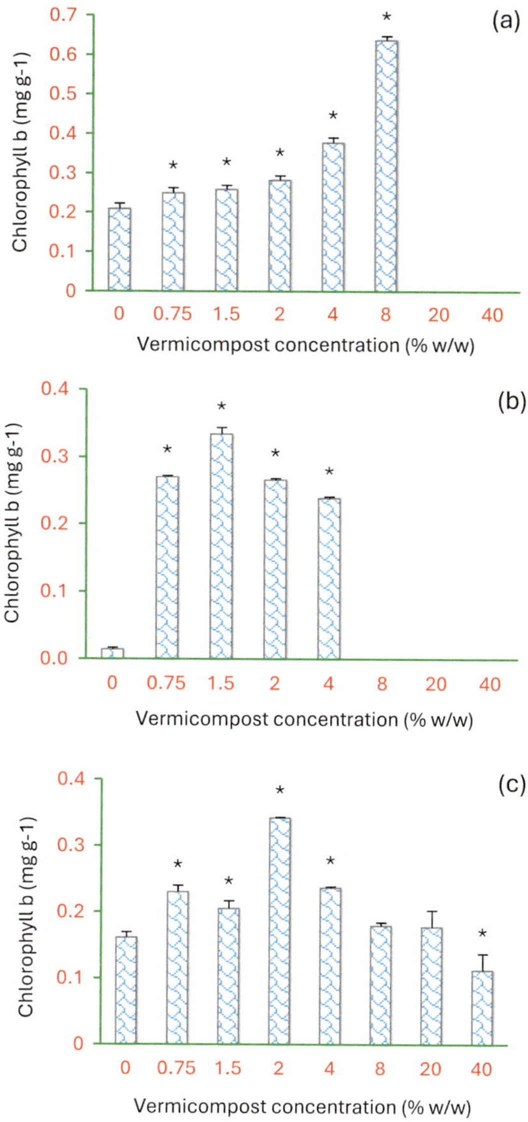

FIGURE 10.15 Effect of LVC on the chlorophyll "b" of (a) green gram, (b) ladies finger, and (c) cucumber. The data on the bars with an asterisk differ significantly ($p \leq 0.05$) from the control (soil not augmented with LVC).

Similarly, *C. sativus* seedlings exhibited greater root length, shoot dry weight, and root dry weight at PVC concentrations between 0.75% and 20%. However, shoot length increased up to 4% PVC and then decreased slightly. Overall, the most consistent growth improvement across all species was observed at PVC concentrations between 0.75% and 2%.

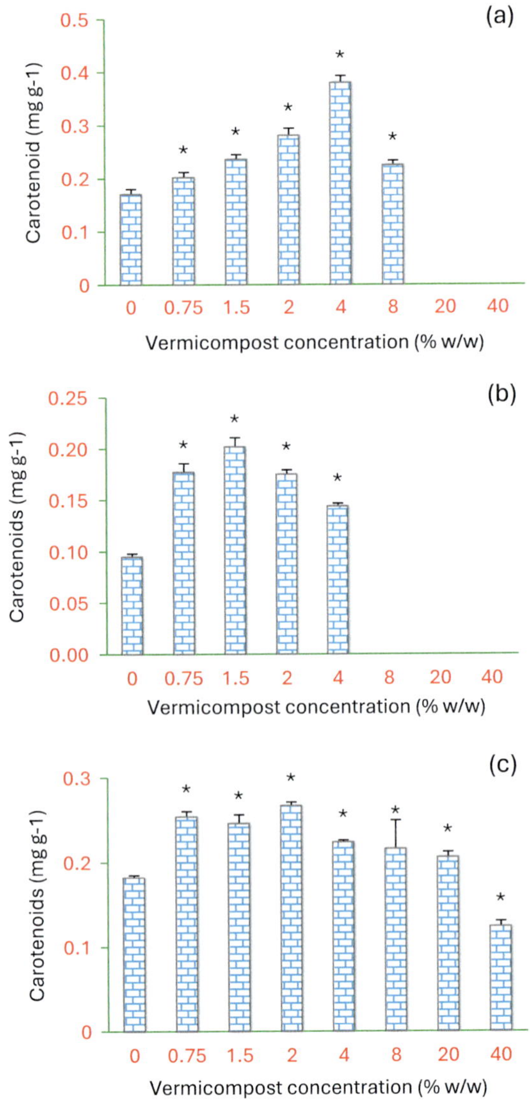

FIGURE 10.16 Effect of LVC on the carotenoids of (a) green gram, (b) ladies finger, and (c) cucumber. The data on the bars with an asterisk differ significantly ($p \leq 0.05$) from the control (soil not augmented with LVC).

Regarding biochemical parameters, chlorophyll-a, chlorophyll-b, carotenoids, and tissue nitrogen levels were higher in PVC treatments ranging from 0.75% to 8%, compared to the control, for all species. Exceptions were the tissue nitrogen content in *A. esculentus* and carotenoid levels in *C. sativus*, which did not follow this trend. While these variables declined at higher PVC concentrations, the drop was less pronounced than the decrease in growth metrics.

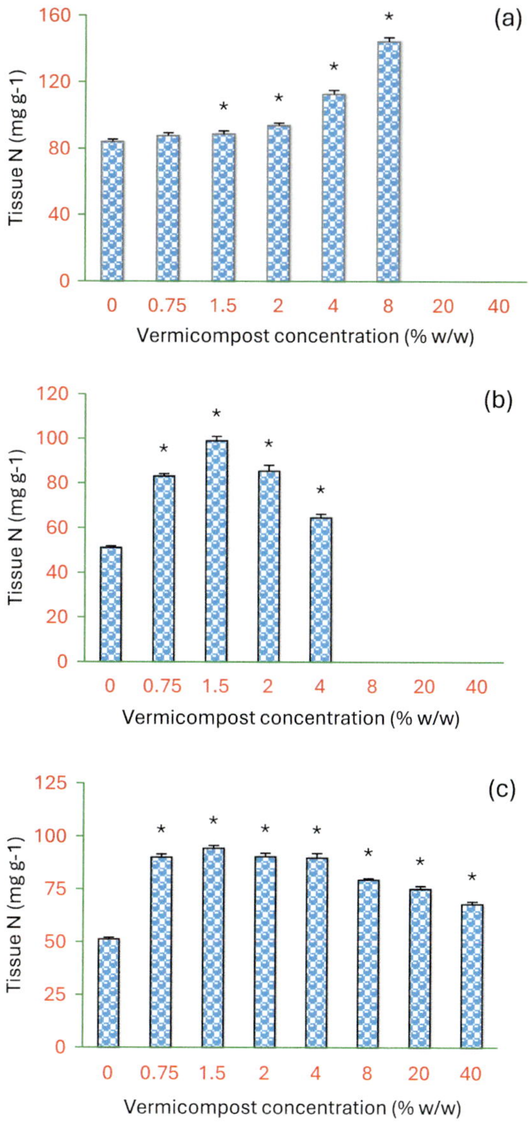

FIGURE 10.17 Effect of LVC on the tissue N levels of (a) green gram, (b) ladies finger, and (c) cucumber. The data on the bars with an asterisk differ significantly ($p \leq 0.05$) from the control (soil not augmented with LVC).

10.4.1.4 Effect of SVC

The results, as summarized in Figure 10.27, show that SVC significantly improved seed germination compared to the control across various concentrations. Germination success was observed within a concentration range of 0.75%–20% in *Abelmoschus esculentus*, 0.75%–8% in *Cucumis sativus*, and 0.75%–4% in *Vigna radiata*.

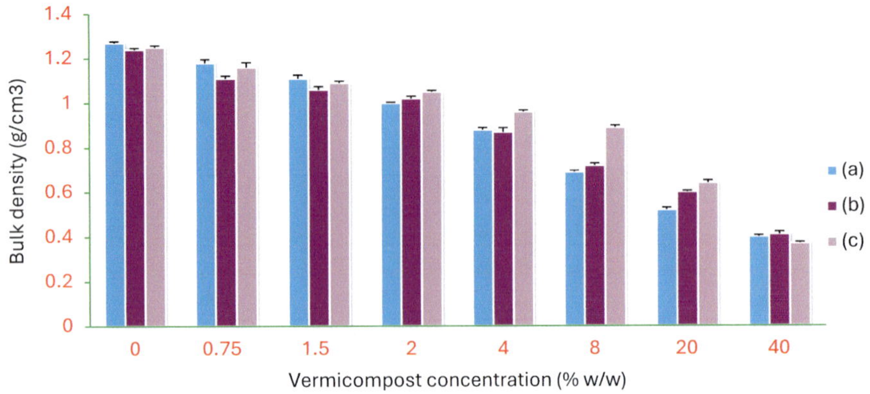

FIGURE 10.18 Bulk density of the soil applied with LVC and used to cultivate (a) green gram, (b) ladies finger, and (c) cucumber plants. Bars topped by an asterisk indicate no significant difference from the control (unaugmented soil) at $p \leq 0.05$.

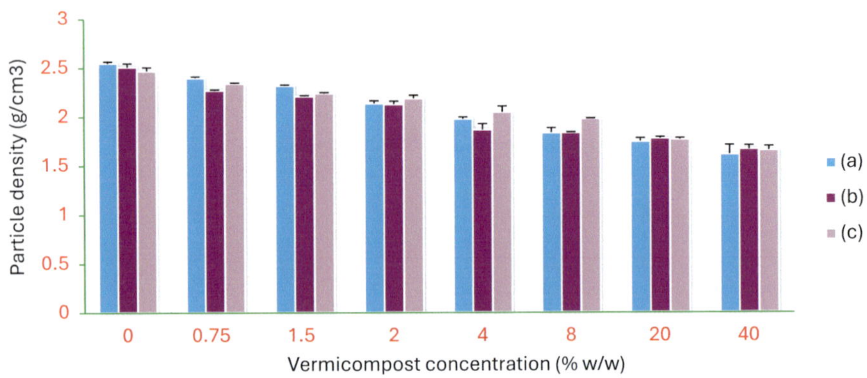

FIGURE 10.19 Particle density of the soil applied with LVC and used to cultivate (a) green gram, (b) ladies finger, and (c) cucumber plants. Bars topped by an asterisk indicate no significant difference from the control (unaugmented soil) at $p \leq 0.05$.

The highest germination rate of 98% was recorded for *C. sativus* at 2% SVC, while *V. radiata* and *A. esculentus* achieved maximum germination rates of 97% at 1.5% and 4% SVC, respectively. However, at SVC concentrations exceeding 4% in *V. radiata*, 8% in *C. sativus*, and 20% in *A. esculentus*, germination rates dropped below those of the control.

Growth stimulation was noted in *A. esculentus* and *C. sativus* across all SVC concentrations, with *A. esculentus* showing the greatest shoot length and shoot dry weight at 20% SVC. Root length and root dry weight, however, were maximized at 8% and 4% SVC, respectively. For *C. sativus*, the most pronounced growth improvements were observed at 1.5%–2% SVC, with diminishing returns at higher SVC concentrations.

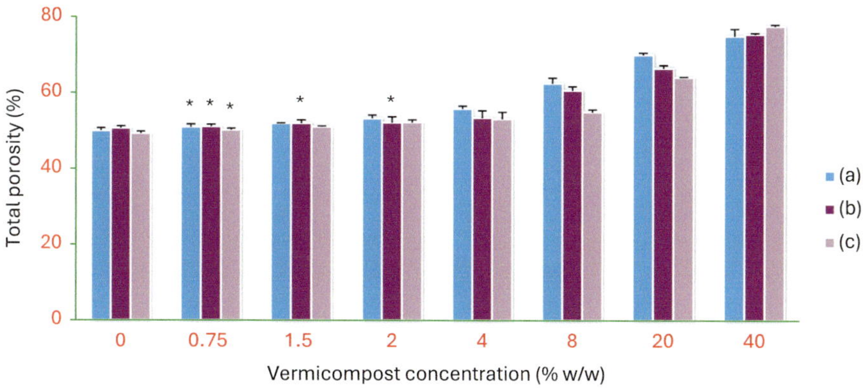

FIGURE 10.20 Total porosity of the soil applied with LVC and used to cultivate (a) green gram, (b) ladies finger, and (c) cucumber plants. Bars topped by an asterisk indicate no significant difference from the control (unaugmented soil) at $p \leq 0.05$.

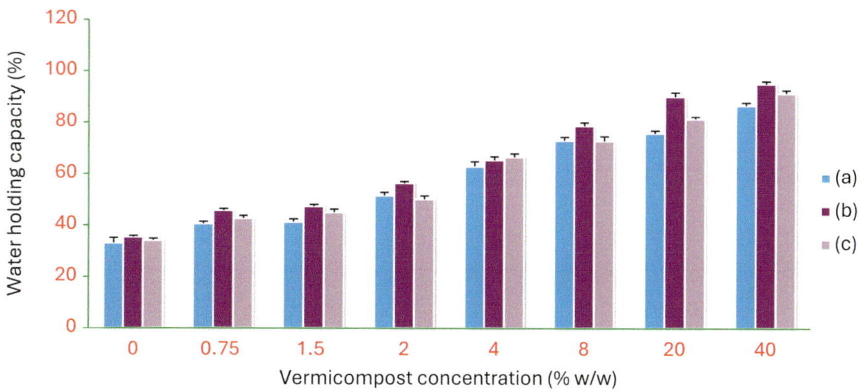

FIGURE 10.21 Water holding capacity of the soil applied with LVC and used to cultivate (a) green gram, (b) ladies finger, and (c) cucumber plants. Bars topped by an asterisk indicate no significant difference from the control (unaugmented soil) at $p \leq 0.05$.

In *V. radiata*, the best growth in terms of shoot length, shoot dry weight, and root dry weight was achieved with 1.5% SVC, followed by 2% SVC. The longest root growth was recorded at 4% SVC, with 8% SVC yielding the second highest root length.

10.4.1.5 Effect on the Physicochemical and Microbial Properties of the Soil

As it was seen in the case of LVC, the IVC, PVC, and SVC also exerted beneficial impacts on the soil to which they were applied. The bulk density and particle density of the soil decreased progressively with increasing concentrations of VC. Conversely, both total porosity and water holding capacity increased. This improvement can be attributed to the enrichment of soil organic matter through the addition of VCs,

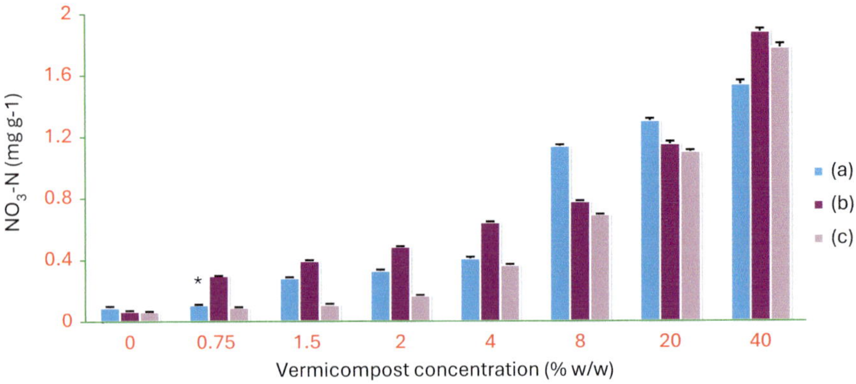

FIGURE 10.22 Nitrate–nitrogen content of the soil applied with LVC and used to cultivate (a) green gram, (b) ladies finger, and (c) cucumber plants. Bars topped by an asterisk indicate no significant difference from the control (unaugmented soil) at $p \leq 0.05$.

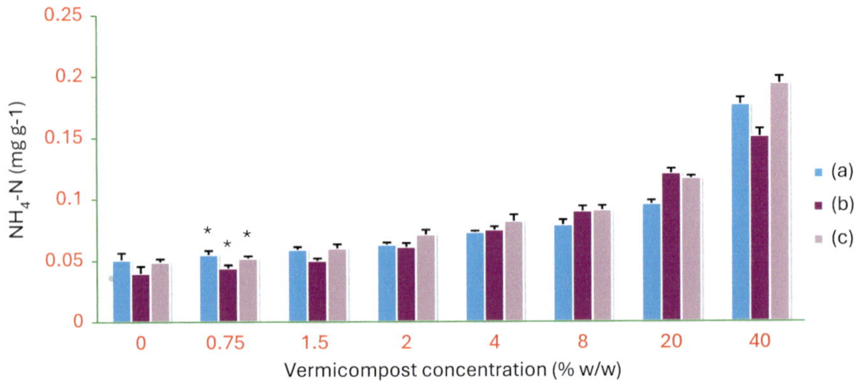

FIGURE 10.23 Ammonium–nitrogen content of the soil applied with LVC and used to cultivate (a) green gram, (b) ladies finger, and (c) cucumber plants. Bars topped by an asterisk indicate no significant difference from the control (unaugmented soil) at $p \leq 0.05$.

which contributed to the formation of soil aggregates. These aggregates enhanced pore space, thereby reducing bulk density (Castellini et al., 2024; Yang et al., 2024).

The incorporation of VC into the soil led to a proportional rise in nitrate and ammonium concentrations, although ammonium levels remained significantly lower than nitrate. It is well-documented that VCs typically contain the majority of their mineral nitrogen in nitrate form (Doklega & Imryed, 2020; Siping et al., 2024). This occurs due to the nitrification process during vermicast formation, which converts most of the ammoniacal nitrogen into nitrate (Yang et al., 2024). Thus, it was not surprising to observe higher nitrate concentrations than ammonium in the VCs derived from these four weeds.

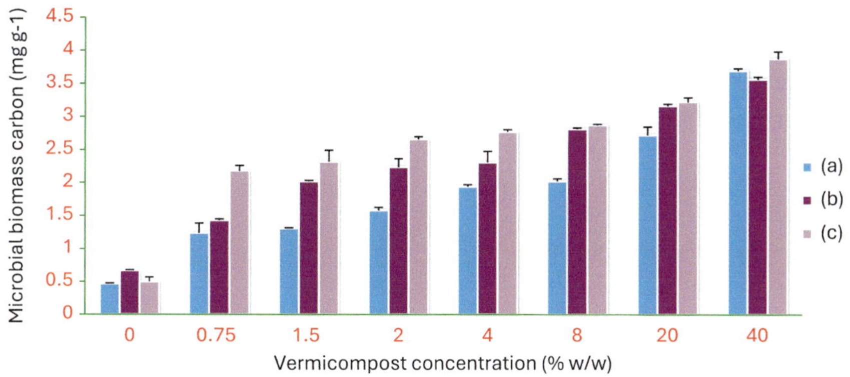

FIGURE 10.24 Microbial biomass of the soil applied with LVC and used to cultivate (a) green gram, (b) ladies finger, and (c) cucumber plants. Bars topped by an asterisk indicate no significant difference from the control (unaugmented soil) at $P \leq 0.05$.

Moreover, all VCs significantly enhanced microbial biomass carbon as their concentration in the soil increased (Figures 10.24 and 10.28–10.30). This suggests that the weed-derived VCs harbour a rich and diverse microbial community similar to that found in VCs produced from animal manure (Chiranjeeb et al., 2020; Lv et al., 2020). The application of VCs effectively enriched the indigenous soil microbial populations, resulting in a substantial increase in microbial biomass carbon compared to the control soil.

10.4.1.6 In Summary...

In summary, VC concentrations of 0.75%–2% boosted germination and all aspects of early growth of the resulting seedlings, independent of the nature of the weed from which the VCs were derived. In some cases, the beneficial impact remained significant at 4% VC levels as well, while in a few exceptional cases it extended to 8% levels and, in rare cases, even higher. But the 0.75%–2% range was common in its beneficial effects across all VC concentrations studied.

Second, even as the four weeds differed from each other in their possession of toxic ingredients as detailed later in Chapter 11, and the extents of negative allelopathy, going through vermicomposting appears to have brought them on a common level of benignness.

Third, even as the positive effect of VC concentration in the 0.75%–2% range encompasses all four VCs and all three species fertilized by them, different species have shown differences in the limits of VC application beyond which beneficial effects begin to cease. This indicates that when agricultural practices are developed based on the use of these and other phytomass-based VCs, they will have to be species-specific in the same way as the practices based on the application of chemical fertilizers are.

An important aspect to note is that when the extent of fertilizer application increased beyond levels at which the benefits were statistically significant, there was a buffer stage when benefits still occurred but less than at highly significant levels.

TABLE 10.1
Summary of Past Work on the Dual Role of Vermicompost in Stimulating or Inhibiting Seed Germination

Source of Vermicompost	Impacted Plant Species	Concentration Range of Vermicompost Which Stimulated Germination	Concentration Range of Vermicompost Which Inhibited Germination (%)	References
Duck waste	Tomatoes (*Solanum lycopersicum*) and Pepper (*Capsicum* sp.)	2%–10% (v/v)	15–20	Wilson and Carlile (1989)
Animal manure and plant wastes	Radish (*Raphanus sativus*)	10% (v/v)	20–100	Buckerfield et al. (1999)
Pig manure	Tomatoes (*Lycopersicon esculentum* Mill)	20%–40% (v/v)	100	Atiyeh et al. (2000)
Organic food and cotton waste	Tomato (*L. esculentum*)	20%–40% (v/v)	60–100	Zaller (2007)
Cow manure	Alfalfa (*Medicago sativa*)	10% (v/v)	20	Zhang et al. (2009)
Cow manure and vegetable waste	Kidney bean (*Phaseolus vulgaris* L)	0%–50%(v/v)	80–100	Rodríguez-Quiroz et al. (2011)
Sheep manure	Radish (*R. sativus*)	10% (v/v)	20–40	Gutierrez-Miceli et al. (2011)
Cow manure	Radish (*R. sativus*) and cabbage (*Brassica oleracea*)	20%	30–100	Ivenish (2011)
Not stated	Groundnut (*Arachis hypogaea*)	1%–2% (w/w)	2.5%	Mathivanan et al. (2012)

Only when the fertilizer concentration was increased still further, did adverse effects begin to manifest. This indicates that in the course of actual field applications, a minor excess in the VC application will not cause any significant harm.

10.4.2 Studies Defined in Section 10.3.1.2

In these studies, the performance of the PWVC as fertilizer in comparison to CD was assessed on four species of food plants: chilli, ladies finger, cucumber, and tomato. The findings on germination success are summarized in Figures 10.31–10.34. Amendment of the soil with CD did not achieve better germination success than the control soil in any of the experiments. But the soil amended with PWVC enabled more seeds to germinate in all trials than the soil amended with CD, barring four exceptions: in 9% PWVC in chilli (*C. annum)* (Figure 10.31) and ladies finger (*A. esculentus)* (Figure 10.32), and in 4.5% PWVC in *A. esculentus* and tomato (*L. esculentum)* (Figures 10.32 and 10.34).

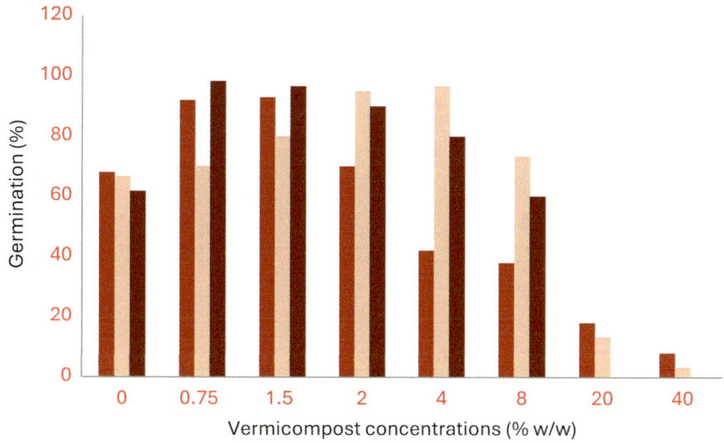

FIGURE 10.25 Effect of supplementation of the soil with different concentrations of IVC on the germination success achieved by (a) green gram ■, (b) ladies finger ■, and (c) cucumber ■.

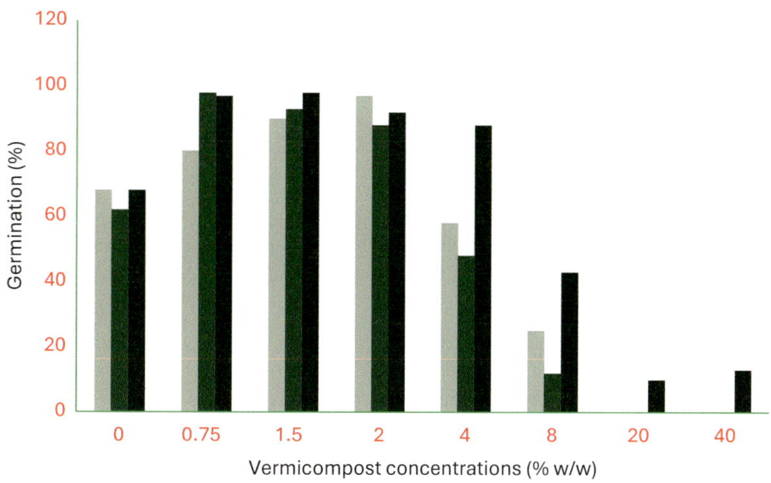

FIGURE 10.26 Effect of supplementation of the soil with different concentrations of PVC on the germination success achieved by (a) ladies finger ■, (b) cucumber ■, and (c) green gram ■.

The PWVC concentration at which maximum beneficial effect occurred in all species was 0.75%. In 1.5% and 2.25% treatments, also, there was greater germination success than in control soil or CD amended soil, but less than at 0.75%.

There was no hostility in evidence of PWVC towards seeds of any of the four species tested in the study. It appears clear that PWVC becomes a benign fertilizer upon vermicomposting, despite containing only 10% by mass of CD. Also, this PWVC is a much more potent fertilizer than 100% CD.

In all the species, the root length and shoot length of the plants were the highest in the soil amended with 1.5% PWVC. The next best results were seen in the soil

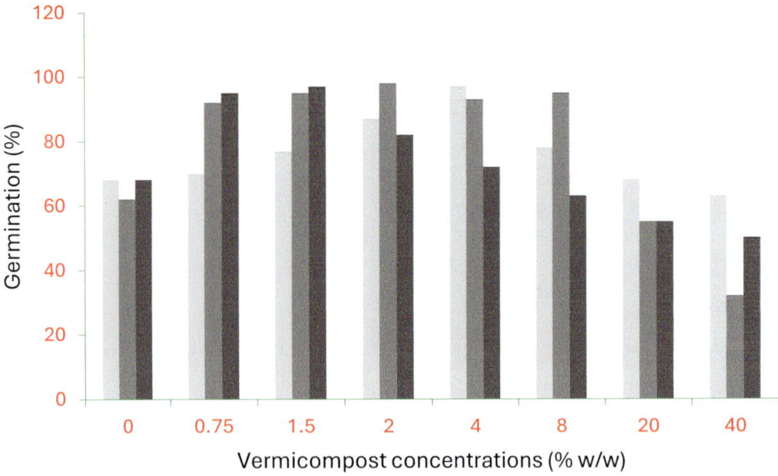

FIGURE 10.27 Effect of supplementation of the soil with different concentrations of SVC on the germination success achieved by (a) ladies finger, (b) cucumber, and (c) green gram.

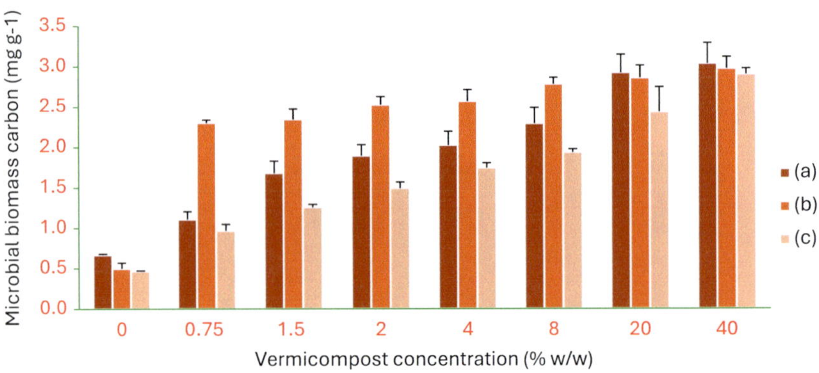

FIGURE 10.28 Effect on the microbial biomass in the soil replenished with IVC and cultivated with (a) ladies finger, (b) cucumber, and (c) green gram plants. Bars topped by an asterisk indicate no significant difference from the control (unfertilized soil) at $p \leq 0.05$.

containing 0.75% PWVC, except in the case of the root length of the *L. esculentum* plants where this trend was in reverse. In all cases, the root and shoot lengths were significantly greater than the ones obtained in control soil or in the corresponding amendments of CD.

The number of leaves and the fresh weight of the seedlings were also maximum in the 1.5% PWVC treatment for all species, the only exception being the fresh weight of *A. esculentus*, which was the highest in the 4.5% PPVC treatment. In each and every set, and barring very few exceptions, the paper waste vermicompost (PWVC) amendment was seen to predominantly exert a beneficial effect compared to the CD treatment. It was statistically significant, too, in most cases.

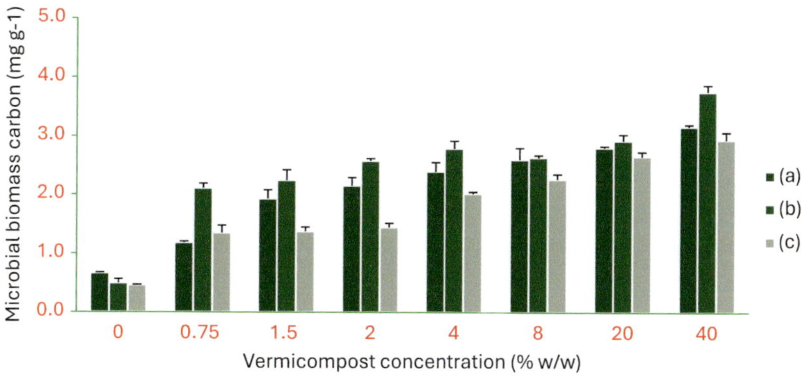

FIGURE 10.29 Effect on the microbial biomass in the soil replenished with PVC and cultivated with (a) ladies finger, (b) cucumber, and (c) green gram plants. Bars topped by an asterisk indicate no significant difference from the control (unfertilized soil) at $p \leq 0.05$.

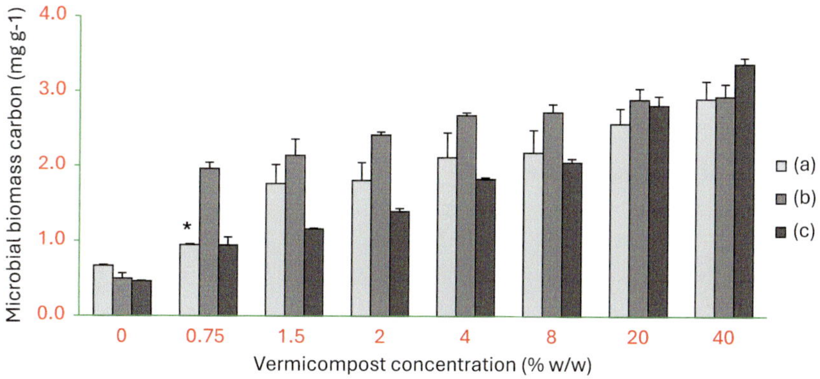

FIGURE 10.30 Effect on the microbial biomass in the soil replenished with SVC and cultivated with (a) ladies finger, (b) cucumber, and (c) green gram plants. Bars topped by an asterisk indicate no significant difference from the control (unfertilized soil) at $p \leq 0.05$.

Depending on the plants, the highest dry weight of the seedlings was recorded either in 4.5% PWVC amended soils or in 1.5% PWVC amended soils. But in all cases, the PWVC amended soils gave significantly better results than the CD-treated soil or the unfertilized soil.

A similar trend was seen in the majority of cases in terms of chlorophyll-*a,* chlorophyll-*b*, and total chlorophyll levels, as typified by Figure 10.35a. The only substantial exception was in chlorophyll-*b* levels of *C. annum,* which were significantly higher in 0.75%, 2.25%, 4.5%, and 9% CD treatments compared to the corresponding PPVC treatments. But the chlorophyll-*a* levels of the *C. annum* plants in PPVC treatments were so much higher than in the corresponding CD treatments that the difference in the chlorophyll-*b* level got muted, and total chlorophyll levels in *C. annum* also followed the trend of other species.

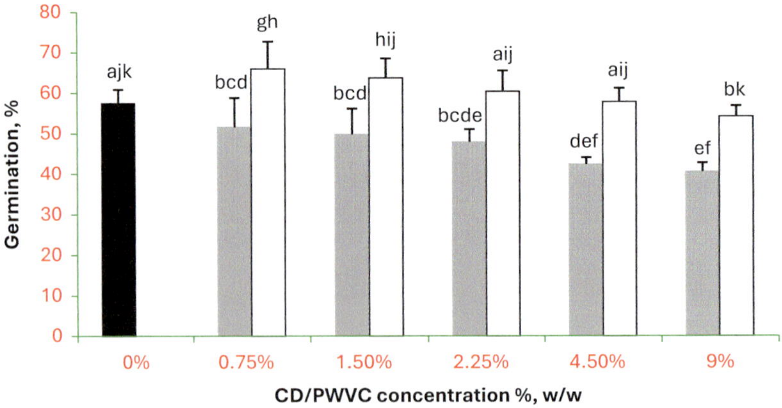

FIGURE 10.31 Germination success of chilli seeds in soils with or without fertilization by CD and PWVC. When pairs of data carry atleast one character in the superscript which is in common significant difference from each other (LSD test; $p < 0.05$), it indicates the absence of ■ control soil; ▨ CD amended soil; □ PVC amended soil.

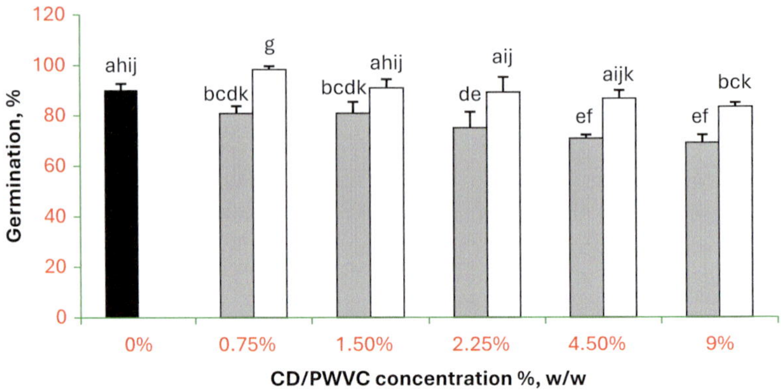

FIGURE 10.32 Germination success of ladies finger seeds in soils with or without fertilization by CD and PWVC. When pairs of data carry atleast one character in the superscript which is in common (LSD test; $p < 0.05$), it indicates the absence of significance difference from each other ■ control soil; ▨ CD amended soil; □ PVC amended soil.

The PWVC-fertilized plants had higher carotene levels than CD-fertilized plants, often significantly so. Figure 10.35b illustrates this fact.

In terms of effect on the soil, the addition of either CD or PWVC improved the properties of the soil by reducing its bulk density (Figure 10.36), pH (Figure 10.37), and enhancing its electrical conductivity (Figure 10.38). These effects were more pronounced in PWVC amendments than in the CD amendments. But CD could not enhance the particle density (Figure 10.39) nor the total porosity (Figure 10.40) of

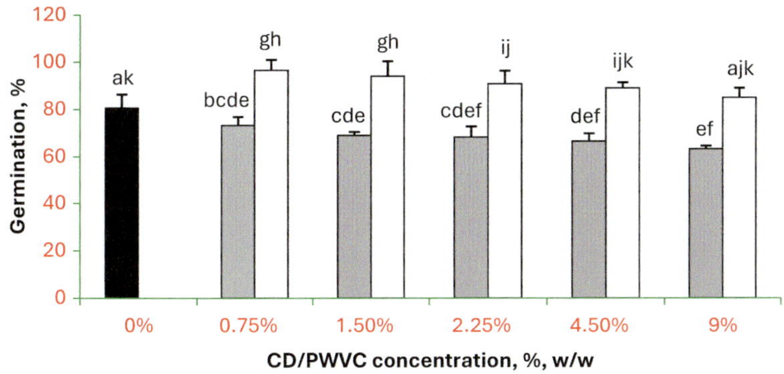

FIGURE 10.33 Germination success of cucumber seeds in soils with or without fertilization by CD and PWVC. When pairs of data carry atleast one character in the superscript which is in common (LSD test; $p < 0.05$), it indicates the absence of significance difference from each other ■ control soil; ▩ CD amended soil; □ PVC amended soil.

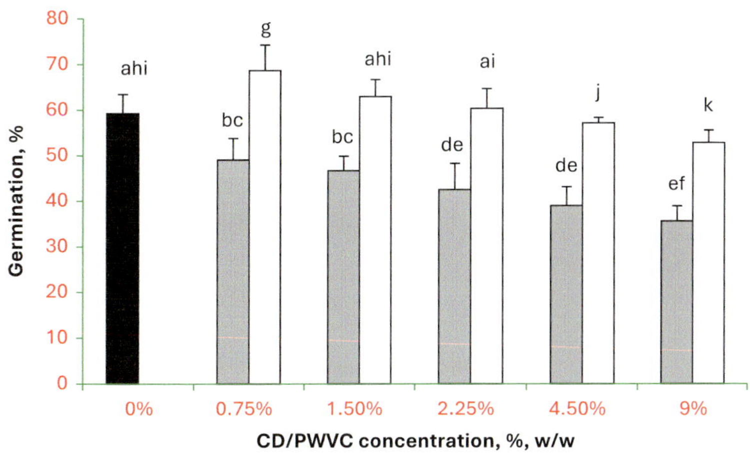

FIGURE 10.34 Germination success of tomato seeds in soils with or without fertilization by CD and PWVC. When pairs of data carry atleast one character in the superscript which is in common (LSD test; $p < 0.05$), it indicates the absence of significance difference from each other ■ control soil; ▩ CD amended soil; □ PVC amended soil.

the soil, which PWVC could. Hence, while in some respects the application of CD improved the soil properties, in some other respects it had an adverse effect. In contrast, PWVC had a uniformly beneficial impact on the soil.

All-in-all, the application of PWVC, especially in concentrations of 0.75% or 1.5%, had a significantly beneficial effect on the germination and early growth of the four species studied. The fact that it was not CD (which had been added to the extent of 10% by weight) that was responsible for these beneficial effects is established by

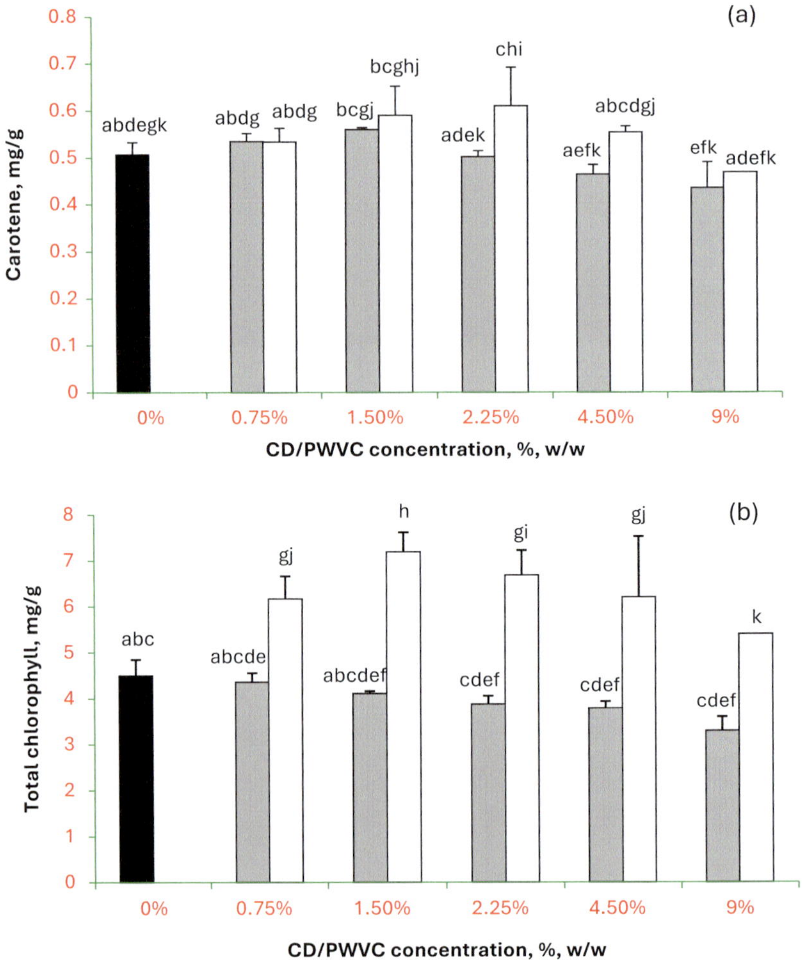

FIGURE 10.35 Carotene (a) and total chlorophyll (b) levels in tomato leaves as affected by fertilization with CD and PWVC. When pairs of data carry atleast one character in the superscript which is in common (LSD test; $p < 0.05$), it indicates the absence of significance difference from each other ■ control soil; ▨ CD amended soil; □ PVC amended soil.

the findings that application of CD alone to the soil had very few of the beneficial effects.

10.4.3 Studies Defined in Section 10.3.1.3

10.4.3.1 Effect on Germination Success

The results, as illustrated in Figure 10.41a, indicate that LVC significantly improved seed germination in *A. esculentus* compared to the control. The highest germination rate of 95% was observed in the 5 t/ha LVC treatment, followed closely by 93% in the 3.75 t/ha treatment. This enhanced germination may be attributed to the increased

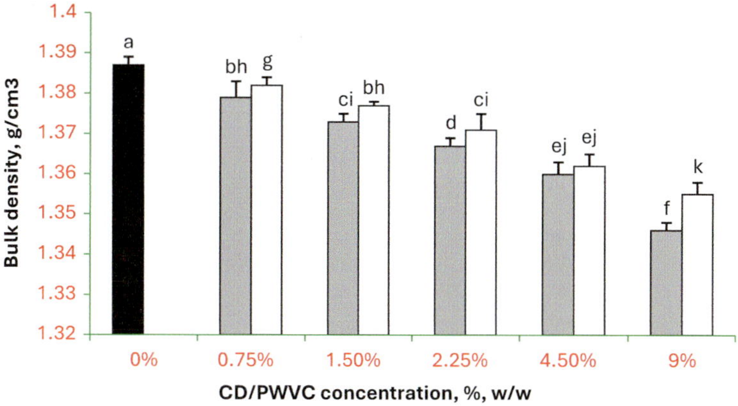

FIGURE 10.36 Bulk density of the container media as affected by fertilization with CD and PWVC. When pairs of data carry atleast one character in the superscript which is in common (LSD test; $p < 0.05$), it indicates the absence of significance difference from each other ■ control soil; ▨ CD amended soil; □ PVC amended soil.

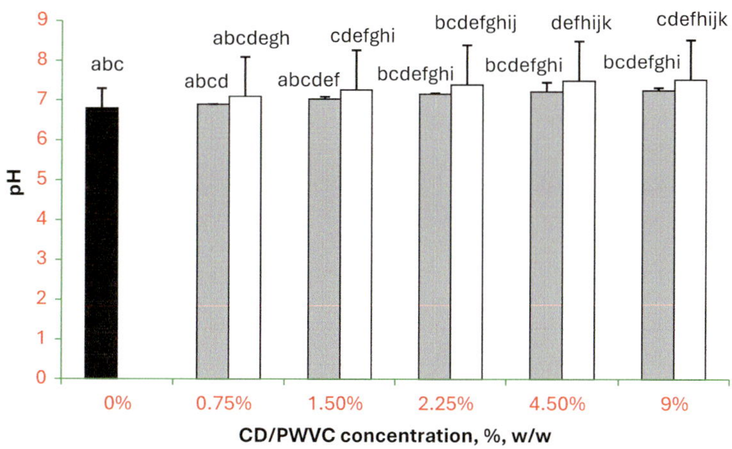

FIGURE 10.37 pH of the container media as affected by fertilization with CD and PWVC. When pairs of data carry atleast one character in the superscript which is in common (LSD test; $p < 0.05$), it indicates the absence of significance difference from each other ■ control soil; ▨ CD amended soil; □ PVC amended soil.

levels of nitrate and ammonium in the LVC-amended soils, both of which are known to break seed dormancy and promote germination (Hilhorst & Karssen, 2000; Yan & Chen, 2020). VCs are also known to contain phenolic acids and plant growth hormones that contribute to seed germination (Ievinsh, 2011).

Similarly, IVC treatments resulted in improved germination rates compared to the control (Figure 10.41b). The highest germination rate of 86% was recorded in the

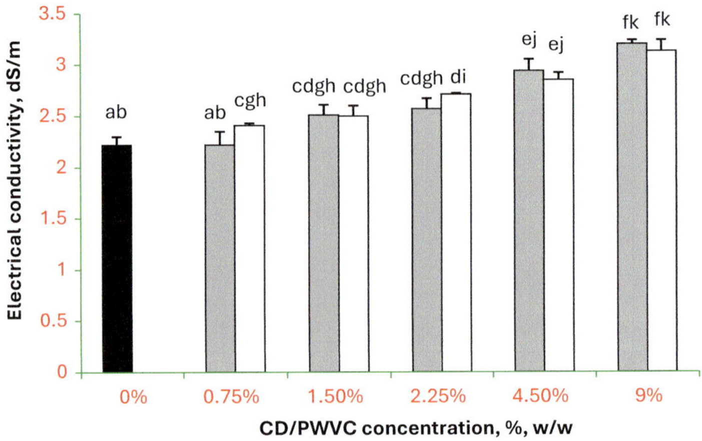

FIGURE 10.38 Electrical conductivity of the container media as affected by fertilization with CD and PWVC. When pairs of data carry atleast one character in the superscript which is in common (LSD test; $p < 0.05$), it indicates the absence of significance difference from each other ■ control soil; ▩ CD amended soil;□ PVC amended soil.

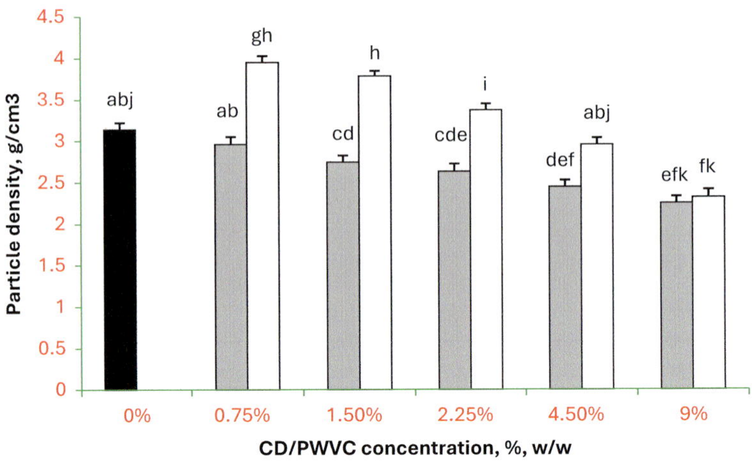

FIGURE 10.39 Particle density of the container media as affected by fertilization with CD and PWVC. When pairs of data carry atleast one character in the superscript which is in common (LSD test; $p < 0.05$), it indicates the absence of significance difference from each other ■ control soil; ▩ CD amended soil; □ PVC amended soil.

3.75 t/ha IVC treatment, while 85% success was observed in both the 2.5 and 5 t/ha treatments. However, no statistically significant differences ($p < 0.05$) were observed among the IVC treatments.

A comparable trend was observed with PVC, where germination success increased with higher concentrations of VC (Figure 10.41c). The maximum germination rate

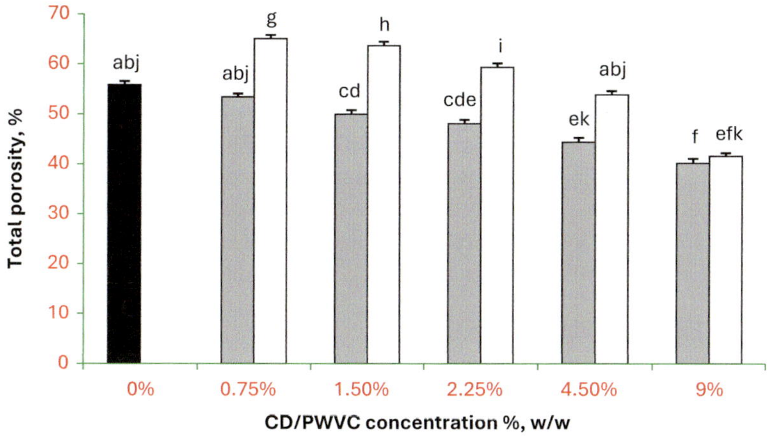

FIGURE 10.40 Total porosity of the container media as affected by fertilization with CD and PWVC. When pairs of data carry atleast one character in the superscript which is in common (LSD test; $p < 0.05$), it indicates the absence of significance difference from each other ■ control soil; ▩ CD amended soil; □ PVC amended soil.

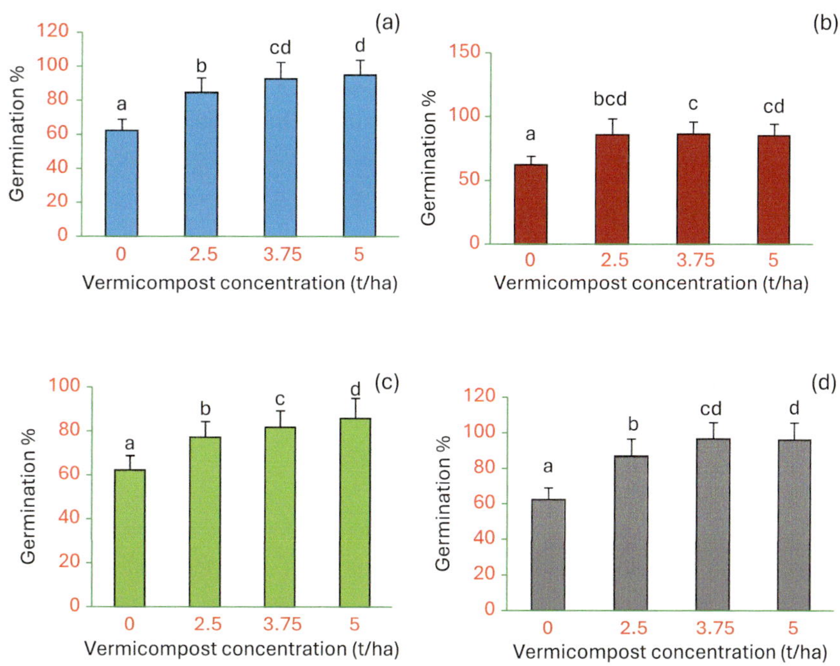

FIGURE 10.41 Effect of vermicomposts derived from (a) lantana; (b) ipomoea; (c) parthenium ; and (d) salvinia on germination. When pairs of bars carry at least one character in the superscript that is common, it indicates that there is no significant difference (LSD test; $p < 0.05$) in the seeds of *A. esculentus*.

of 86% was achieved in the 5 t/ha PVC treatment, whereas the control showed the lowest germination rate of 62%. All PVC treatments showed significantly higher germination success ($p < 0.05$) compared to the control.

For SVC, the highest germination rate of 97% was recorded in the 3.75 t/ha treatment, followed by 96% in the 5 t/ha and 87% in the 2.5 t/ha treatments (Figure 10.41d). The control group exhibited the lowest germination success. Germination rates in all SVC treatments were significantly higher ($p < 0.05$) than in the controls. All-in-all, each of the four weed VCs enhanced germination success at all levels of application studied, and the improvement was statistically significant in the majority of cases.

10.4.3.2 Effect on Plant Growth

A. esculentus plants grown in soils amended with LVC exhibited enhanced growth across all evaluated parameters (Figure 10.42). A positive correlation was observed between the rate of LVC application and plant growth, with the hierarchy of effectiveness being 5 t/ha > 3.75 t/ha > 2.5 t/ha > control (no LVC). Remarkably, even at the lowest LVC concentration of 2.5 t/ha, there was a substantial increase in growth compared to the control group, except for root length, which did not show a significant change. These findings suggest that the application of LVC, even at modest rates, can significantly promote the growth of *A. esculentus* in container cultivation.

As seen with the LVC, in the IVC treatments also the growth of *A. esculentus* in terms of all the variables studied was higher than in controls (Figure 10.43). The same trends were seen when PVC or SVC were applied (Figures 10.44 and 10.45). All treatments had shown significantly greater growth than the controls. The positive impact extended to chlorophyll, carotenoids, carbohydrates, and protein content of the four VCs (Figures 10.46–10.49)

10.4.3.3 Effect on Flowering and Yield

All four VCs induced earlier flowering (Figure 10.50) and significantly increased ($p < 0.05$) the number of flowers in *A. esculentus* plants compared to the control group. Additionally, yield-related parameters, including the number and weight, as well as the length and diameter of the pods, were also significantly enhanced in plants treated with VCs, each compared to controls (Figures 10.50–10.53). These results demonstrate the positive impact of VC applications on both flowering (Figure 10.54) and fruit yield (Figure 10.55) in *A. esculentus.*

10.4.3.4 Effect on Biochemistry

The application of VCs significantly increased the levels of chlorophyll and carotenoids in the leaves of *A. esculentus* plants, with pigment concentrations in VC-treated plants being notably higher than those in the control group. In addition to the enhanced pigment levels, total solids and ash (mineral) content were also significantly elevated in VC-treated plants compared to controls (Figures 10.46–10.49). Furthermore, there was a marked increase in the protein and carbohydrate concentrations in plants grown in VC-amended soils, further highlighting the positive impact of VC on the nutritional quality of the plants.

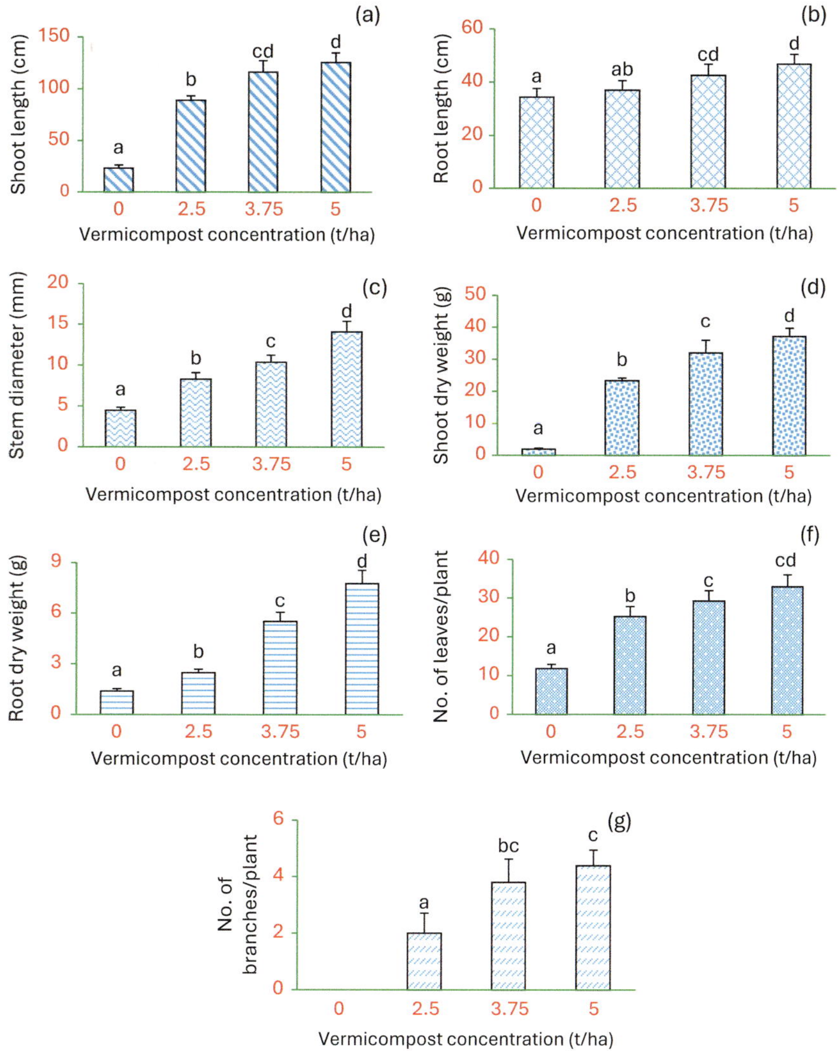

FIGURE 10.42 Effect of LVC on (a) shoot length; (b) root length; (c) stem diameter; (d) shoot dry weight; (e) root dry weight; (f) number of leaves; and (g) number of branches, of *A. esculentus* plants. When pairs of bars carry at least one character in the superscript that is common, it indicates that there is no significant difference (LSD test; $p < 0.05$) in the data represented by these bars.

10.4.3.5 Pest-Repellant and Anti-Pathogen Attributes

Typically, diseased leaves and pods are shown in Figures 10.56 and 10.57. These effects were much less prevalent and intense in plants fertilized by the VCs of the four weeds than in the controls (Figures 10.58 and 10.59). In effect, VCs derived from

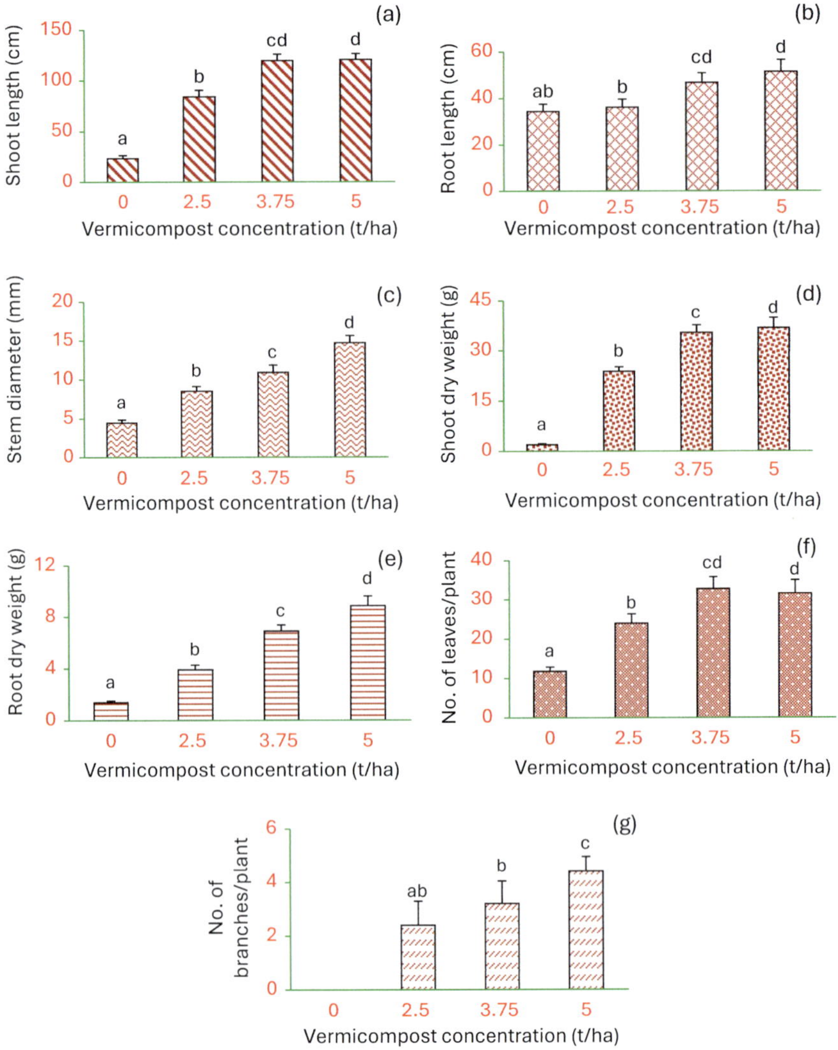

FIGURE 10.43 Effect of IVC on (a) shoot length; (b) root length; (c) stem diameter; (d) shoot dry weight; (e) root dry weight; (f) number of leaves; and (g) number of branches, of *A. esculentus* plants. When pairs of bars carry at least one character in the superscript that is common, it indicates that there is no significant difference (LSD test; $p < 0.05$) in the data represented by these bars.

all four weed species effectively induced resistance in the test plants against pests and pathogens. A significant reduction in fruit borer infestation was observed, with the severity of infestation following the trend: 0 t/ha (control) > 2.5 t/ha > 3.75 t/ha > 5 t/ha. Additionally, during the early stages of plant growth, seedlings experienced heavy infestation by leaf miners. However, as plant growth progressed, the incidence

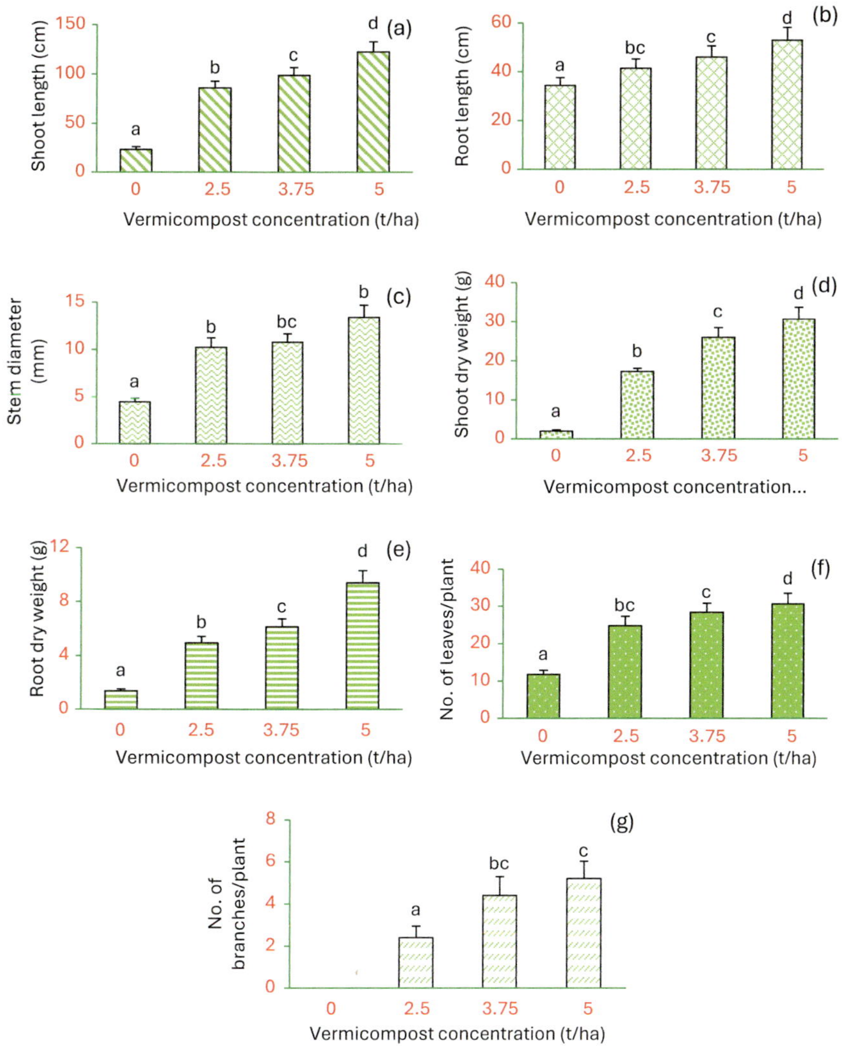

FIGURE 10.44 Effect of PVC on (a) shoot length; (b) root length; (c) stem diameter; (d) shoot dry weight; (e) root dry weight; (f) number of leaves; and (g) number of branches, of *A. esculentus* plants. When pairs of bars carry at least one character in the superscript that is common, it indicates that there is no significant difference (LSD test; $p < 0.05$) in the data represented by these bars.

of infestation diminished proportionally, indicating an enhanced ability of the plants to resist pests as they matured under VC treatment.

10.4.3.6 Possible Reasons behind the Findings

In several past reports, which are predominantly focused on manure-based VCs, the possible reasons behind the effectiveness of the VCs as fertilizers have been itemized.

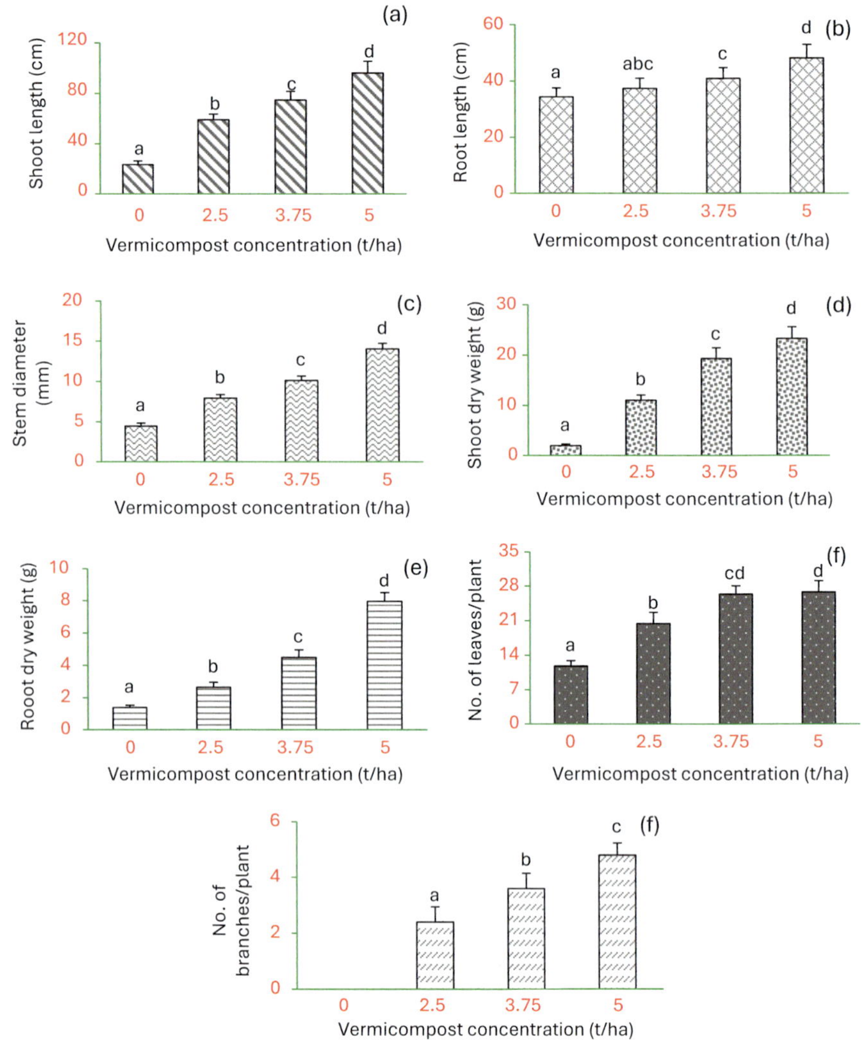

FIGURE 10.45 Effect of SVC on (a) shoot length; (b) root length; (c) stem diameter; (d) shoot dry weight; (e) root dry weight; (f) number of leaves; and (g) number of branches, of *A. esculentus* plants. When pairs of bars carry at least one character in the superscript that is common, it indicates that there is no significant difference (LSD test; $p < 0.05$) in the data represented by these bars.

The gist is presented here. It is likely that similar factors are operative in the case of the LVC/IVC/PVC/SVC applications as well.

In addition to improving nutrient bioavailability compared to the parent substrate, vermicomposting introduces several biologically active substances, including fulvic acids, humic acids, and phytohormones, into the substrate (Ievinsh 2020: Wong et al., 2020). These components likely contribute to the superior growth observed in VC-treated plants over the control group (Karthikeyan et al., 2014; Hussain et al.,

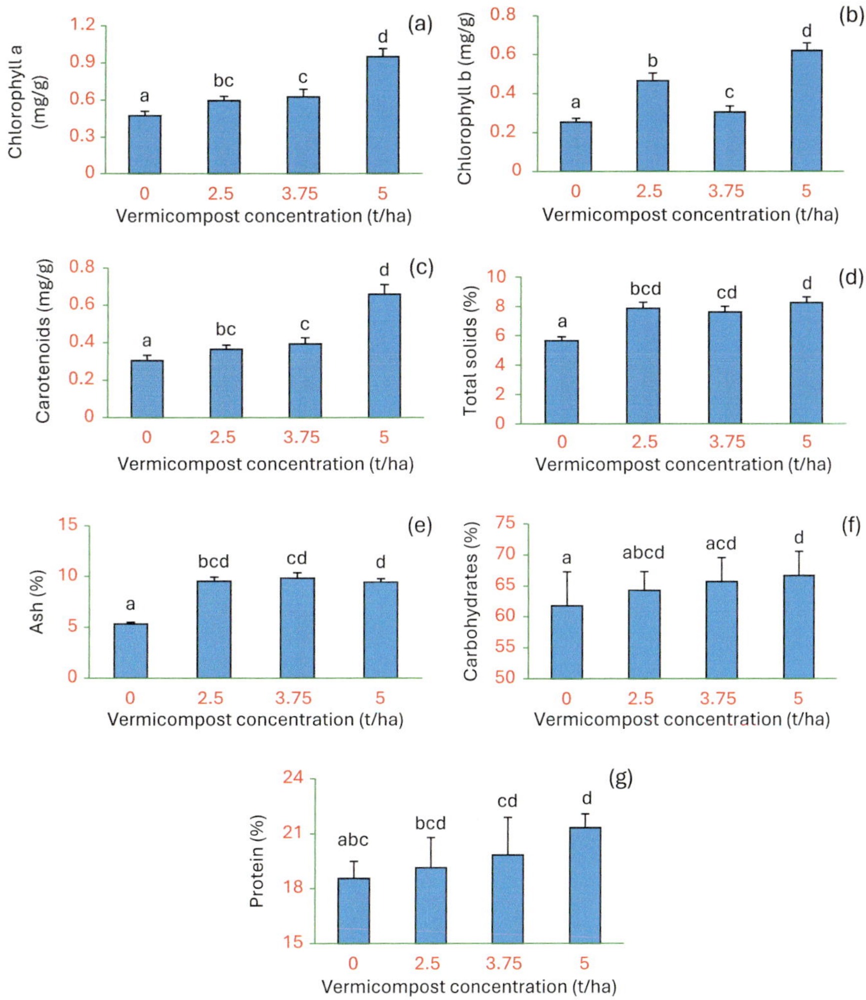

FIGURE 10.46 Effect of LVC on (a) chlorophyll "a"; (b) chlorophyll "b"; (c) carotenoid content of ladies finger leaves and (d) total solids; (e) ash; (f) carbohydrates and (g) protein content, of *A. esculentus* fruits. When pairs of bars carry at least one character in the superscript that is common, it indicates that there is no significant difference (LSD test; $p < 0.05$) in the data represented by these bars.

2015; Rehman et al., 2023). Canellas et al. (2002) found that humic acids extracted from VCs stimulate root elongation, lateral root development, and the activity of H+-ATPase in maize roots. These biomolecules are likely produced due to the diverse microbial community present in the earthworm gut, which plays a crucial role in the vermicomposting process (Parthasarathi et al., 2008; Zandonadi & Busato, 2012; Zhang et al., 2024). Actinomycetes and fungi in the soil are thought to release phytohormones that further support plant growth (Frankenberger & Arshad,

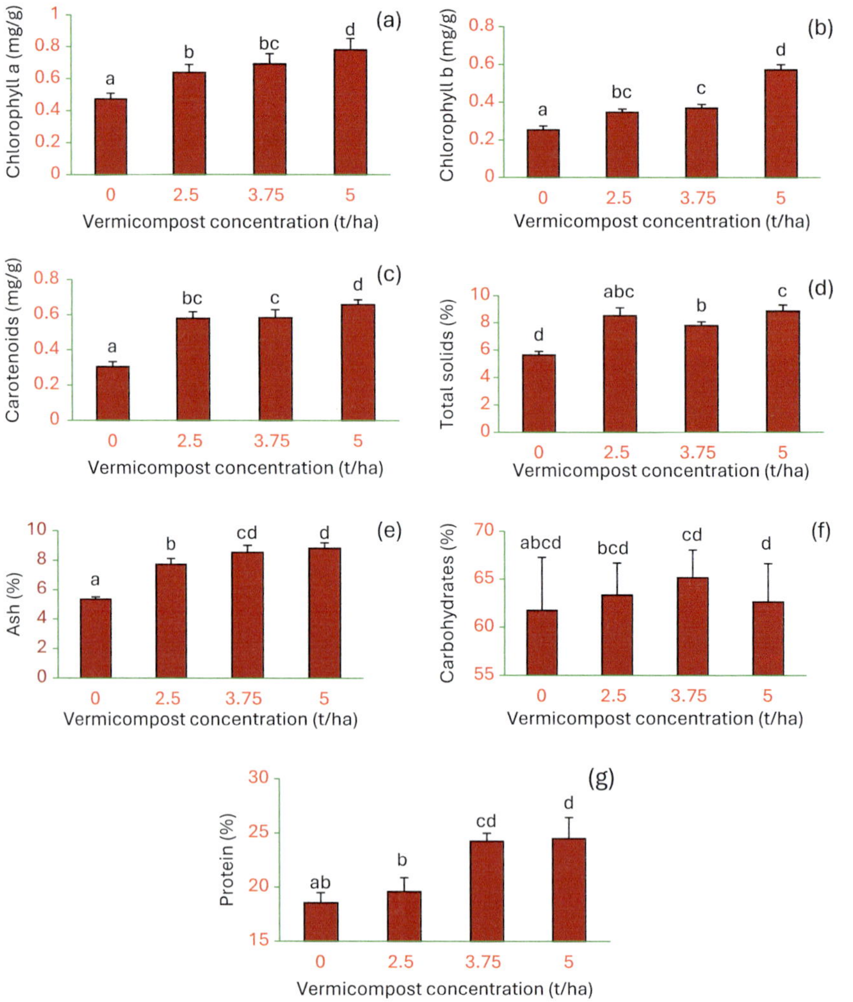

FIGURE 10.47 Effect of IVC on (a) chlorophyll "a"; (b) chlorophyll "b"; (c) carotenoid content of ladies finger leaves and (d) total solids; (e) ash; (f) carbohydrates and (g) protein content, of *A. esculentus* fruits. When pairs of bars carry at least one character in the superscript that is common, it indicates that there is no significant difference (LSD test; $p < 0.05$) in the data represented by these bars.

1995; Canellas et al., 2020). The improved soil physical properties, such as increased aeration and water holding capacity, as demonstrated in previous sections, may also have contributed to enhanced plant productivity in VC-treated soils (Le Bayon et al., 2021; Edwards et al., 2022).

Additionally, previous research has shown that VCs possess pest repellent properties (Edwards et al., 2010; Cardoza and Buhler, 2012; Khiew et al., 2020), and the current study suggests that weed-derived VCs may share similar characteristics. Studies

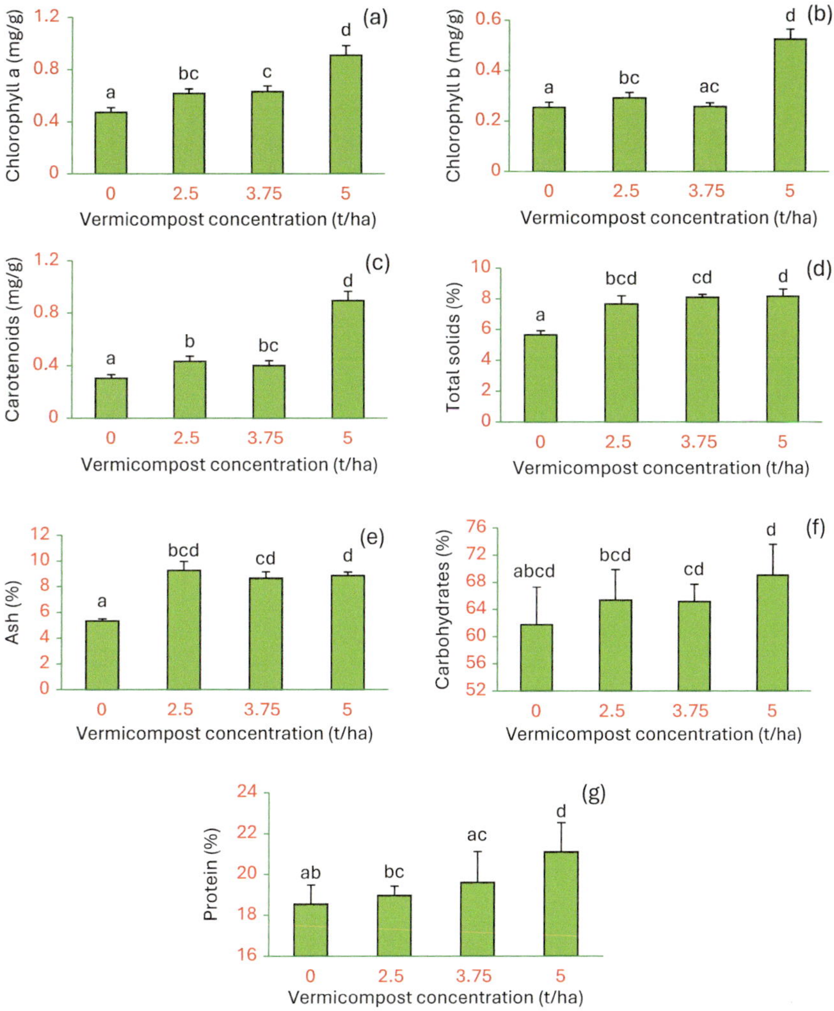

FIGURE 10.48 Effect of PVC on (a) chlorophyll "a"; (b) chlorophyll "b"; (c) carotenoid content of ladies finger leaves and (d) total solids; (e) ash; (f) carbohydrates and (g) protein content, of *A. esculentus* fruits. When pairs of bars carry at least one character in the superscript that is common, it indicates that there is no significant difference (LSD test; $p < 0.05$) in the data represented by these bars.

on manure-based VCs indicate that improved nutrient availability and the presence of antimicrobial compounds, such as flavonoids, phenolics, and humic acids, likely promote pathogen resistance in plants (Edwards et al., 2010; Cimen et al., 2023). Cardoza (2011) further suggests that the microbial flora, rather than the chemical constituents, in VC may play a key role in conferring resistance to pests and diseases.

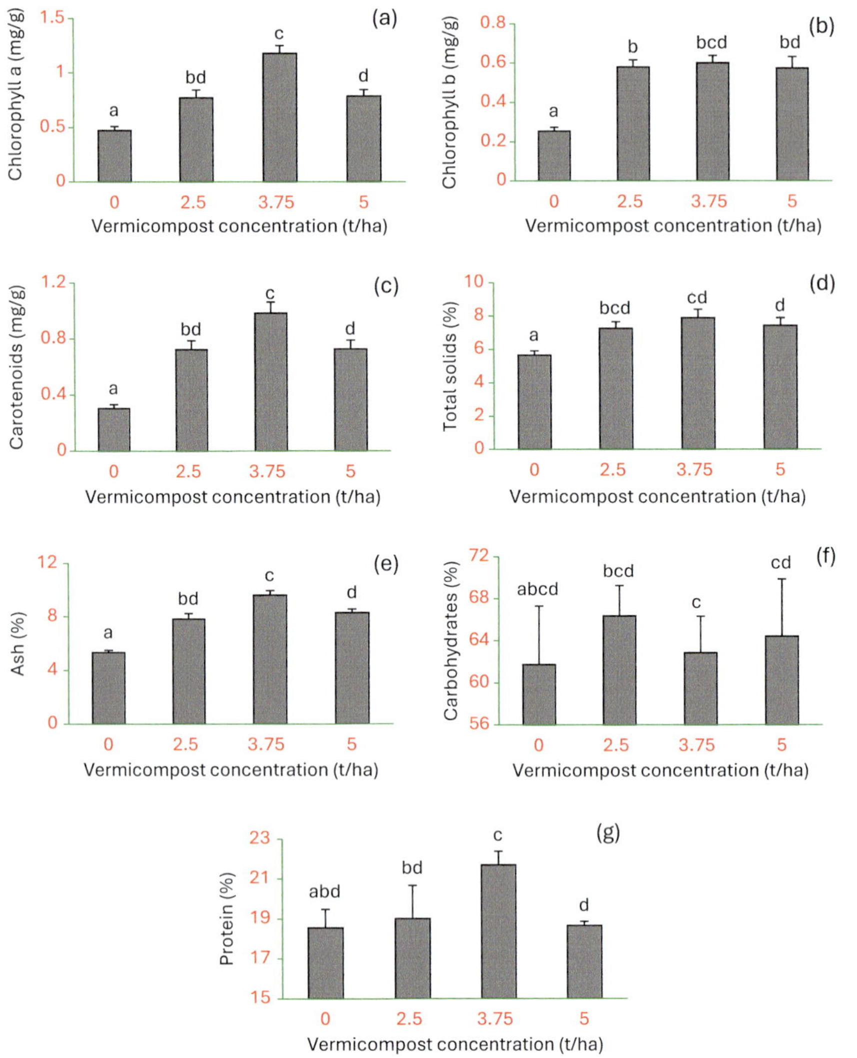

FIGURE 10.49 Effect of SVC on (a) chlorophyll "a"; (b) chlorophyll "b"; (c) carotenoid content of ladies finger leaves and (d) total solids; (e) ash; (f) carbohydrates and (g) protein content, of *A. esculentus* fruits. When pairs of bars carry at least one character in the superscript that is common, it indicates that there is no significant difference (LSD test; $p < 0.05$) in the data represented by these bars.

10.4.4 Studies Defined in Section 10.3.1.4

To assess the effectiveness of LVC compared to SCF, application rates were determined based on equivalent nutrient content (Karthikeyan et al., 2014). SCF was applied to cluster bean (*Cyamopsis tetragonoloba*) in concentrations corresponding to the nutrient levels present in the respective LVC treatments (Figures 10.60 and 10.61).

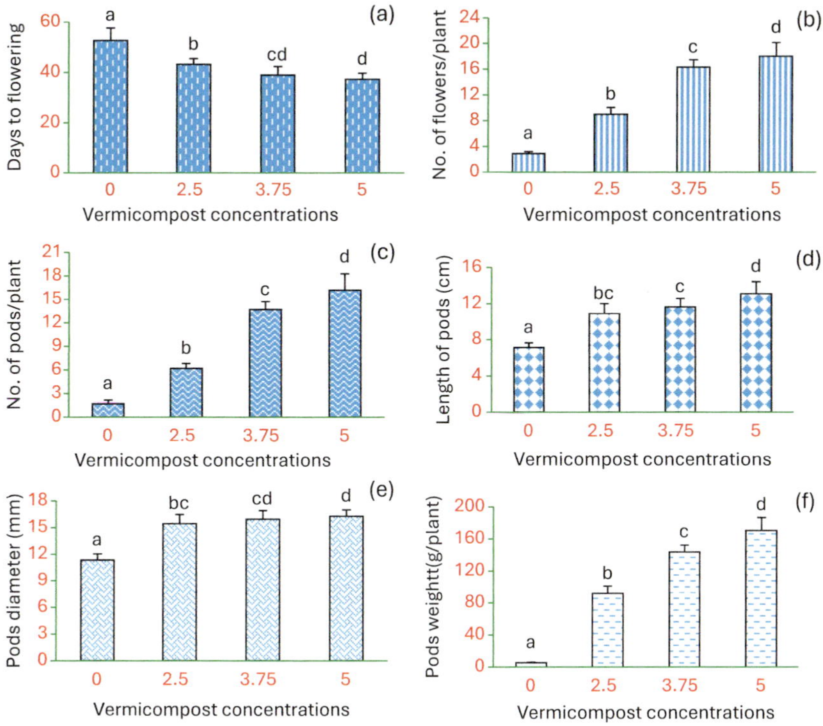

FIGURE 10.50 Effect of LVC on (a) early flowering; (b) number of flowers per plant; (c) number of pods per plant; (d) length of pods; (e) diameter of pods; and (f) weight of pods per plant, of *A. esculentus*. When pairs of bars carry at least one character in the superscript that is common, it indicates that there is no significant difference (LSD test; $p < 0.05$) in the data represented by these bars.

A control group, without any fortification of LVC or SCF, was also included. The results showed that seed germination success in *tetragonoloba* was up to 51.5% higher in LVC-treated soils compared to those treated with SCF (Figure 10.62). LVC also promoted superior plant growth, reflected in greater stem diameter, shoot length, shoot biomass, leaf number, and leaf pigment levels (Figure 10.63). These positive effects extended to increased fruit yield (Figure 10.64). Additionally, LVC application enhanced root nodule formation, reduced disease incidence, and led to fewer stunted plants. These findings suggest that the allelopathic compounds in lantana were likely neutralized during the vermicomposting process, enabling LVC to outperform SCF in supporting germination, growth, and fruit yield.

Yet-to-be-reported studies along similar lines by the authors, in which IVC was deployed instead of LVC, have shown that use of IVC led to witnessing of positive effects similar to the ones seen with LVC. Even though ipomoea is more toxic than lantana, its VC seemed to be as benign, plant friendly, and soil friendly as LVC. And its performance was better than that of SCF (Tables 10.2–10.4)

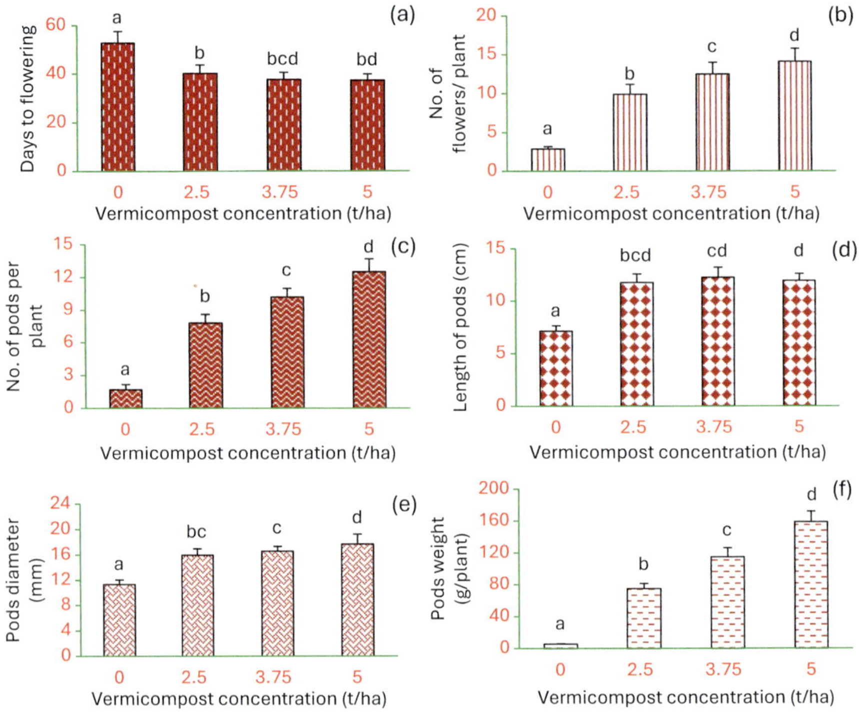

FIGURE 10.51 Effect of IVC on (a) early flowering; (b) number of flowers per plant; (c) number of pods per plant; (d) length of pods; (e) diameter of pods; and (f) weight of pods per plant, of *A. esculentus*. When pairs of bars carry at least one character in the superscript that is common, it indicates that there is no significant difference (LSD test; $p < 0.05$) in the data represented by these bars.

10.4.5 Studies Defined in Section 10.3.1.5

These studies are perhaps the most significant of all described in this chapter. The reason is that these studies conclusively show that phytomass-based VCs – even the ones generated from toxic weeds – are as potent and benign fertilizers as CD-based VC (CDVC) is known to be. In some respects, phytomass-based VCs are seen to perform better than CDVC. These studies, together with the ones described in the previous section, which show that LVC and IVC are superior as fertilizers to SCF, should remove the last doubts on the fertilizer value of phytomass/phytowaste and paper waste.

10.4.5.1 Seed Germination

VC treatments significantly improved seed germination compared to the controls (Figure 10.65a). However, no statistically significant differences were observed between the effects of CDVC and LVC treatments. The highest germination success rate, 95%, was recorded in the 5 t/ha LVC treatment, closely followed by a 94% germination rate in the 3.75 t/ha CDVC treatment.

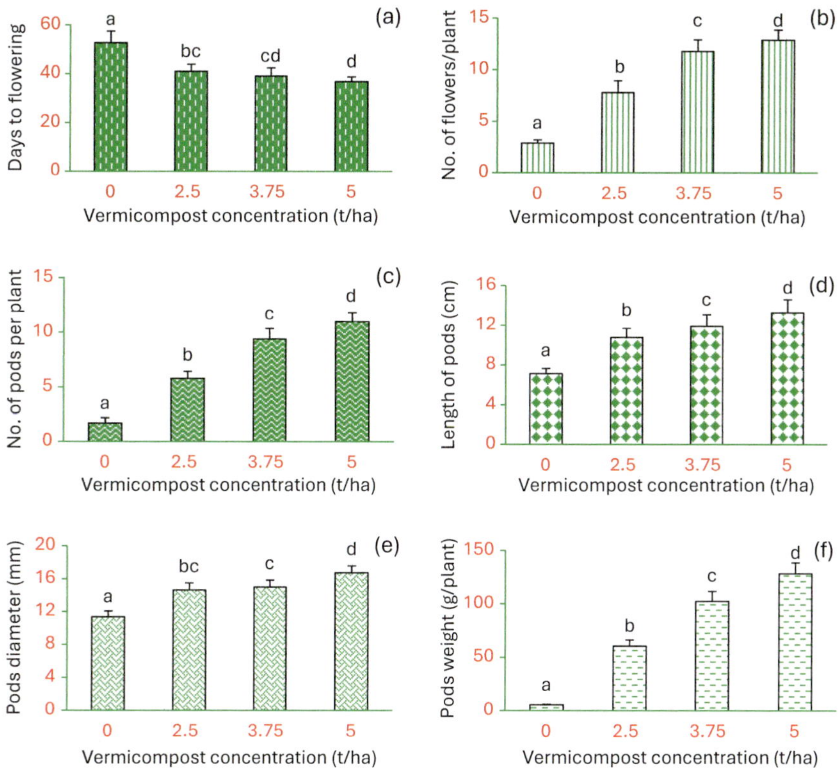

FIGURE 10.52 Effect of PVC on (a) early flowering; (b) number of flowers per plant; (c) number of pods per plant; (d) length of pods; (e) diameter of pods; and (f) weight of pods per plant, of *A. esculentus*. When pairs of bars carry at least one character in the superscript that is common, it indicates that there is no significant difference (LSD test; $p < 0.05$) in the data represented by these bars.

10.4.5.2 Plant Growth

Both CDVC and LVC stimulated the growth of *A. esculentus* plants to a similar extent (Figure 10.65b–g). Within the range of VC concentrations tested, the general trend was that the greater the application rate, the more pronounced the growth benefits. With the exception of leaf number in CDVC treatments, all growth parameters followed the pattern: 5 t/ha > 3.75 t/ha > 2.5 t/ha > control. Notably, apart from root length, plant growth increased significantly even with a marginal rise in VC concentration from zero to 2.5 t/ha. A similar trend was observed for flowering, where higher LVC treatments led to a greater number of flowers and earlier flowering compared to the controls and lower LVC treatments. For CDVC, the 3.75 t/ha treatment outperformed other concentrations in promoting flowering (Table 10.2).

Compared to CDVC, *A. esculentus* plants grown in LVC-amended soil exhibited significantly greater shoot length and plant biomass. However, no statistically significant differences were observed in shoot diameter or the number of branches between the two VC treatments.

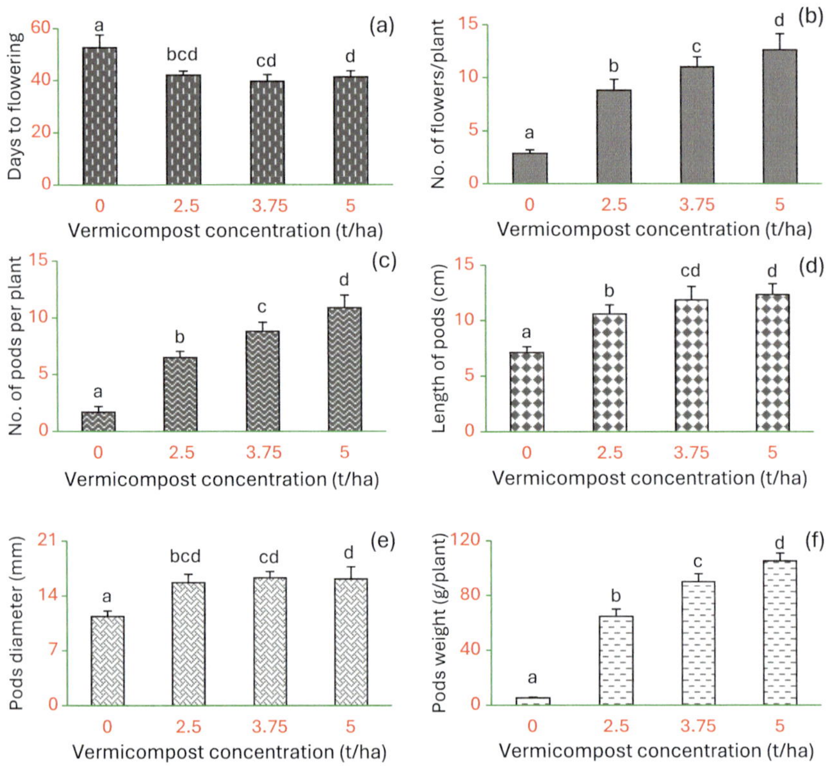

FIGURE 10.53 Effect of SVC on (a) early flowering; (b) number of flowers per plant; (c) number of pods per plant; (d) length of pods; (e) diameter of pods; and (f) weight of pods per plant, of *A. esculentus*. When pairs of bars carry at least one character in the superscript that is common, it indicates that there is no significant difference (LSD test; $p < 0.05$) in the data represented by these bars.

10.4.5.3 Yield and Biochemical Aspects

Even as both VCs significantly enhanced the yield of the *A. esculentus* pods in comparison to the unfertilized plants (Figure 10.65), LVC induced a significantly higher number of pods per plant than CDVC did. It also led to pods of higher average weight. In the matter of boosting chlorophyll, carotenoid, total solids, and protein content, however, CDVC performed as well as LVC, and no statistically significant difference was seen between the two (Figure 10.66).

10.4.5.4 Disease Incidence

Both VCs effectively induced disease resistance in the test plants (Figure 10.65h, Table 10.2). While LVC marginally outperformed CDVC in reducing disease incidence, the difference was not statistically significant. In contrast, the proportion of infected fruits was lower in the 3.75 and 5 t/ha CDVC treatments compared to the

FIGURE 10.54 An okra (*A. esculentus*) flower, with an already formed pod.

FIGURE 10.55 Harvests of the okra (*A. esculentus*) pods.

FIGURE 10.56 Typically diseased leaves of ladies finger.

FIGURE 10.57 Typically diseased pods of ladies finger.

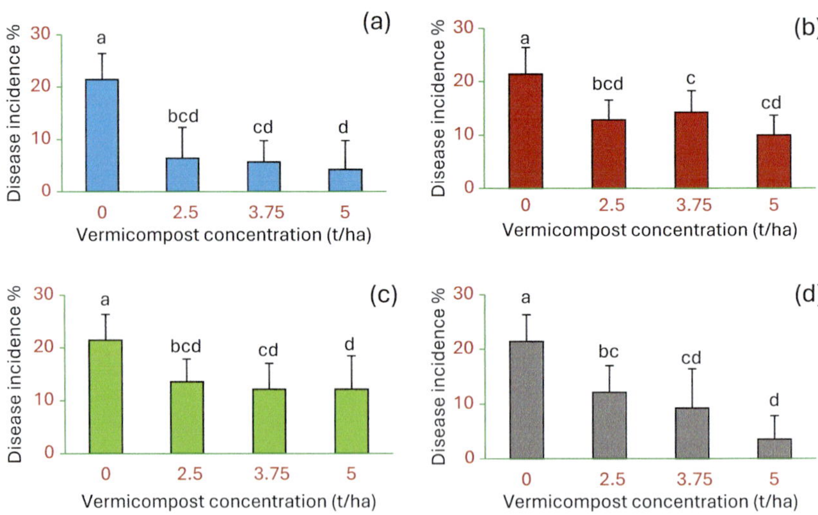

FIGURE 10.58 Influence of the vermicomposts derived from (a) lantana; (b) ipomoea; (c) parthenium, and (d) salvinia on disease incidence in *A. esculentus* plants. When pairs of bars carry at least one character in the superscript that is common, it indicates that there is no significant difference (LSD test; $p < 0.05$) in the data represented by these bars.

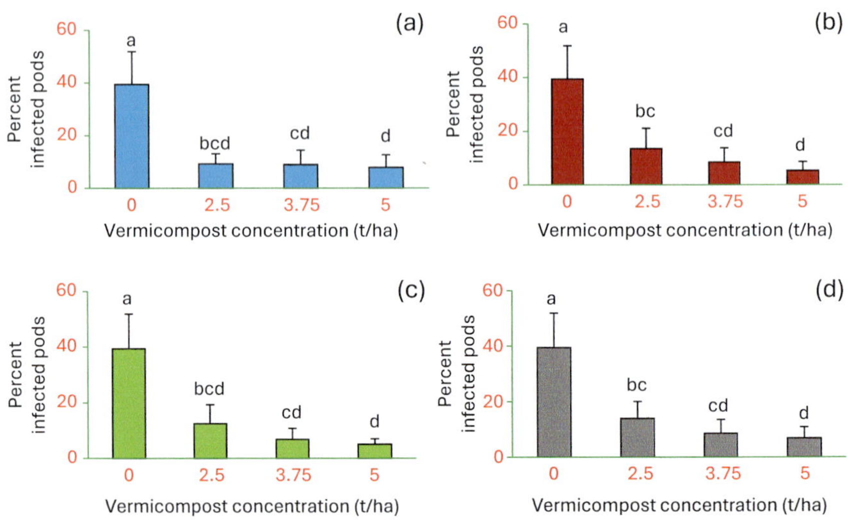

FIGURE 10.59 Influence of the vermicomposts derived from (a) lantana; (b) ipomoea; (c) parthenium; and (d) salvinia on infected fruits (%). When pairs of bars carry at least one character in the superscript that is common, it indicates that there is no significant difference (LSD test; $p < 0.05$) in the data represented by these bars.

FIGURE 10.60 Early growth experiments on cluster bean fertilized either by LVC or equivalent SCF.

FIGURE 10.61 Cluster bean plants laden with fruit in plots fertilized by LVC.

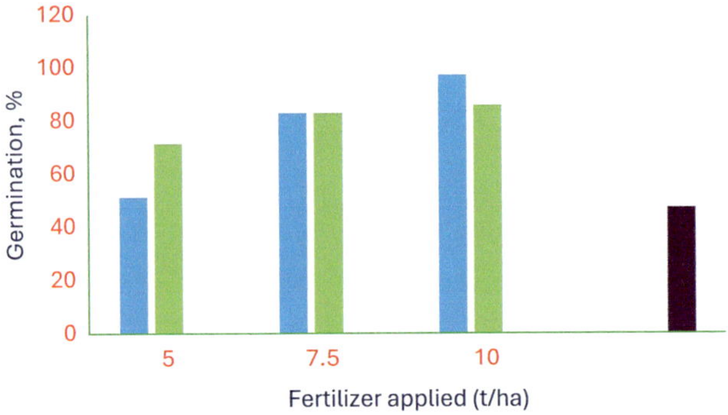

FIGURE 10.62 Effect of fertilizer applied (t/ha) on germination success of cluster bean LVC ■, SCF ■, and control ■.

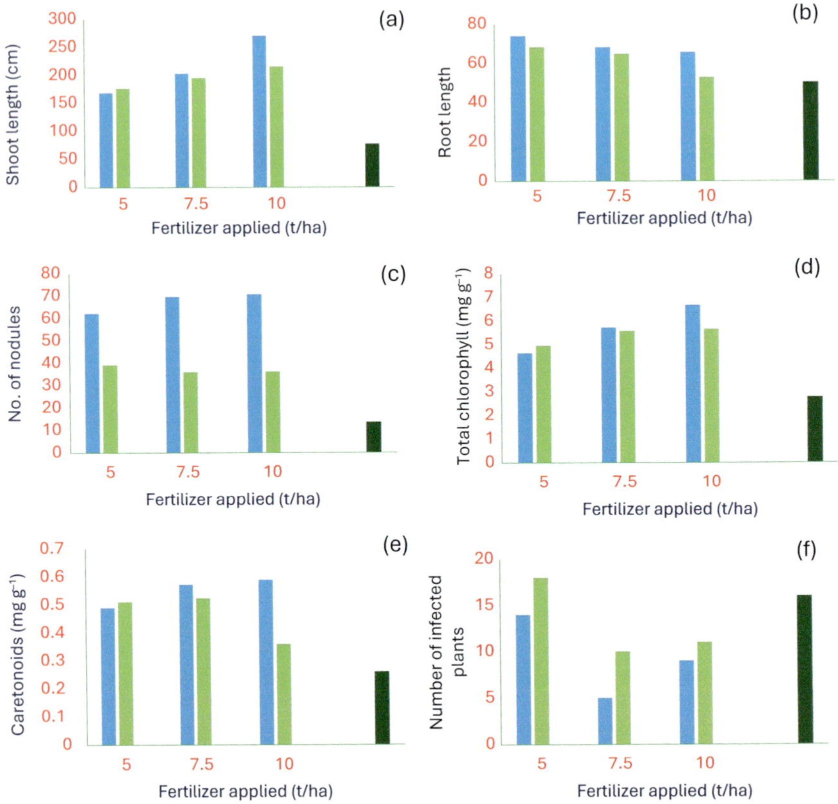

FIGURE 10.63 Growth of the cluster bean plants when fertilized by LVC ■, SCF ■, or neither ■.

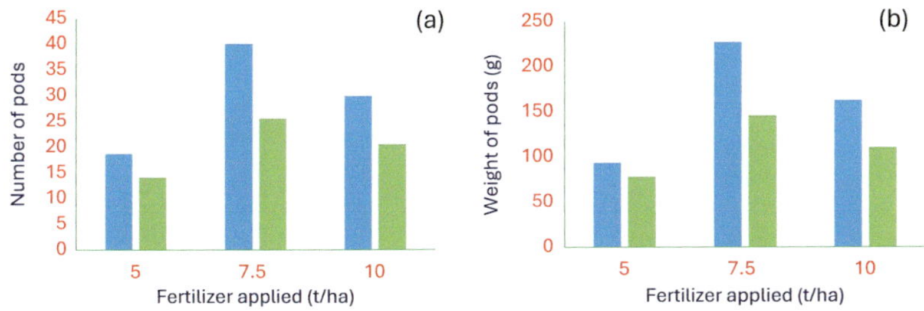

FIGURE 10.64 Number (a) and weight (b) of cluster bean pods (fruits) per plant in plots fertilized by LVC ■ and SCF ■.

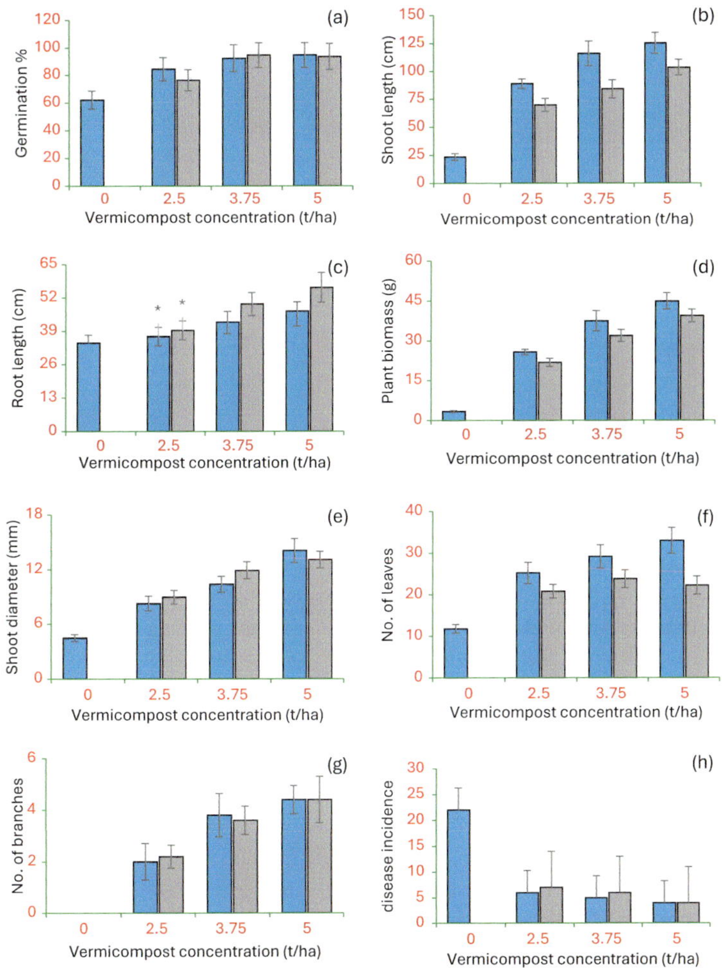

FIGURE 10.65 Effect of LVC ▭ and CDVC ▭ on *A. esculentus* in terms of (a) germination success; (b) length of shoots; (c) length of roots; d) plant biomass; (e) shoot diameter; (f) number of leaves; (g) number of branches; and (h) disease incidence. Bars topped with an asterisk indicate that the corresponding numbers do not differ significantly from the controls at $P \leq 0.05$.

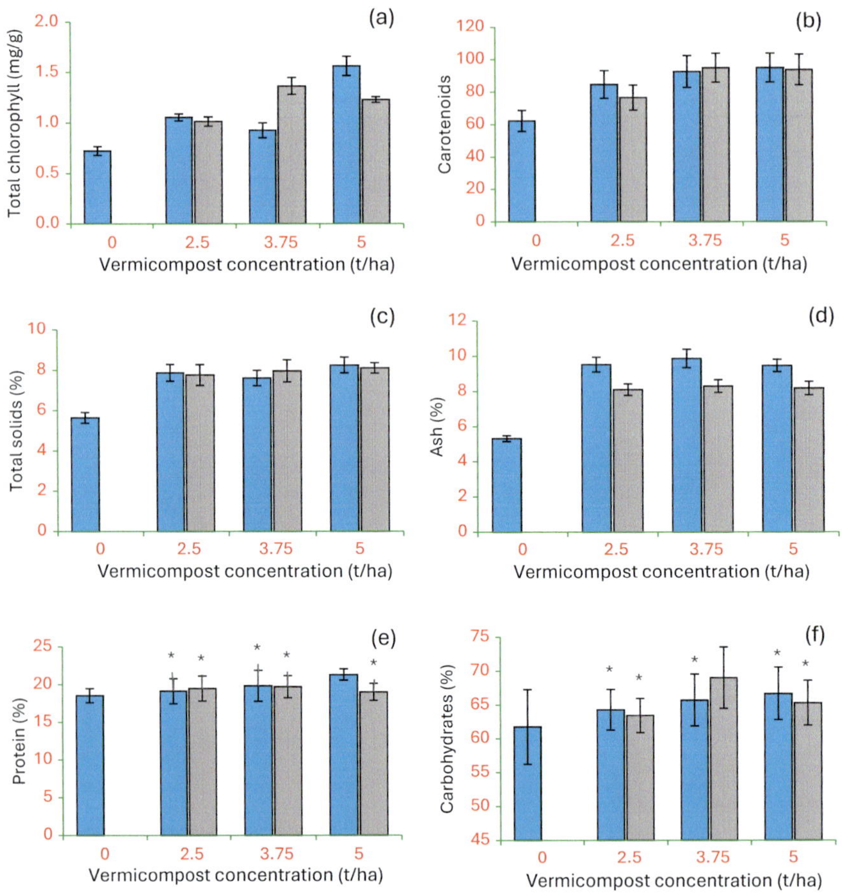

FIGURE 10.66 Effect of LVC and CDVC on *A. esculentus* in terms of (a) total chlorophyll in the leaves; (b) carotenoids in the leaves; (c) total solids; (d) ash; (e) protein in pods; and (f) carbohydrates in pods. Bars topped with an asterisk indicate that the corresponding numbers do not differ significantly from the controls at $p \leq 0.05$.

corresponding LVC applications, though this difference was also not statistically significant.

All-in-all, the studies reveal that in some aspects LVC has outperformed CDVC, while in some other aspects no significant difference was seen between the two. This makes it evident that lantana loses its toxic and allelopathic constituents during its vermicomposting. Equally significant is the finding that the positive influence matches — at times even surpasses — that of CDVC.

TABLE 10.2
Germination Value (GV) and Germination Percentage (GP) of the Seeds of Cluster bean as Influenced by Ipomoea Vermicast and Inorganic Fertilizers. Results That Do Not Differ Significantly (LSD Test; $p < 0.05$) Carry at Least One Character (a/b/c/d) in the Superscript That is Common

Treatment	Amount	Fourth Day		Fifth Day		Sixth Day		Seventh Day		Eighth Day	
		GV	GP	GV	GP	GV	GP	GV	GP	GV	GP
Vermicompost	5 t/ha[a]	24.70	62.86	17.27	65.71	13.06	68.57	13.06	80.00	10.00	80.00
	7.5 t/ha[ac]	22.50	60.00	15.81	62.86	11.99	65.71	10.41	71.43	9.30	77.14
	10 t/ha[b]	14.74	48.57	13.06	57.14	10.00	60.00	7.35	60.00	6.17	62.86
Inorganic fertilizer equivalent to the VC	5 t/ha[cb]	16.53	51.43	13.06	57.14	11.99	65.71	8.81	65.71	7.35	68.57
	7.5 t/ha[a]	31.89	71.43	22.08	74.29	15.33	74.29	11.26	74.29	8.62	74.29
	10 t/ha[a]	29.39	68.57	20.41	71.43	15.33	74.29	12.14	77.14	9.65	78.57
Control	Nil[d]	2.50	20.00	2.64	25.71	2.74	31.43	4.26	45.71	3.47	47.14

TABLE 10.3
Plant Growth, Leaf Pigments, Flowering, Disease Incidence, and Retarded Growth Stunned Growth/Death in Cluster Bean Plants as Impacted by Ipomoea Vermicast or Inorganic Fertilizers. Results That Do Not Differ Significantly (LSD Test; p <0.05) Carry at Least One Character (a/b/c/d/e/f) in the Superscript That is Common

Parameters	Vermicompost at Dose			Inorganic Fertilizers at Dose			Control	F-value	
	5 t/ha	7.5 t/ha	10 t/ha	5 t/ha	7.5 t/ha	10 t/ha		Type of Fertilizer	Amount
Plant Growth									
Stem diameter (mm)	16.9 ± 1.5^{a}	22.6 ± 4.2^{b}	24.1 ± 2.5^{b}	16.8 ± 1.4^{a}	18.4 ± 1.2^{a}	22.9 ± 2.0^{b}	11.2 ± 1.7^{c}	36.70***	70.94***
Shoot length (cm)	144.5 ± 5.0^{a}	197.6 ± 9.8^{b}	231.6 ± 10.5^{c}	138.6 ± 5.7^{a}	162.7 ± 7.3^{d}	191.4 ± 6.9^{b}	76.8 ± 5.9^{e}	70.73***	202.3***
Root length (cm)	55.0 ± 8.0^{a}	41.4 ± 4.1^{b}	38.3 ± 4.8^{b}	47.7 ± 5.2^{c}	47.2 ± 4.3^{c}	39.6 ± 3.9^{b}	50.2 ± 8.4^{c}	2.364$^{n.s}$	18.33***
Number of leaves	98.4 ± 5.5^{a}	120.3 ± 7.7^{b}	133.8 ± 5.6^{c}	92.3 ± 3.8^{d}	108.3 ± 6.0^{e}	112.2 ± 6.6^{e}	42.0 ± 5.3^{f}	162.1***	236.9***
Number of nodules	61.3 ± 10.8^{a}	90.3 ± 15.4^{b}	129.3 ± 23.2^{c}	31.9 ± 7.2^{d}	50.0 ± 11.9^{e}	37.1 ± 7.3^{d}	13.5 ± 6.0^{f}	76.96***	15.41***
Shoot dry weight (g)	51.5 ± 5.6^{a}	89.3 ± 6.6^{b}	102.6 ± 10.8^{c}	42.9 ± 4.6^{d}	59.6 ± 5.2^{e}	82.8 ± 6.6^{f}	45.3 ± 5.2^{d}	18.64***	74.71***
Root dry weight (g)	5.02 ± 0.22^{a}	6.25 ± 0.60^{b}	11.90 ± 2.14^{c}	4.60 ± 0.74^{a}	8.10 ± 1.18^{d}	6.84 ± 0.93^{b}	3.21 ± 0.58^{f}	15.21***	40.90***
Leaf Pigments									
Chlorophyll-a (mg g^{-1})	2.17 ± 0.14^{a}	3.21 ± 0.25^{b}	3.45 ± 0.11^{c}	1.99 ± 0.09^{d}	2.69 ± 0.10^{e}	2.72 ± 0.06^{e}	1.32 ± 0.05^{f}	58.35***	145.0***
Chlorophyll-b (mg g^{-1})	2.21 ± 0.05^{a}	2.30 ± 0.06^{b}	2.47 ± 0.08^{c}	2.10 ± 0.14^{d}	2.28 ± 0.10^{ba}	2.57 ± 0.08^{e}	1.45 ± 0.04^{f}	137.5***	370.8***
Total chlorophyll (mg g^{-1})	4.38 ± 0.14^{a}	5.51 ± 0.24^{b}	5.92 ± 0.16^{c}	4.09 ± 0.08^{d}	4.97 ± 0.12^{e}	5.29 ± 0.10^{f}	2.77 ± 0.04^{g}	88.05***	329.7***
Carotenoids (mg g^{-1})	0.42 ± 0.04^{a}	0.51 ± 0.05^{b}	0.55 ± 0.08^{b}	0.36 ± 0.04^{c}	0.41 ± 0.05^{a}	0.53 ± 0.07^{b}	0.26 ± 0.02^{d}	38.36***	59.70***
Flowering									
Number of flowers	27.4 ± 7.0^{a}	37.6 ± 12.4^{b}	31.7 ± 11.1^{cd}	15.7 ± 4.2^{e}	27.9 ± 8.0^{ad}	18.1 ± 8.7^{ce}	0.7 ± 1.0^{f}	158.0***	84.87***
Disease Incident, Plant Death, and Stunted Plants									
Number of infected plant	13^{a}	3^{a}	8^{a}	17^{a}	9^{a}	15^{a}	16^{a}	1.201$^{n.s}$	1.193$^{n.s}$
Number of plant died	Nil^{a}	1^{a}	Nil^{a}	2^{a}	2^{a}	1^{a}	Nil^{a}	2.873$^{n.s}$	0.874$^{n.s}$
Number of stunted plant	Nil^{a}	1^{a}	1^{a}	1^{a}	5^{a}	6^{a}	11^{a}	2.380$^{n.s}$	1.384$^{n.s}$

$^{*}p < 0.05$, $^{**}p < 0.01$, $^{***}p < 0.001$, n.s. not significant.

TABLE 10.4
Harvest Index and Yield Attributes of Plants as Impacted by Ipomoea Vermicast and Equivalent Inorganic Fertilizers (Mean ± SD). Results That Do Not Differ Significantly (LSD Test; $p < 0.05$) Carry at Least One Character in the Superscript That is Common

Parameters, Average Value	5 t/ha	7.5 t/ha	10 t/ha	5 t/ha	7.5 t/ha	10 t/ha	Type of Fertilizer	Amount
Number of pods per plant	20.1 ± 8.1^{a}	36.3 ± 10.0^{b}	28.3 ± 7.2^{c}	19.5 ± 6.7^{a}	30.9 ± 7.8^{c}	22.4 ± 7.5^{a}	103.4^{***}	91.39^{***}
Weight of the pods (g)	110.1 ± 43.8^{a}	201.2 ± 55.9^{b}	159.3 ± 42.4^{c}	103.7 ± 35.7^{a}	161.4 ± 42.7^{cd}	120.8 ± 41.2^{a}	134.8^{***}	147.8^{***}
Pod length (cm)	14.6 ± 1.8^{a}	15.8 ± 1.6^{b}	15.5 ± 1.6^{b}	14.1 ± 1.6^{a}	14.7 ± 1.9^{a}	14.3 ± 1.7^{a}	22.24^{***}	6.851^{**}
Pod width (cm)	10.2 ± 1.0^{a}	10.8 ± 0.9^{bd}	16.0 ± 1.3^{c}	10.2 ± 1.3^{ade}	10.4 ± 0.9^{be}	10.8 ± 1.5^{b}	53.32^{***}	83.730^{***}
Pod thickness (cm)	6.39 ± 0.85^{a}	6.01 ± 0.79^{b}	6.79 ± 0.58^{c}	5.80 ± 0.54^{b}	5.99 ± 1.04^{b}	5.74 ± 0.74^{b}	36.21^{***}	$2.604^{n.s}$
Number of seeds per pod	9.24 ± 0.92^{a}	9.50 ± 1.16^{a}	9.50 ± 1.16^{a}	9.18 ± 1.16^{bc}	9.30 ± 1.28^{a}	9.18 ± 1.12^{a}	$2.175^{n.s}$	$0.728^{n.s}$
Seed diameter (mm)	5.59 ± 1.01^{a}	6.01 ± 0.86^{b}	6.23 ± 1.69^{c}	5.61 ± 0.95^{a}	5.96 ± 0.89^{b}	6.51 ± 0.72^{d}	5.565^{*}	170.9^{***}
Seed thickness (mm)	2.30 ± 0.57^{a}	1.90 ± 0.59^{b}	2.43 ± 0.59^{c}	2.14 ± 0.63^{d}	2.15 ± 0.46^{d}	2.51 ± 0.49^{e}	5.705^{***}	153.3^{***}
Harvest index %	65.2 ± 40.1^{abcf}	86.9 ± 67.8^{bdg}	49.7 ± 34.3^{ceh}	73.4 ± 51.1^{ade}	87.8 ± 61.4^{agh}	47.0 ± 39.2^{cf}	$0.292^{n.s}$	7.713^{**}

$^{*}p < 0.05$, $^{**}p < 0.01$, $^{***}p < 0.001$, n.s. not significant.

REFERENCES

Abbasi, S.A., Hussain, N., Tauseef, S.M., Abbasi, T. 2018. A novel flippable units vermireactor train system – FLUVTS – for rapidly vermicomposting paper waste to an organic fertilizer, *Journal of Cleaner Production, 198*, pp.917–930. doi.org/10.1016/j.jclepro.2018.07.040.

Abbasi, S.A., Nayeem-Shah, M., and Abbasi, T. 2015. Vermicomposting of phytomass: limitations of the past approaches and the emerging directions. *Journal of Cleaner Production, 93*, pp.103–114.

Ahmad, A., Aslam, Z., Bellitürk, K., Iqbal, N., Naeem, S., Idrees, M., ... and Kamal, A. 2022. Vermicomposting methods from different wastes: an environment friendly, economically viable and socially acceptable approach for crop nutrition: a review. *International Journal of Food Science and Agriculture, 5*(1), pp.58–68.

Ahmed, M.M.M., 2000. Studies on the control of insect pests in vegetables (okra, Tomato, and Onion) in Sudan with special reference to neem preparations. PhD thesis submitted to University of Göttingen, Germany.

Amenduini, M., D'Amico, M., Colella, C., and Cirulli, M., 2003. Severe outbreaks of Alternaria leaf spot on kiwi in Southern Italy. *FergusonBasilicata, 53*(11), pp.39–43.

Anandyawati, A., Muktamar, Z., Prameswari, W., Amrullah, A.H.K., and Anggoro, A. 2023. Comparison of the quality of animal manure compost conventional methods with vermicompost animal manure from Lumbricus rubellus. *International Journal of Agricultural Technology, 19*, pp.1435–1446.

Antonijevic, D., Fakultet, P., Zemun, B., and Mitrovic, M., 2007. Leaf spot of oilseed rape Biljni lekar (Serbia), *Plant Doctor, 35*, pp.443–449.

AOAC (Association of Official Agricultural Chemists), 2012. *Official methods of analysis of AOAC International*, 19th ed. AOAC International, Gaithersburg.

Arain, A.R., Mithal Jiskani, M., Wagan, K.H., Khuhro, S.N., and Khaskheli, M.I., 2012. Incidence and chemical control of okra leaf spot disease. *Pakistan Journal of Botany, 44*, pp.1769–1774.

Arif, N., Yadav, V., Singh, S., Singh, S., Ahmad, P., Mishra, R.K., ... and Chauhan, D.K. 2016. Influence of high and low levels of plant-beneficial heavy metal ions on plant growth and development. *Frontiers in Environmental Science, 4*, p.69.

Atiyeh, R.M., Arancon, N., Edwards, C.A. and Metzger, J.D., 2000. Influence of earthworm-processed pig manure on the growth and yield of greenhouse tomatoes. *Bioresource Technology, 75*(3), pp.175–180.

Bashour Issam, I. and Sayegh Antoine, H., 2007. *Methods of Analysis for Soils of Arid and Semi-arid Regions*. Food and Agriculture Organization of the United Nations, Rome. 128pp. ISBN 978-92-5-105661-5.

Buckerfield, J.C., Flavel, T.C., Lee, K.E. and Webster, K.A., 1999. Vermicompost in solid and liquid forms as a plant-growth promoter. *Pedobiologia, 4*, pp.753–759.

Canellas, L.P., Canellas, N.O., da S. Irineu, L.E.S., Olivares, F.L., and Piccolo, A. 2020. Plant chemical priming by humic acids. *Chemical and Biological Technologies in Agriculture, 7*, pp.1–17.

Canellas, L.P., Olivares, F.L., Okorokova-Facanha, A.L., Facanha, A.R. 2002. Humic acids isolated from earthworm compost enhance root elongation, lateral root emergence, and plasma membrane H+ -ATPase activity in maize roots. *Plant Physiology, 130*, pp.1951–1957.

Canihos, Y., Peever, T.L., and Timmer, L.W., 1999. Temperature, leaf wetness and isolate effects on infection of Minneola tangelo leaves by Alternaria sp. *Plant Diseases, 83*, pp.429–433.

Cardoza, Y.J. 2011. *Arabidopsis thaliana* resistance to insects, mediated by an earthworm-produced organic soil amendment. *Pest Management Science, 67*(2), pp.233–238.

Cardoza, Y.J., and Buhler, W.G., 2012. Soil organic amendment impacts on corn resistance to *Helicoverpa zea*: Constitutive or induced? *Pedobiologia*, *55*, pp.343–347.

Carter, M.R., and Gregorich, E.G. 2007. *Soil Sampling and Methods of Analysis*. CRC Press, Boca Raton. https://doi.org/10.1201/9781420005271.

Castellini, M., Bondì, C., Giglio, L., and Iovino, M. 2024. Impact of vermicompost addition on water availability of differently textured soils. *Heliyon*, *10*(15), e35699.

Chatterjee, R., Debnath, A., and Mishra, S. 2020. Vermicompost and soil health. In: Giri, B., Varma, A. (eds) *Soil Health,* Springer, Cham, pp. 69–88.

Chiranjeeb, K., Prasad, S.S., Singh, S.P., Bharati, V., and Jha, S. 2020. Effect of household vermicompost and fertilizer on soil microbial biomass carbon, biomass phosphorus and biomass nitrogen in incubation experiment. *International Journal of Current Microbiology and Applied Sciences*, *9*(2), 1508–1516.

Cho, J.T. and Moon, B.J., 1980. The occurrence of strawberry black leaf spot caused by Alternaria alternate (Fr.) Keissler in Korea. *Korean Journal of Applied Entomology*, *19*(4), pp.221–227.

Cimen, A., Baba, Y., Yildirim, A.B., and Turker, A.U. 2023. Do vermicompost applications improve growth performance, pharmaceutically important alkaloids, phenolic content, free radical scavenging potency and defense enzyme activities in summer snowflake (Leucojum aestivum L.)?. *Chemistry & Biodiversity*, *20*(11), e202301074.

Clewer, A.G. and Scarisbrick, D.H., 2001. *Practical Statistics and Experimental Design for Plant and Crop Sciences*, John Wiley & Sons, Ltd, England. p.10.

Doklega, S., and Imryed, Y.F.E. 2020. Effect of vermicompost and nitrogen levels fertilization on yield and quality of head lettuce. *Journal of Plant Production*, *11*(12, 1495–1499.

Edwards, C.A., Arancon, N.Q., Bennett, M.V., Askar, A., Keeney, G., Little, B., 2010. Suppression of green peach aphid (*Myzus persicae*) (Sulz.), citrus mealybug (*Planococcus citri*), and two spotted spider mite (*Tetranychus urticae*) (Koch.) attacks on tomatoes and cucumbers by aqueous extracts from vermicomposts, *Crop Protection*, *29*, pp.80–93.

Edwards, C.A., Arancon, N.Q., Bohlen, P.J., and Hendrix, P. 2022. *Biology and Ecology of Earthworms* (Vol. 567). Springer, New York.

Field Andy, 2009. *Discovering Statistics Using SPSS*, In SAGE Publications London.

Frankenberger, W.T., Jr., and Arshad, M. 1995. *Phytohormones in Soils: Microbial Production and Function*. Marcel and Deckker, New York, p. 503.

Ganeshkumar, T., Premalatha, M., Gajalakshmi, S., and Abbasi, S.A. 2014. A new process for the rapid and direct vermicomposting of the aquatic weed salvinia (Salvinia molesta). *Bioresources and Bioprocessing*, *1*, 1–5.

Guptha, J., and Tayde, A.R. 2022. Efficacy of selected biopesticides and chemicals against okra shoot and fruit borer, Earias vittella (fabricius.) infesting Okra. *Annals of Plant Protection Sciences*, *30*(2), 132–135.

Gutiérrez-Miceli, F.A., Llaven, M.A.O., Nazar, P.M., Sesma, B.R., Álvarez-Solís, J.D. and Dendooven, L., 2011. Optimization of vermicompost and worm-bed leachate for the organic cultivation of radish. *Journal of Plant Nutrition*, *34*(11), pp.1642–1653.

He, H.Q. and Lin, W.X., 2001. Studies on allelopathic physiobiochemical characteristics of rice. *Chinese Journal of Eco-Agriculture*, *9*, 56–57.

Hilhorst, H.W.M. and Karssen, C.M., 2000. Effect of chemical environment on seed germination. In: Fenner, M. (ed.) *Seeds: The Ecology of Regeneration in Plant Communities.* CABI publishing, Wallingford, UK.

Hussain, N., Abbasi, T., and Abbasi, S.A. 2015. Vermicomposting eliminates the toxicity of Lantana (Lantana camara) and turns it into a plant friendly organic fertilizer. *Journal of Hazardous Materials*, *298*, 46–57.

Hussain, N., and Abbasi, S.A. 2018. Efficacy of the vermicomposts of different organic wastes as "clean" fertilizers: state-of-the-art. *Sustainability*, *10*(4), 1205.

ICAR, 2011. *Handbook of Agriculture*. Indian Counsel of Agriculture Research, Pusa, New Delhi.

Ievinsh, G. 2020. Review on physiological effects of vermicomposts on plants. *Biology of Composts*, 63–86.
Ievinsh, G., 2011. Vermicompost treatment differentially affects seed germination, seedling growth and physiological status of vegetable crop species. *Plant Growth Regulation*, *65*, pp.169–181.
Iqbal, A., Hussain, Q., Mo, Z., Hua, T., Mustafa, A.E.Z.M., and Tang, X. 2024. Vermicompost supply enhances fragrant-Rice yield by improving soil fertility and eukaryotic microbial community composition under environmental stress conditions. *Microorganisms*, *12*(6), p.1252.
Jenkinson, D.S., 1988. Determination of microbial biomass carbon and nitrogen in soil. In: Wilson, J.R. (ed.) *Advances in Nitrogen Cycling in Agricultural Ecosystems*, Wallingford: CABI. pp.368–386.
Jones Jr, J.B. 2012. *Plant Nutrition and Soil Fertility Manual*. CRC Press, Boca Raton. https://doi.org/10.1201/b11577.
Jones, J.B. Jr., 2001. *Laboratory Guide for Conducting Soil Tests and Plant Analysis*, CRC Press, Washington.
Kadam, D., and Pathade, G. 2014. Effect of tendu (Diospyros melanoxylon RoxB.) leaf vermicompost on growth and yield of French bean (*Phaseolus vulgaris* L.). *International Journal of Recycling of Organic Waste in Agriculture*, *3*, pp.1–7.
Karthikeyan, M., Hussain, N., Gajalakshmi, S., and Abbasi, S.A. 2014. Effect of vermicast generated from an allelopathic weed lantana (Lantana camara) on seed germination, plant growth, and yield of cluster bean (Cyamopsis tetragonoloba). *Environmental Science and Pollution Research*, *21*, pp.12539–12548.
Khiew, A.A.M., Lew, J.H., Shamsuddin, M.R., Mustapa, N.I., and Narasimha, M.M. 2020. Study on humic acid derived from EFB based vermicompost as biopesticide. In *IOP Conference Series: Materials Science and Engineering*, USA. (Vol. 736, No. 7, p. 072012). IOP Publishing, England. doi:10.1088/1757-899X/736/7/072012.
Kinigopoulou, V., Hatzigiannakis, E., Stefanou, S., Guitonas, A., and Oikonomou, E.K. 2023. Utilization of vermicompost sludge instead of peat in olive tree nurseries in the frame of circular economy and sustainable development. *AgriEngineering*, *5*(3), pp.1630–1643.
Le Bayon, R.C., Bullinger, G., Schomburg, A., Turberg, P., Brunner, P., Schlaepfer, R., and Guenat, C. 2021. Earthworms, plants, and soils. In: Hunt, A., Egli, M., Faybishenko, B. (Eds). *Hydrogeology, Chemical Weathering, and Soil Formation*, Washington, pp.81–103. https://doi.org/10.1002/9781119563952.ch4
Leslie, C.A.; Romani, R.J. 1998. Inhibition of ethylene biosynthesis by salicylic acid. *Plant Physiology, 88*, pp.833–837.
Lv, M., Li, J., Zhang, W., Zhou, B., Dai, J., and Zhang, C. 2020. Microbial activity was greater in soils added with herb residue vermicompost than chemical fertilizer. *Soil Ecology Letters*, *2*, 209–219.
Manivannan, K., and Selvamani, P. 2014. Influence of organic inputs on the yield and quality of fruits in banana cultivar 'Poovan' (syn. Mysore AAB). *Acta Horticult.* 1018, 139–148.
Margesin, R., and Schinner, F. (Eds.. 2005. *Manual for Soil Analysis-Monitoring and Assessing Soil Bioremediation* (Vol. 5). Springer Berlin, Heidelberg, pp XVI, 366. https://doi.org/10.1007/3-540-28904-6.
Mathivanan, S., Chidambaram, A.L., Sundaramoorthy, P. and Kalaikandhan, R., 2012. Effect of vermicompost on germination and biochemical constituents of groundnut (Arachis hypogaea L.) seedling. *International Journal of Biological Sciences, 2*(2), pp.54–59.
Mendoza-Hernández, D., Fornes, F., and Belda, R.M. 2014. Compost and vermicompost of horticultural waste as substrates for cutting rooting and growth of rosemary. *Scientia Horticulturae*, *178*, pp.192–202.
Moran, R., Porath, D., 1980. Chlorophyll determination in intact tissues using N, N dimethyl formamide. *Plant Physiology*, *65*, pp.478–479.

Munoz-Ucros, J., Panke-Buisse, K., and Robe, J. 2020. Bacterial community composition of vermicompost-treated tomato rhizospheres. *PLoS One*, *15*(4), e0230577.

Nayeem-Shah, M., Gajalakshmi, S., Abbasi, S.A., 2015. Direct, rapid and sustainable vermicomposting of the leaf litter of neem (*Azadirachta indica*), *Applied Biochemistry and Biotechnology*, *175*, pp.792–801.

Nelson, S. Suzanne, 2009. *Food Analysis*, 4th edition, Springer, USA.

Oyege, I., and Balaji Bhaskar, M.S. 2023. Effects of vermicompost on soil and plant health and promoting sustainable agriculture. *Soil Systems*, *7*(4), p.101.

Parthasarathi, K., Balamurugan, M., and Ranganathan, L.S. 2008. Influence of vermicompost on the physico-chemical and biological properties in different types of soil along with yield and quality of the pulse cropblack gram. *Journal of Environmental Health Science and Engineering*, *5*(1), pp.51–58.

Patra, R.K., Behera, D., Mohapatra, K.K., Sethi, D., Mandal, M., Patra, A.K., and Ravindran, B. 2022. Juxtaposing the quality of compost and vermicompost produced from organic wastes amended with cow dung. *Environmental Research*, *214*, p.114119.

Patterson, D.T. 1981. Effects of allelopathic chemicals on growth and physiological responses of soybean (Glycine max). *Weed Science*, *29*(1), 53–59. pp.293–310.

Pratyusha, S. 2022. Phenolic compounds in the plant development and defense: an overview. In Hasanuzzaman, M. and Nahar, K. (Eds), *Plant Stress Physiology-Perspectives in Agriculture*, IntechOpen, London, UK. pp.125–140. DOI: 10.5772/intechopen.102873

Ramos, R.F., Santana, N.A., de Andrade, N., Romagna, I.S., Tirloni, B., de Oliveira Silveira, A., ... and Jacques, R.J.S. 2022. Vermicomposting of cow manure: Effect of time on earthworm biomass and chemical, physical, and biological properties of vermicompost. *Bioresource Technology*, *345*, p.126572.

Raza, S.T., Wu, J., Rene, E.R., Ali, Z., and Chen, Z. 2022. Reuse of agricultural wastes, manure, and biochar as an organic amendment: a review on its implications for vermicomposting technology. *Journal of Cleaner Production*, *360*, p.132200.

Rehman, S.U., De Castro, F., Aprile, A., Benedetti, M., and Fanizzi, F.P. 2023. Vermicompost: enhancing plant growth and combating abiotic and biotic stress. *Agronomy*, *13*(4), p.1134.

Rodriguez-Quiroz, G., Valenzuela-Quiñonez, W. and Nava-Pérez, E., 2011. Vermicomposting as a nitrogen source in germinating kidney bean in trays. *Journal of Plant Nutrition*, *34*(10), pp.1418–1423.

Rostaei, M., Fallah, S., Carrubba, A., and Lorigooini, Z. 2024. Organic manures enhance biomass and improve content, chemical compounds of essential oil and antioxidant capacity of medicinal plants: A review. *Heliyon*, *10*, p.e36693.

Silva, J.A., and Uchida, R.S. 2000. *Plant Nutrient Management in Hawaii's Soils: Approaches for Tropical and Subtropical Agriculture*. College of Tropical Agriculture and Human Resources, University of Hawaii at Manoa, 158pp.

Singh, A., Kumar, V., Verma, S., Majumdar, M., and Sarkar, S. 2020. Significance of vermicompost on crop and soil productivity: A review. *IJCS*, *8*(5), pp.1529–1534.

Siping, L I., Chong, W.A.N.G., HUANG, H., Jia, C.A.O., Rui, X.U.E., and Binglei, W.A.N.G. 2024. Vermicompost maintains fertility of topsoil by reducing NH3 volatilization and improving 15N/NO_3-retention in a saline-alkali soil. *Pedosphere*.

Smith, J.L. and Paul, E.A., 1990. The significance of soil microbial biomass estimations. In: Bollag, J.M. and Stotzky, G. (eds.) *Soil Biochemistry*, 6, Taylor & Francis, New York, pp. 357–396. https://doi.org/10.1201/9780203739389

Sun, L., Wang, Y., and Miao, M. 2020. Inhibition of α-amylase by polyphenolic compounds: Substrate digestion, binding interactions and nutritional intervention. *Trends in Food Science & Technology*, *104*, 190–207.

Tanase, C., Bujor, O.C., and Popa, V.I. 2019. Phenolic natural compounds and their influence on physiological processes in plants. In Watson, R.R. (Ed). *Polyphenols in Plants*. Academic Press, London and USA, pp. 45–58.

Thakur, A., Kumar, A., Kumar, C.V., Kiran, B.S., Kumar, S., and Athokpam, V.A.R.U.N. 2021. A review on vermicomposting: by-products and its importance. *Plant Cell Biotechnology and Molecular Biology*, *22*, pp.156–164.

Tohyama, A., Hayashi, K., Taniguchi, N., and Tsuda, M., 1995. Infection sources and causal agents of *Alternaria* rot of okra pods, *Annals of the Phytopathological Society of Japan (Japan)*, 61, pp.346–349.

Uçarlı, C. 2020. Effects of salinity on seed germination and early seedling stage. In Fahad, S., Saud, S., Chen, Y., Wu, C., and Wang, D. (Eds). *Abiotic Stress in Plants*, IntechOpen, London, UK. *211* DOI: 10.5772/intechopen.93647

Vibhuti, C.S., Bargali, K., and Bargali, S.S. 2015. Seed germination and seedling growth parameters of rice (*Oryza sativa* L.) varieties as affected by salt and water stress. *Indian Journal of Agricultural Sciences*, *85*(1), pp.102–108.

Villarreal, A.P.B., Tumipamba, D.E.G., and Sarzosa, F.V.C. 2021. Vermicomposting: Production of Humus and Biol. In *Communication, Smart Technologies and Innovation for Society: Proceedings of CITIS 2021*. Springer Singapore, Singapore, pp. 591–600.

Wellburn, A.R., 1994. The spectral determination of chlorophylls a and b as well as total carotenoids, using various solvents with spectrophotometers of different resolutions, *Journal of Plant Physiology*, *144*, pp.307–313.

Werner, M., 1987. Necrotic leaf spot of apple caused by fungi of the genus Alternaria. *Ochrona Roslin (Poland)*, *31*(10). pp. 6–7

Wilson, D.P. and Carlile, W.R., 1988. Plant growth in potting media containing worm-worked duck waste. In M.J. Maher (Ed). *Symposium on Substrates in Horticulture other than Soils in situ 238*. ISHS, Dublin, Ireland pp. 205–220.

Wong, W.S., Zhong, H.T., Cross, A.T., and Yong, J.W.H. 2020. Plant biostimulants in vermicomposts: Characteristics and plausible mechanisms. In Geelen, D., Xu, L., (Eds) *The Chemical Biology of Plant Biostimulants*, John Wiley & Sons Ltd, US. pp.155–180. https://doi.org/10.1002/9781119357254.ch6

Yan, A. and Chen, Z., 2020. The control of seed dormancy and germination by temperature, light and nitrate. *The Botanical Review*, *86*(1), pp.39–75.

Yang, Z., Luo, Y., Chen, H., Zhang, Y., Wu, S., Yang, M., ... and Zhou, Y. 2024. Vermicompost addition improved soil aggregate stability, enzyme activity, and soil available nutrients. *Journal of Soil Science and Plant Nutrition*, *24*, pp.6760–6774.

Zaller, J.G., 2007. Vermicompost as a substitute for peat in potting media: Effects on germination, biomass allocation, yields and fruit quality of three tomato varieties. *Scientia Horticulturae*, *112*(2), pp.191–199.

Zandonadi, D.B. and Busato, J.G. 2012. Vermicompost humic substances: technology for converting pollution into plant growth regulators, *International Journal of Environmental Science and Engineering Research, 3*, pp.73–84.

Zeng, R.S., Luo, S.M. and Shi, Y.H., 2001. Physiological and biochemical mechanism of allelopathy of secalonic acid on higher plants. *Agronomy Journal*, *93*, pp.72–79.

Zhang, J., Xu, Y. and Li, F., 2009. Influence of cow manure vermicompost on plant growth and microbes in rhizosphere on iron tailing. In *2009 3rd International Conference on Bioinformatics and Biomedical Engineering*. IEEE, Beijing, China, pp. 1–4.

Zhang, Y., Yuan, H., Peng, S., Wang, Z., Cai, S., Chen, Z., ... and Zhang, W. 2024. Vermicomposting preferably alters fungal communities in wasted activated sludge and promotes the production of plant growth-promoting biostimulants in the vermicompost. *Chemical Engineering Journal*, *495*, p.153232.

11 Transformations Caused by Vermicomposting Which Convert Even Toxic and Allelopathic Weeds into Benign Organic Fertilizers[1]

11.1 INTRODUCTION

The body of work presented in the preceding chapter, which is further bolstered by the studies on the fertilizer value of a xerophyte's VC as described in Chapter 13, puts the following two deductions beyond doubt:

a. Vermicompost (VC of phytomass is as good, if not better, a fertilizer as manure-based VCs are known to be.
b. Even strongly allelopathic and toxic species of plants get transformed into very benign plant-friendly and soil-friendly fertilizers when subjected to direct vermicomposting as per the HRVP.

But what are the transformations that achieve this kind of metamorphosis? And what factors cause/influence those transformations? How and why do the weeds shed their harmful attributes as they undergo earthworm-mediated zoocomposting? This chapter is devoted to addressing these questions.

11.2 GUIDING CONSIDERATIONS

The following parameters appeared most relevant for assessing the changes that occur when a substrate is turned into vermicast:

i. The C:N ratio – because the lower this ratio in a fertilizer, the more suitable it is deemed to be. A C:N ratio of 15 or lower is considered highly desirable for a fertilizer (Morais & Queda, 2003; Deka et al., 2011).

[1] This chapter has been co-authored with Naseer Hussain, School of Life Sciences, B.S. Abdur Rahman Crescent Institute of Science and Technology, Chennai 600048, India.

DOI: 10.1201/9780429331886-13

ii. The degree of humifications as quantified by humification index (HI) – because fall in the HI to 5 or lesser reflects the transformation of a substrate into a highly humidified and stabilized entity which is known to support agriculture (Zbytniewski & Buszewski, 2005; Ravindran et al., 2013).
iii. The change in the concentration of phenols, fatty acids, lipids, ketones etc – as these are identified with the allelopathy and toxicity of the parent substrates (Hussain & Abbasi, 2018).
iv. The extent of mineralization – as reflected in the thermogravimetric (TG) and differential scanning calorimetric (DSC) profiles of the substrate in comparison to its VC – because mineralization makes those nutrients bioavailable, which are otherwise locked in the matrices of organic carbon.
v. The molecular rearrangements that occur in terms of number of chemicals, their types, and concentrations in a VC relative to its parent substrate – as assessed with gas chromatography–mass spectrometry (GC-MS) – because such rearrangements reflect transition from toxicity to beneficial entities.
vi. Changes in the surface characteristics – because they can potentially influence bioavailability of nutrients – as reflected in scanning electron microscopy.

Details of methodology are available in Hussain et al. (2016a–d). *Eisenia fetida* was the earthworm species used to vermicompost all four weeds.

11.3 THE TRANSFORMATIONS

11.3.1 C:N Ratio

As seen from Figure 11.1, lantana, ipomoea, parthenium, and salvinia have C:N ratios 22.7, 26.1, 23.9, and 53.9, respectively. Upon getting vermicomposted, their C:N ratios fall sharply to 8.1, 9.9, 9.1, and 9.35, respectively. In other words, the proportion of nitrogen in the VC of three of the weeds is about 2.7 times higher than it was in the parent substrates, while in the case of salvinia the nitrogen content relative to its carbon content has gone up 5-fold. In all cases, the C:N has reached levels known to be "highly desirable" in a fertilizer.

There are two reasons behind this turnabout. First is the mineralization of 50%–60% of the organic carbon of each substrate that occurs during vermicomposting. Feeding by the bacteria and their metabolic activity converts this much-bound organic carbon into mobile CO_2 gas. The gas escapes to the atmosphere, and only 40%–50% of the original carbon remains in the VC. Since vermicomposting is a fully aerobic process, no denitrification – which needs anaerobic conditions – takes place. For this reason, there is little loss of nitrogen. Rather, there is some addition of nitrogen contributed in the form of earthworm mucus and other exudates.

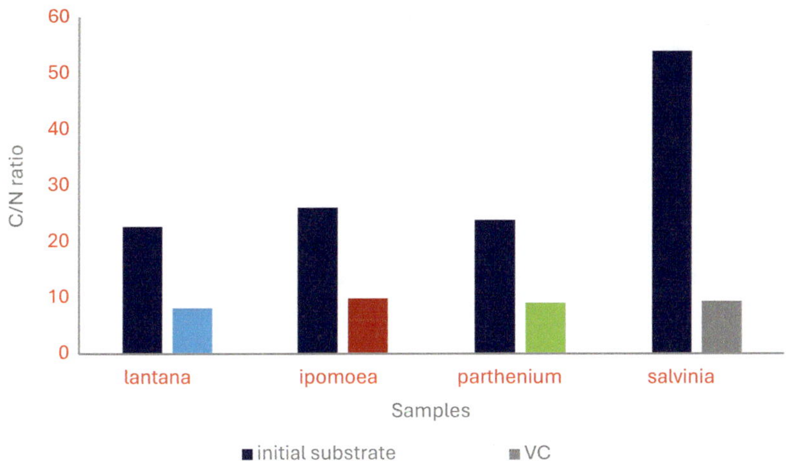

FIGURE 11.1 Carbon-to-nitrogen (C:N) ratios of parent weeds ■ and their VCs.

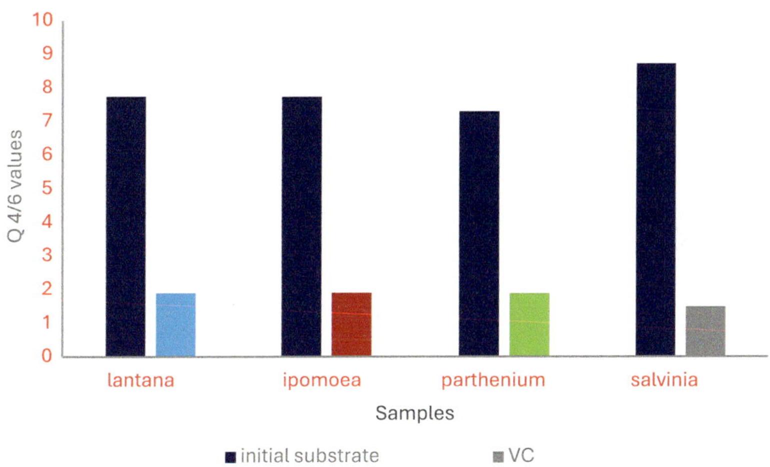

FIGURE 11.2 HI values of parent weeds ■ and their VCs.

11.3.2 HI

The HI – which is also referred to as $Q_{4/6}$ – values of lantana, ipomoea, parthenium, and salvinia were found to be 8.4, 7.4, 7.3, and 7.7, respectively. Their VCs displayed sharply reduced HIs: 2.0, 1.9, 1.9, and 1.9, respectively (Figure 11.2). The values were 2.5 times less than the level of HI 5, which is known to indicate adequately high extents of humification. The crossing of this limit by 2.5 times reflects a very high degree of humification and stabilization occurring in the course of vermicomposting. As recounted in Chapter 3, after ingesting a substrate, earthworms comminute it with their gizzards into fine particles. These particles acquire increasingly rich and diverse microorganisms as they pass through the earthworm gut (Edwards, 1998; Arancon et al., 2005). This happening boosts the processes of aeration, decomposition, and the resultant stabilization of the comminuted substrate into stable humus (Albanell et al., 1988; Masciandaro et al., 2010). When VC is applied to the soil, its humus content plays a decisive role in the efficacy of the VC as a fertilizer. The humus serves as a reservoir of nutrients, applying them according to the plant's needs. Humus also improves the physical proportion of soil immensely. The HI values are reflective of the high efficiency with which earthworm-mediated zoocomposting humifies the phytomass.

11.3.3 The Fate of Allelochemicals and Other Harmful Biomolecules

To identify the problematic biomolecules in the four weeds and see what happens to them when the weeds are vermicomposted, we took recourse to Fourier transform infrared spectroscopy (FTIR). The pointers that emerged from the studies of the FTIR spectra of the parent weeds, in contrast to the FTIR spectra of the corresponding VCs, are illustrated with the example of lantana (Figure 11.3). As detailed in Hussain et al. (2015), to quote,

> Lantana has a very prominent peak at 3,448 cm^{-1}, which is caused by the O–H stretching frequencies of alcohol and phenol (Pavia et al., 2001). Next to it are peaks at 2925 and 2852 cm^{-1} due to aliphatic C–H stretching which is assigned to fatty acids and lipids (Teh et al., 2014), followed by a peak at 1733 cm^{-1} due to unsaturated sesquiterpene lactones (Iranshahi et al., 2007; Xu et al., 2012). The sesquiterpene lactones are known to be responsible for the allelopathic effect of Lantana and several other plants (Putnam, 1988; Rizvi & Rizvi, 1992), while high phenol concentration as indicated by 3448 cm^{-1} peak is also a likely reason for the allelopathy and toxicity of Lantana. The peak at 1661 cm^{-1} is due to the aromatic C–C vibrations followed by the peak at 1515 cm^{-1} which is known to be due to aromatic skeletal vibrations of the cellulosic or lignocellulosic materials (Baddi et al., 2004; Mochochoko et al., 2013). There is also a peak at 1447 cm^{-1} due to the OCH_3 stretching of lignin (Bykov, 2008; Ravindran et al., 2013).
>
> Further up, a peak at 1256 cm^{-1} corresponds to C–O stretch of phenols or C–O and O–H of COOH groups (Wang et al., 2004). The peak at 1159 cm^{-1} is identified with C–O–C stretching skeletal vibration of carbohydrates. It is followed by a prominent peak at 1015 cm^{-1} due to C – O stretch of polysaccharides, cellulose or hemicellulose (Huang et al., 2006; Ravindran et al., 2013). There is a small peak at 836 cm^{-1} which can be attributed to N–H wag vibrations (Smidt & Meissl, 2007).
>
> In comparison, the FTIR spectra of vermicompost show a marked reduction in the band at 3,448 cm^{-1}, reflecting that a drastic reduction in the phenolic content accompanies the conversion of Lantana to vermicompost. The peaks at 2925, 2852 and 1661 cm^{-1} are also much shallower, indicating a decrease in CH_2, CH_3 and CH groups, due to the possible degradation of aliphatic and aromatic compounds contained in Lantana in the course of vermicomposting. The peak at 1733 cm^{-1} which was characteristic of sesquiterpene lactone functional group in the substrate is totally absent in the vermicompost, which shows that the vermicomposting process has completely degraded the chemicals that were specifically responsible for the allelopathy in Lantana.

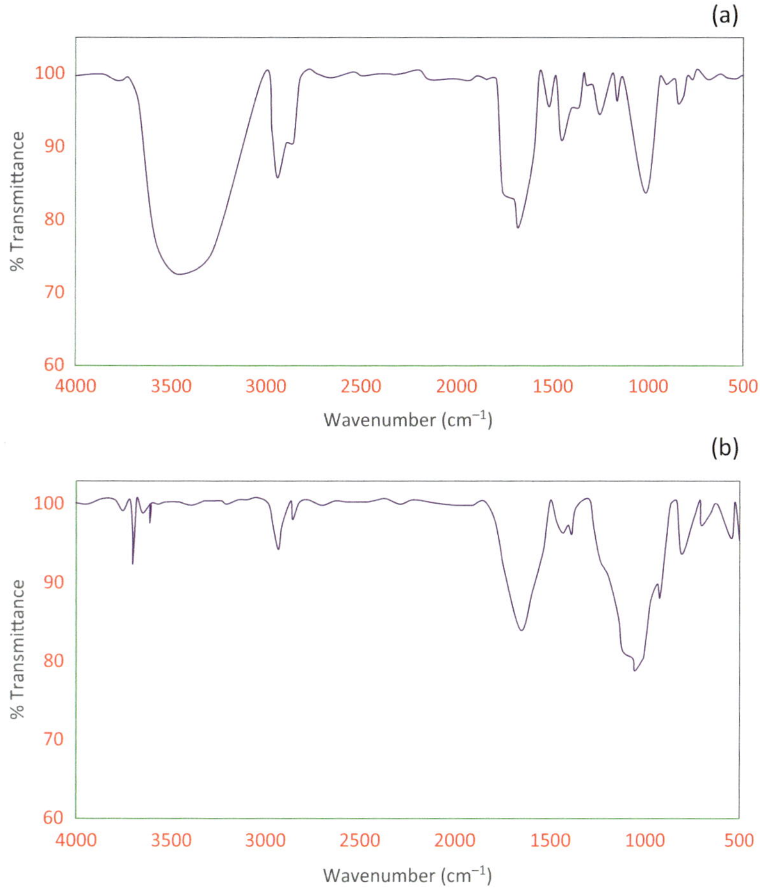

FIGURE 11.3 FTIR spectra of (a) lantana leaves, and (b) VC derived from lantana leaves.

While vermicomposting has evidently caused degradation of lignocellulose, lignin, and carbohydrates contained in Lantana, there is formation of polysaccharides as indicated by an increase in the intensities of the peaks in the 1000–1100 cm^{-1} range and emergence of a peak at 914 cm^{-1} (Smith & Paul, 2017; Ravindran et al., 2013). A sharp peak at 1384 cm^{-1} is also seen which is attributable to N–O stretching frequencies (Kumar et al., 2013) and at 794 cm^{-1} which is known to be due to the out of plane bend of primary amine group (Pavia et al., 2001). It suggests that a rise in the concentration of nitrogenous compounds occurred during the course of vermicomposting. This observation is consistent with the findings presented in Section 11.3.1.

The FTIR spectra of the VCs of ipomoea, parthenium, and salvinia displayed a similar pattern of change relative to the spectra of their original substrates as was seen with lantana (Hussain et al., 2016e, 2017, 2018). The concentrations of the respective allelochemicals and of the constituents responsible for different forms of toxicity were drastically reduced in the course of the vermicomposting of the other three weeds as well.

11.3.4 Extent of Mineralization as Shown by TG and Differential Scanning Coulometric Studies

The TG curves of ipomoea leaves and that of the ipomoea vermicompost (IVC) are shown in Figure 11.4. The curves of the three other weeds have displayed a similar pattern. Figure 11.4 indicates that ipomoea leaves lost water in the temperature range 60°C–150°C. At about 200°C, the weed's decomposition started, leading to a mass loss of 78% by the time the assessment was completed at 800°C. In the corresponding period, and within the same temperature range of 200°C–800°C, the IVC lost only 49% of its original mass.

In case of lantana, parthenium, and salvinia, the weeds lost 94%, 79%, and 70% mass, respectively, while the corresponding VCs lost only 53%, 55%, and 59%, respectively, in the 200°C–800°C temperature range. This shows that all the VCs were substantially more mineralized than the parent substrates.

The differential coulometric scans of the four weeds and their VCs, as typified by Figure 11.5 (pertaining to salvinia), have three common features:

a. All the weeds yield two endothermic peaks in the temperature range 100°C–200°C, signalling dehydration and dihydroxylation occurring in that range (Critter & Airoldi, 2006; Dell'Abate et al., 2000; Hussain et al.,

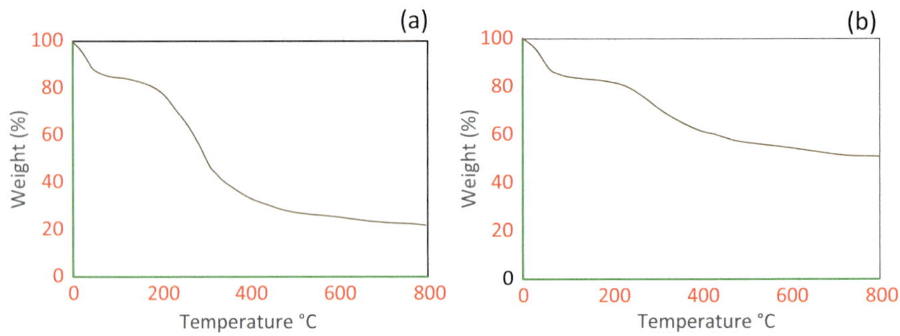

FIGURE 11.4 TG curves of (a) ipomoea leaves, and (b) VC, derived from ipomoea leaves.

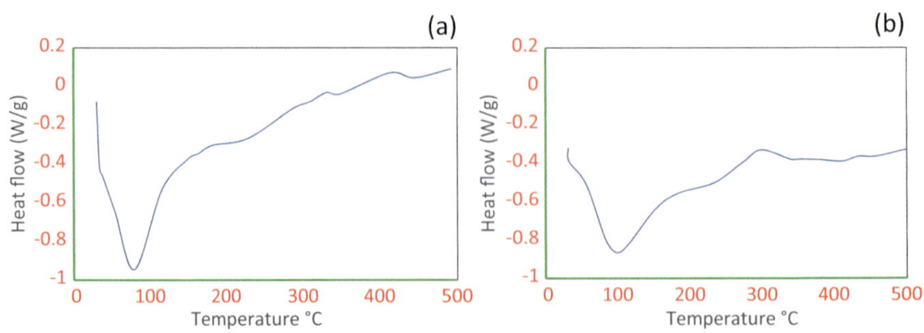

FIGURE 11.5 DSC curves of (a) salvinia leaves, and (b) VC, derived from salvinia leaves.

2016a). This is followed by two exothermic peaks in the temperature ranges 250°C–350°C and 400°C–500°C. The first can be attributed to the decomposition of simpler carbohydrates, lipids, amino acids, and other less recalcitrant compounds (Ravindran et al., 2013; Hussain et al., 2016b). The second peak is likely due to the breakdown of more complex aromatics, including polyphenols and lignins (Pietro & Paola, 2004; Hussain et al., 2016c).

b. All the corresponding VCs also display the endothermic peaks at 100°C–200°C seen in the case of the parent weeds and which are of similar depth as seen with the parent weeds. But the exothermic peaks of the VCs are much shallower, indicating that in the process of VC formation, several complex and problematic constituents of the parent weeds have already been converted to simpler, and likely benign, compounds (Hussain et al., 2016d).

11.3.5 Molecular Rearrangements as Revealed by GC-MS Analysis

GC-MS analysis showed that lantana, ipomoea, parthenium, and salvinia contained 23, 24, 20, and 25 organic compounds, respectively, of whose concentration in terms of the area of their respective GC-MS peaks was 0.1% or higher (Tables 11.1–11.4). In comparison, the number of constituents of the respective VCs was 86, 85, 41, and 37, respectively (Tables 11.5–11.8). Moreover, as seen from Tables 11.1–11.4, certain constituents were present in very high concentrations in the parent weeds – for instance, ethenone, 1-(2,4,5-triethylphenyl) and pyrido[3,4-d]pyridine-4(3+1)-one are present in lantana leaves to the extent of 28.5% and 25.6% of the total mass, while 2-propynoic acid's methyl ester has 35.9% of the mass of ipomoea's chemicals. The most striking dominance is seen in the form of 4-pyridinol-1-oxide, of which 58.3% is present in parthenium leaves. In salvinia, the chemical profile is dominated by dibutyl phthalate, which constitutes 23.4% of salvinia's chemical mass.

But the chemical profiles of the corresponding VCs (Table 11.2) show either drastic reduction, or elimination, of these and similarly dominant chemicals that were present in the parent weeds. Indeed, the substance present in most of the chemical mass of LVC, IVC, and PVC is phytol – to the extents of 9.6%, 16.4%, and 8.1%, respectively. Phytol is known to possess metabolism-modulating, antioxidant, antinociceptive, anti-inflammatory, and immune-modulating attributes (Islam et al., 2018).

TABLE 11.1
List of Compounds Identified in the Methanolic Extract of Lantana Leaves

Peaks #	Retention Time	Compound Identified	Peak Area (%)
1	33.731	Ethanone, 1-(2,4,5-triethylphenyl)	28.49
2	32.929	Pyrido[3,4-d]pyrimidin-4(3H)-one, 3,6,8-trimethyl-	25.6
3	20.827	Phytol	9.64
4	34.392	Silane, triethyl(2-phenylethenyl)-	4.36
5	33.508	Gamma.-Sitosterol	3.91
6	28.939	Pyrene, 1,6-bis(1,1-dimethylethyl)	2.75

(*Continued*)

TABLE 11.1 (*Continued*)
List of Compounds Identified in the Methanolic Extract of Lantana Leaves

Peaks #	Retention Time	Compound Identified	Peak Area (%)
7	19.349	*n*-Hexadecanoic acid	2.74
8	17.945	Bicyclo[3.1.1]heptane, 2,6,6-trimethyl-, (1.alpha.,2.beta.,5.alpha.)	2.69
9	20.716	9,12,15-Octadecatrienoic acid, methyl ester, (Z,Z,Z)-	2.48
10	18.903	Hexadecanoic acid, methyl ester	2.16
11	15.107	Levomenol	1.61
12	26.897	Squalene	1.5
13	20.649	9,12-Octadecadienoic acid, methyl ester	1.44
14	15.635	Butanoic acid, 2,2-dimethyl-, methyl ester	1.25
15	17.002	Phenol, 5-(1,5-dimethyl-4-hexenyl) -2-methyl-, (R)-	1.24
16	22.328	Acetic acid, (phenylmethoxy)-, ethyl ester	1.2
17	17.373	3-Carene, 4-acetyl-	1.03
18	15.026	1H-Cycloprop[e]azulen-7-ol, decahydro-1,1,7-trimethyl-4-methylene-,[1ar-(1a.alpha.,4a. alpha.,7.beta., 7a.beta.,7b.alpha.)]-	1.02
19	15.464	Naphthalene, decahydro-, cis-	0.82
20	14.179	[1,2,4] Triazolo[1,5-a] pyrimidine-6-carboxylic acid, 4,7-dihydro-7-imino-, ethyl ester	0.7
21	18.005	Sclaral (sclareolide lactol)	0.69
22	22.261	11,13-Dimethyl-12-tetradecen-1-ol acetate	0.6
23	18.421	Phytol	0.21

TABLE 11.2
List of Compounds Identified in the Methanolic Extract of Ipomoea Leaves

# Peaks	Retention Time	Compound Name	Peak Area (%)
1	24.757	2-Propynoic acid, 3-(1-hydroxycyclohexyl)-, methyl ester	35.87
2	24.869	Pyrido[3,4-d]pyrimidin-4(3H)-one, 2,6,8-trimethyl-	14.06
3	11.408	Propanoic acid, 2,3-dihydroxy-	11.72
4	25.961	2,5-Dimethyl-4-phenyl-5,6-dihydro- 4H-[1,3]oxazine	7.93
5	15.939	Butanoic acid, pentyl ester	5.75
6	24.118	Spiro[4.5]decane-2,6-dione	4.77
7	25.292	2,3-Pyrazinedicarboxamide	3.33
8	15.991	Butanoic acid, pentyl ester	2.88
9	20.82	Phytol	2.34
10	32.877	1H-Indole-2-carboxylic acid, 1-methyl-, methyl ester	1.84
11	22.075	1-Naphthalenecarbonitrile	1.83
12	33.671	Silane, (4-bromo-3-buten-1-ynyl)trimethyl-, (E)-	1.72

(*Continued*)

TABLE 11.2 (*Continued*)
List of Compounds Identified in the Methanolic Extract of Ipomoea Leaves

# Peaks	Retention Time	Compound Name	Peak Area (%)
13	8.407	4H-Pyran-4-one, 2,3-dihydro-3,5-dihydroxy-6-methyl-	1.5
14	24.245	Dihydronopol	0.82
15	21.162	2,3-Dimethylhydroquinone	0.68
16	8.956	Carbamic acid, methyl-, ethyl ester	0.62
17	23.145	Sebacic acid, isohexyl 2,3,5-trichlorophenyl ester	0.52
18	16.927	4-((1E)-3-Hydroxy-1-propenyl)-2-methoxyphenol	0.51
19	22.462	S-[2-[*N*,*N*-Dimethylamino]ethyl]*N*,*N*-dimethylcarbamoyl thiocarbo hydroximate	0.46
20	17.38	2-Cyclopenten-1-one, 4-hydroxy-3-methyl-2-(2,4-pentadienyl)-, (Z)-(+)-	0.39
21	22.254	Aciphyllyl alcohol	0.31
22	17.945	Bicyclo[10.8.0]eicosane, cis-	0.29
23	20.716	9,12,15-Octadecatrienoic acid, methyl ester, (Z,Z,Z)-	0.29
24	19.275	Dibutyl phthalate	0.13

TABLE 11.3
List of Compounds Identified in the Methanolic Extract of Parthenium Leaves

# Peaks	Retention Time	Compound Name	Peak Area (%)
1	24.898	4-Pyridinol-1-oxide	58.32
2	25.411	Azuleno[4,5-b]furan-2,9-dione, decahydro-6a-hydroxy-6,9a-dimethyl-3-methylene-, [3aS-(3a.alpha.,6.beta .,6a.alpha., 9a.beta., 9b.alpha.)]-	10.15
3	24.155	2-Aminopyrimidine-1-oxide	7.03
4	11.259	Propanoic acid, ethyl(dimethyl)silyl ester	5.15
5	20.827	Phytol	3.81
6	24.274	Isoaromadendrene epoxide	3.28
7	23.658	2H-Benzocyclohepten-2-one, 1,4a,5,6,7,8,9,9a-octahydro-4a-methyl-, trans-	1.69
8	33.664	1,3-Cyclohexadiene, 2,6,6-trimethy l-1-(3-methyl-1,3-butadienyl)-	1.42
9	21.176	Benzoic acid, 2,5-dichloro-3,6-dimethoxy-, methyl ester	1.28
10	22.075	3a,9-Dimethyldodecahydrocyclohepta [d]inden-3-one	1.04
11	22.87	Alloaromadendrene oxide-(1)	1
12	23.45	Cyclohexylamine, *N*-(2-chlorocyclop entylidene)-, *N*-oxide	0.95
13	21.63	2,4-Dimethyl-3-nitrobicyclo[3.2.1] octan-8-one	0.91
14	18.903	Hexadecanoic acid, methyl ester	0.69
15	20.716	9,12,15-Octadecatrienoic acid, methyl ester, (Z,Z,Z)-	0.63

(*Continued*)

TABLE 11.3 (*Continued*)
List of Compounds Identified in the Methanolic Extract of Parthenium Leaves

# Peaks	Retention Time	Compound Name	Peak Area (%)
16	17.945	Bicyclo [3.1.1] heptane, 2,6,6-trimethyl-, (1.alpha.,2.beta., 5.alpha.)	0.62
17	22.261	11,13-Dimethyl-12-tetradecen-1-ol acetate	0.59
18	18.004	2-Pentadecanone, 6,10,14-trimethyl	0.51
19	22.469	S-[2-[N,N-Dimethylamino]ethyl]N,N-dimethylcarbamoyl thiocarbo hydroximate	0.51
20	20.656	9,12-Octadecadienoic acid (Z,Z)-, methyl ester	0.42

TABLE 11.4
List of Compounds Identified in the Methanolic Extract of Salvinia Leaves

# Peak	Retention Time	Compound Name	Peak Area (%)
1	19.409	*n*-Hexadecanoic acid	23.37
2	20.828	Phytol	7.93
3	9.848	4-Mercaptophenol	7.36
4	18.904	Hexadecanoic acid, methyl ester	6.73
5	25.953	2,2,4,7-Tetramethyl-3,6,9-trioxa-2 -silatridecane	5.53
6	32.409	Stigmasterol	4.46
7	17.945	Bicyclo[3.1.1]heptane, 2,6,6-trimethyl-, (1.alpha.,2.beta.,5.alpha.)	4.36
8	17.098	Tetradecanoic acid	4.31
9	33.486	Gamma.-Sitosterol	3.95
10	20.716	11-Octadecenoic acid, methyl ester	3.47
11	19.126	Cis-9-Hexadecenoic acid	3.21
12	24.497	Hexadecanoic acid, 2-hydroxy-1-(hydroxymethyl)ethyl ester	2.79
13	8.429	4H-Pyran-4-one, 2,3-dihydro-3,5-dihydroxy-6-methyl-	2.6
14	32	Campesterol	2.59
15	20.657	9,12-Octadecadienoic acid (Z,Z)-, methyl ester	2.38
16	18.213	Pentadecanoic acid	2.25
17	30.336	Vitamin E	2.2
18	18.005	2-Pentadecanone, 6,10,14-trimethyl	1.61
19	30.44	26-Nor-5-cholesten-3.beta.-ol-25-one	1.58
20	29.786	Rishitin	1.53
21	18.421	Phytol, acetate	1.31
22	20.961	Methyl stearate	1.28
23	17.381	2-Benzothiazolamine, 4,6-dimethyl-	1.17
24	19.275	Dibutyl phthalate	1.02
25	17.804	Pentadecanoic acid	1

TABLE 11.5
List of Compounds Identified in the Methanolic Extract of Lantana VC

Peaks #	Retention Time	Compound Identified	Peak Area (%)
1	20.828	Phytol	9.58
2	34.749	2-Ethylacridine	8.64
3	19.312	*n*-Hexadecanoic acid	7.1
4	17.938	Phytol, acetate	5.26
5	18.421	Citronellyl butyrate	4.37
6	34.452	Benzo[h]quinoline, 2,4-dimethyl-	4.35
7	33.478	Gamma.-Sitosterol	3.89
8	34.273	Benzo[h]quinoline, 2,4-dimethyl-	3.75
9	34.073	1,4-Bis(trimethylsilyl)benzene	3.44
10	20.716	9,12,15-Octadecatrienoic acid, methyl ester, (Z,Z,Z)-	3.32
11	18.213	Levo-menthoxyacetic acid	3.14
12	18.904	Hexadecanoic acid, methyl ester	2.52
13	33.768	Arsenous acid, tris(trimethylsilyl) ester	2.02
14	33.635	4-Dehydroxy-*N*-(4,5-methylenedioxy- 2-nitrobenzylidene) tyramine	1.71
15	32.409	Stigmasterol	1.52
16	26.429	17-(1,5-Dimethylhexyl)-10,13-dimethyl-4-vinylhexadecahydro cyclopenta [a]phenanthren-3-ol	1.44
17	25.953	9,19-Cycloergost-24(28)-en-3-ol, 4,14-dimethyl-, acetate, (3.beta., 4.alpha.,5.alpha.)-	1.34
18	20.649	9,12-Octadecadienoic acid (Z,Z)-, methyl ester	1.26
19	30.054	Cholest-5-en-3-ol (3.beta.)-, carbonochloridate	1.26
20	20.768	9-Octadecenoic acid, methyl ester, (E)-	1.14
21	25.478	4H-Pyran-3-carboxylic acid, 2-amino-5-cyano-6-ethyl-4-(3-pyridinyl)- , methyl ester	1.14
22	28.576	4-Dehydroxy-*N*-(4,5-methylenedioxy- 2-nitrobenzylidene) tyramine	1.06
23	17.069	Tetradecanoic acid	1.01
24	21.102	6-Octadecenoic acid	0.93
25	29.318	Carbonic acid, monoamide, *N*-(2-ethylphenyl)-, propyl ester	0.93
26	17.373	Pentadecanoic acid, methyl ester	0.9
27	21.801	6-Octadecenoic acid	0.89
28	30.329	DI-.alpha.-Tocopherol	0.88
29	26.184	1H-Indole, 5-methyl-2-phenyl-	0.77
30	26.897	Squalene	0.77
31	20.961	Heptadecanoic acid, 16-methyl-, methyl ester	0.72
32	19.572	Hexadecanoic acid, 15-methyl-, methyl ester	0.67
33	17.789	Pentadecanoic acid	0.65
34	29.058	Anthracene, 9,10-dihydro-9,9,10-trimethyl-	0.64
35	25.775	Tricyclo[4.3.0.0(7,9)]non-3-ene, 2,2,5,5,8,8-hexamethyl-, (1.alpha., 6.beta.,7.alpha.,9.alpha.)-	0.63

(Continued)

TABLE 11.5 (*Continued*)
List of Compounds Identified in the Methanolic Extract of Lantana VC

Peaks #	Retention Time	Compound Identified	Peak Area (%)
36	18.681	9-Hexadecenoic acid, methyl ester,	0.59
37	32.877	2-Ethylacridine	0.59
38	20.568	1-Octadecene	0.56
39	21.318	Oleic acid	0.56
40	28.122	Cyclotrisiloxane, hexamethyl-	0.56
41	32	4-Dehydroxy-*N*-(4,5-methylenedioxy- 2-nitrobenzylidene) tyramine	0.54
42	29.786	4-Quinolinecarboxylic acid, 2-chloro-	0.53
43	18.005	Tetradecane, 4,11-dimethyl-	0.52
44	18.502	Pentadecanoic acid, 14-methyl-, methyl ester	0.51
45	19.951	Heptadecanoic acid, methyl ester	0.51
46	20.322	Valeric acid, undec-2-enyl ester	0.51
47	18.785	9-Hexadecenoic acid, methyl ester, 117513 001120-25-8 99(Z)-	0.49
48	30.433	2-Ethylacridine	0.48
49	20.087	Thunbergol	0.45
50	26.087	2-Ethylacridine	0.45
51	19.178	Palmitoleic acid	0.44
52	19.089	Palmitoleic acid	0.43
53	24.631	Phthalic acid, 6-ethyloct-3-yl 2-ethylhexyl ester	0.43
54	16.281	Butanamide, *N*-(4-methoxyphenyl)-	0.42
55	19.802	Cis-Vaccenic acid	0.4
56	19.661	Hexadecanoic acid, 14-methyl-, methyl ester	0.39
57	31.592	1,2-Bis(trimethylsilyl)benzene	0.39
58	32.104	Indole-2-one, 2,3-dihydro-*N*-hydroxy-4-methoxy-3,3-dimethyl-	0.39
59	17.47	Tetradecanoic acid, 12-methyl-, methyl ester	0.37
60	19.141	Succinic acid, dodec-2-enyl 3-methylbutyl ester	0.33
61	19.728	Cis-10-Heptadecenoic acid	0.33
62	25.612	Indole-2-one, 2,3-dihydro-*N*-hydroxy-4-methoxy-3,3-dimethyl-	0.3
63	18.844	Cis-10-Heptadecenoic acid	0.27
64	17.878	*n*-Hexadecanoic acid	0.26
65	25.277	4H-Pyran-3-carboxylic acid, 2-amino-5-cyano-6-ethyl-4-(3-pyridinyl)- , methyl ester	0.26
66	21.006	Cis-Vaccenic acid	0.24
67	23.019	1-Docosene	0.23
68	24.274	Mecamylamine	0.23
69	20.025	7-Thiabicyclo[4.2.1]nonane	0.22
70	22.87	2,2-Dimethylpropanoic acid, 2,6-dimethylnon-1-en-3-yn-5-yl ester	0.21
71	23.598	9-Octadecenoic acid (Z)-, 2,3-dihydroxypropyl ester	0.2
72	16.638	Tridecanoic acid, 12-methyl-, methyl ester	0.18
73	24.542	2-Methyl-Z,Z-3,13-octadecadienol	0.18

(*Continued*)

TABLE 11.5 (*Continued*)
List of Compounds Identified in the Methanolic Extract of Lantana VC

Peaks #	Retention Time	Compound Identified	Peak Area (%)
74	22.232	Methyl 6,9,12-hexadecatrienoate	0.17
75	23.955	Culmorin	0.16
76	22.833	Oleic acid	0.15
77	14.632	3-Buten-2-one, 4-(dimethylamino)-4-ethoxy-	0.14
78	23.093	9-Octadecenoic acid (Z)-, 2,3-dihydroxypropyl ester	0.14
79	23.13	2-Methyl-Z,Z-3,13-octadecadienol	0.14
80	15.226	4-Phosphatricyclo[6.1.1.0(2,6)]dec-2(6)-ene, 4,9,9-trimethyl-	0.13
81	22.15	*n*-Hexyltrichlorosilane	0.12
82	24.572	Cyclopropane carboxamide, 2-cyclopropyl-2-methyl-N-(1-cyclopropylethyl)-	0.12
83	16.682	3-Tetradecene, (Z)-	0.1
84	22.514	Cis-10-Nonadecenoic acid	0.08
85	12.909	2-Pyridinemethanamine, .alpha.-methyl-	0.07
86	22.566	10,13-Eicosadienoic acid, methyl ester	0.07

TABLE 11.6
List of Compounds Identified in the Methanolic Extract of Ipomoea VC

# Peaks	Retention Time	Compounds Identified	Peak Area (%)
1	20.85	Phytol	16.44
2	34.741	4,22-Stigmastadiene-3-one	7.15
3	18.904	Hexadecanoic acid, methyl ester	4.56
4	17.938	Bicyclo[3.1.1]heptane, 2,6,6-trimethyl-, (1.alpha.,2.beta.,5.alpha.)	3.66
5	32.527	Stigmasterol	3.6
6	19.29	*n*-Hexadecanoic acid	3.08
7	18.42	5-Nonadecen-1-ol	2.98
8	20.775	cis-13-Octadecenoic acid, methyl ester	2.98
9	34.555	Benzo[h]quinoline, 2,4-dimethyl-	2.79
10	20.649	Methyl 10-trans,12-cis-octadecadie noate	2.5
11	20.716	9-Octadecenoic acid, methyl ester, (E)-	2.45
12	34.466	N-Methyl-1-adamantaneacetamide	2.25
13	20.961	Methyl stearate	2.04
14	18.212	Phytol, acetate	1.89
15	28.404	2-Ethylacridine	1.64
16	21.8	Cyclopropaneoctanoic acid, 2-octyl-, methyl ester, cis-	1.61
17	18.784	11-Hexadecenoic acid, methyl ester	1.58
18	23.019	Stigmast-4-en-3-one	1.51
19	33.649	1H-Indole, 5-methyl-2-phenyl-	1.45

(*Continued*)

TABLE 11.6 (*Continued*)
List of Compounds Identified in the Methanolic Extract of Ipomoea VC

# Peaks	Retention Time	Compounds Identified	Peak Area (%)
20	21.132	Cyclopentanone, 2-(5-oxohexyl)-	1.28
21	18.68	9-Hexadecenoic acid, methyl ester, (Z)-	1.07
22	23.086	Methyl 2-hydroxy-pentacosanoate	1.05
23	27.543	4-Dehydroxy-N-(4,5-methylenedioxy- 2-nitrobenzylidene) tyramine	1.03
24	34.08	1,2,5-Oxadiazol-3-amine, 4-(4-methoxyphenoxy)-	0.99
25	18.502	Pentadecanoic acid, 14-methyl-, methyl ester	0.95
26	27.223	t-Butyldiphenyl(prop-2-ynyloxy)silane	0.93
27	18.004	2-Pentadecanone, 6,10,14-trimethyl	0.91
28	22.848	Eicosanoic acid, methyl ester	0.9
29	26.02	17-(1,5-Dimethylhexyl)-10,13-dimethyl-4-vinylhexadecahydro cyclo penta [a]phenanthren-3-ol	0.89
30	17.38	Pentadecanoic acid, methyl ester	0.85
31	27.119	1,2-Bis(trimethylsilyl)benzene	0.85
32	17.47	Tetradecanoic acid, 12-methyl-, methyl ester	0.83
33	19.572	Hexadecanoic acid, 15-methyl-, methyl ester	0.83
34	25.916	2-Ethylacridine	0.76
35	30.054	Carbonic acid, monoamide, N-(2-ethylphenyl)-, propyl ester	0.73
36	21.006	4-n-Pentylthiane, S,S-dioxide	0.69
37	19.661	Methyl 10-methyl-hexadecanoate	0.68
38	25.79	1-Docosene	0.65
39	29.883	Tris(tert-butyldimethylsilyloxy)arsane	0.65
40	26.644	2-Ethylacridine	0.63
41	29.318	2-Ethylacridine	0.62
42	21.325	Octadecanoic acid, 10-methyl-, methyl ester	0.6
43	33.367	1,2-Bis(trimethylsilyl)benzene	0.6
44	29.073	Benzene, 2-[(tert-butyldimethylsilyl)oxy]-1-isopropyl-4-methyl-	0.57
45	30.551	Benzo[h]quinoline, 2,4-dimethyl-	0.56
46	30.343	DI-.alpha.-Tocopherol	0.53
47	19.735	Ethyl pentadecyl ether	0.52
48	19.951	Hexadecanoic acid, 15-methyl-, methyl ester	0.51
49	24.534	4H-Chromen-4-one, 3-(2-thenoyl)-	0.51
50	29.794	1,1,1,3,5,5,5-Heptamethyltrisiloxane	0.51
51	19.802	11-Hexadecenoic acid, 15-methyl-, methyl ester	0.5
52	33.805	N-Methyl-1-adamantaneacetamide	0.47
53	21.927	Z,E-2-Methyl-3,13-octadecadien-1-ol	0.46
54	24.631	2-(Heptyloxycarbonyl)benzoic acid	0.44
55	22.618	cis-11-Eicosenoic acid, methyl ester	0.43
56	17.068	Tetradecanoic acid	0.41
57	17.796	Methyl 9-methyltetradecanoate	0.4
58	16.355	Tricyclo[4.3.1.1(3,8)]undecan-1-amine	0.37

(*Continued*)

TABLE 11.6 (*Continued*)
List of Compounds Identified in the Methanolic Extract of Ipomoea VC

# Peaks	Retention Time	Compounds Identified	Peak Area (%)
59	19.148	Phytol	0.37
60	22.462	6-Isopropenyl-4,8a-dimethyl-1,2,3, 5,6,7,8,8a-octahydro naphthalene-2,3-diol	0.36
61	22.684	2-Pentadecanone, 6,10,14-trimethyl	0.35
62	20.322	2H-Pyran, 2-[(2-furanylmethoxy)methyl]tetrahydro-	0.34
63	24.571	Docosanoic acid, methyl ester	0.33
64	19.082	Cis-13-Octadecenoic acid	0.32
65	21.748	Stearic acid hydrazide	0.32
66	21.563	Cyclopropaneoctanoic acid, 2-octyl -	0.31
67	20.567	9,12-Octadecadienoic acid (Z,Z)-,	0.3
68	18.101	1,1-Dichloro-2-methyl-3-(4,4-diformyl-1,3-butadien-1-yl) cyclo propane	0.26
69	32.119	2-Ethylacridine	0.26
70	22.573	Cis-11,14-Eicosadienoic acid, methyl ester	0.25
71	23.591	8-Hexadecenal, 14-methyl-, (Z)-	0.24
72	25.322	N-methyl-N-acetyl-3,4-methylenedioxybenzylamine	0.22
73	17.752	Octadecanal	0.21
74	19.23	Spiro[2.5]octane, 5,5-dimethyl-4-(3-oxobutyl)-	0.2
75	16.199	Ethanamine, 2-phenoxy-	0.19
76	24.252	3-Tetradecanynoic acid	0.19
77	17.336	Spiro[4.5]decan-7-one, 1,8-dimethyl-8,9-epoxy-4-isopropyl-	0.18
78	18.725	9-Hexadecenoic acid, methyl ester, (Z)-	0.17
79	23.724	Heneicosanoic acid, methyl ester	0.17
80	18.836	3-Methyl-2-(3,7,11-trimethyldodecyl) furan	0.16
81	20.478	Methyl 5,9,12-octadecatrienoate	0.16
82	20.538	9,12-Octadecadienoic acid, methyl ester, (E,E)-	0.15
83	24.044	Z-8-Methyl-9-tetradecen-1-ol acetate	0.14
84	16.638	Methyl tetradecanoate	0.13
85	22.231	Methyl 10-trans,12-cis-octadecadie noate	0.12

TABLE 11.7
List of Compounds Identified in the Methanolic Extract of Parthenium VC

# Peaks	Retention Time	Compounds Identified	Peak Area (%)
1	20.835	Phytol	8.14
2	32.461	1,2-Bis(trimethylsilyl)benzene	7.66
3	26.436	Benzo[h]quinoline, 2,4-dimethyl-	5.38

(*Continued*)

TABLE 11.7 (*Continued*)
List of Compounds Identified in the Methanolic Extract of Parthenium VC

# Peaks	Retention Time	Compounds Identified	Peak Area (%)
4	34.749	5-Methyl-2-trimethylsilyloxy-acetophenone	5.31
5	27.513	1H-Indole, 1-methyl-2-phenyl-	4.32
6	24.527	Cyclohexane-1,3-dione, 2-allylaminomethylene-5,5-dimethyl-	4.13
7	18.904	Hexadecanoic acid, methyl ester	3.46
8	33.538	Cyclotrisiloxane, hexamethyl-	3.4
9	20.597	Octadecane, 1-(ethenyloxy)-	3.33
10	20.768	9-Octadecenoic acid (Z)-, methyl ester	3.28
11	34.437	Anthracene, 9,10-dihydro-9,9,10-trimethyl-	3.1
12	29.772	Tetrasiloxane, decamethyl-	2.85
13	17.373	4-methoxy-3-hydroxyamphetamine	2.84
14	30.054	Silane, trimethyl[5-methyl-2-(1-methylethyl)phenoxy]-	2.84
15	24.631	4H-Pyran-3-carboxylic acid, 2-amino-5-cyano-6-ethyl-4-(3-pyridinyl)-, methyl ester	2.83
16	22.96	1-Tetradecanamine	2.79
17	20.716	11-Octadecenoic acid, methyl ester	2.5
18	27.164	Arsenous acid, tris(trimethylsilyl) ester	2.32
19	18.488	Diepicedrene-1-oxide	2.26
20	29.311	Cyclotrisiloxane, hexamethyl-	2.17
21	24.037	Benzo[h]quinoline, 2,4-dimethyl-	2.01
22	20.961	Methyl stearate	1.77
23	20.649	9,12-Octadecadienoic acid (Z,Z)-, methyl ester	1.75
24	21.801	Methyl hexadec-9-enoate	1.71
25	17.938	3-Decanynoic acid	1.69
26	15.226	6-(3,3-Dimethyl-oxiran-2-ylidene)- 5,5-dimethyl-hex-3-en-2-one	1.45
27	19.275	Benzoic acid, 2-(1-oxopropyl)-	1.38
28	23.086	Benzo[h]quinoline, 2,4-dimethyl-	1.38
29	33.642	5-Methyl-2-phenylindolizine	1.36
30	19.944	Z,Z-3,13-Octadecedien-1-ol	1.18
31	23.026	3-Cyclohexene-1-carboxaldehyde, 4- methyl-	1.13
32	18.005	1,3-Adamantanediacetamide	1.12
33	19.164	Cyclotetradecane	1.12
34	23.962	Tricyclo[10.2.2.2(5,8)]octadeca-5,7,12,14,15,17-hexaene, 6-nitro-	0.96
35	24.193	2-Ethylacridine	0.94
36	22.997	Sarcosine, N-valeryl-, hexyl ester	0.9
37	25.196	1H-Indole, 5-methyl-2-phenyl-	0.85
38	18.421	1,3-dicyclohexylpropene	0.76
39	16.289	2-Pyrimidinamine, 4,6-dimethyl-	0.65
40	19.572	Tetradecanoic acid, 5,9,13-trimethyl-, methyl ester	0.58
41	26.035	2-Ethylacridine	0.38

TABLE 11.8
List of Compounds Identified in the Methanolic Extract of Salvinia VC

S. No	Retention Time	Compounds Identified	Peak Area (%)
1	19.282	*n*-Hexadecanoic acid	11.3
2	22.945	1-Hydroxypyrene	8.93
3	23.101	Pyrene, hexadecahydro-	5.82
4	18.903	Hexadecanoic acid, methyl ester	5.55
5	22.826	Pyrene, hexadecahydro-	4.78
6	20.716	6-Octadecenoic acid, methyl ester, (Z)-	3.79
7	21.801	Cyclopropanedecanoic acid, 2-hexyl -, methyl ester	3.47
8	30.054	1,2-Bis(trimethylsilyl)benzene	3.42
9	20.768	9-Octadecenoic acid, methyl ester, (E)-	3.28
10	33.657	2-Ethylacridine	3.25
11	34.414	1,2-Bis(trimethylsilyl)benzene	2.93
12	22.521	Ethanamine, 2-phenoxy-	2.66
13	22.699	Pyrene, hexadecahydro-	2.59
14	24.631	Bis(2-ethylhexyl) phthalate	2.54
15	20.649	9,12-Octadecadienoic acid, methyl ester, (E,E)-	2.44
16	22.737	1H-Pyrazole-4-carboxylic acid, 5-amino-3-(2-fluorophenyl amino)-, amide	2.39
17	29.311	1,2-Bis(trimethylsilyl)benzene	2.19
18	29.779	t-Butyldiphenyl(prop-2-ynyloxy)silane	2.19
19	24.297	Cyclopropaneoctanal, 2-octyl-	2.08
20	18.421	1,4-Eicosadiene	1.77
21	17.782	Pentadecanoic acid	1.75
22	23.977	2-Pyridinamine, N-(4,5-dihydro-5-methyl-2-thiazolyl)-3-methyl-	1.75
23	29.445	1,2-Bis(trimethylsilyl)benzene	1.71
24	20.961	Heptadecanoic acid, 14-methyl-, methyl ester	1.66
25	18.205	Butanoic acid, 3-methyl-, 3,7-dimethyl-6-octenyl ester	1.63
26	28.122	N-Methyl-1-adamantaneacetamide	1.55
27	28.724	Arsenous acid, tris(trimethylsilyl) ester	1.52
28	17.938	Phytol	1.48
29	22.135	Cyclopropaneoctanal, 2-octyl-	1.34
30	22.573	2,2-Dimethylpropanoic acid, 2,6-dimethylnon-1-en-3-yn-5-yl ester	1.34
31	17.054	Tetradecanoic acid	1.3
32	22.625	1-Hydroxypyrene	1.27
33	18.785	Methyl 11-hexadecenoate	1.21
34	28.442	1,2-Bis(trimethylsilyl)benzene	0.89
35	19.943	Tridecane	0.88
36	19.802	cis-10-Heptadecenoic acid, methyl ester	0.78
37	26.124	2-Ethylacridine	0.57

Its degradation products serve as biogeochemical tracers in aquatic environments (Rontani & Volkman, 2003). Even though phytol's pharmaceutical properties have been studied with reference to animals, including humans, it is possible that it plays a role in boosting soil and/or plant health.

The chemical perhaps enhances the fertilizer value and pest-resistant attributes of the weeds' VCs. The chemical seen most abundant (11.3%) in SVC – *n*-hexadecenoic acid – is also anti-inflammatory (Aparna et al., 2012) and might be beneficial to soil or plants.

11.3.6 Surface Characteristics

The scanning electron micrographs (SEMs) of the surfaces of all four weeds displayed robust and contiguous features, indicating their formation by lignin-containing fibres. A representative SEM is shown in Figure 11.6. But the surfaces of the corresponding VCs had a withered, disaggregated, and frequented ambiance, as typified by Figure 11.7. This change is consistent with the explanation given earlier by us, as follows:

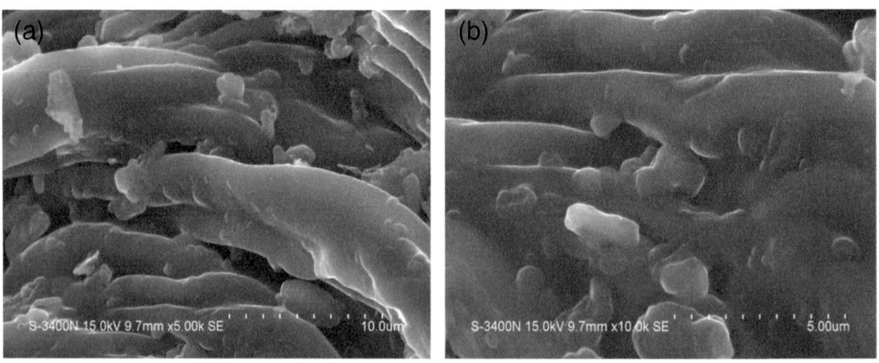

FIGURE 11.6 SEM images (a, b) of ipomoea leaves, at different resolutions.

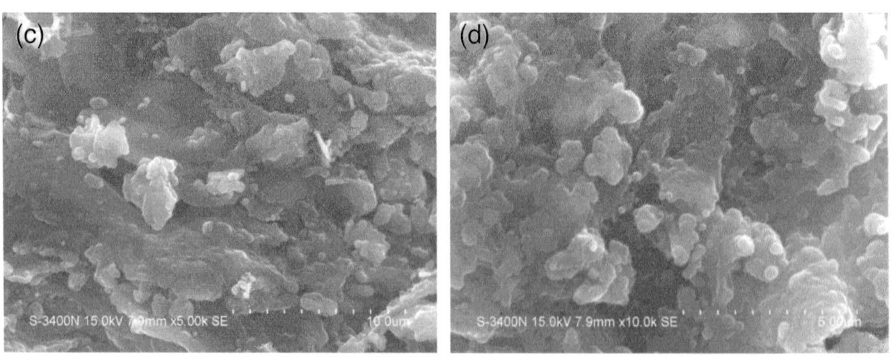

FIGURE 11.7 SEM images (c, d) of VC derived from lantana leaves, at different resolutions.

a. After ingesting any substrate, earthworms comminute it with their gizzard into fine particles.
b. As these particles move through the earthworm gut, they are acted upon by the microflora, getting progressively biodegraded (or zoocomposted).
c. There is a drastic reduction of organic carbon as the carbon is mineralized as CO_2 due to the metabolic activity of the earthworms and the microflora.
d. There is a high degree of humification.

Due to these and other factors, the surface of the exited vermicast (VC) appears much more rough and uneven in a SEM than the SEM of the parent weed. This roughness and unevenness enable the nutrients present in the vermicast to be much more accessible than the nutrients present in the parent weeds.

11.4 ROLE OF THE EARTHWORM SPECIES

When studies that eventually led us to the HRVP were going on, we had explored several species of earthworms for their suitability as vermicomposters. They included *Eudrilus eugeniae* (Gajalakshmi & Abbasi 2004), *Lampito mauritii* (Ganesh et al., 2009), *Perionyx excavatus* (Gajalakshmi & Abbasi, 2003), *Eisenia fetida* (Godson et al., 2015), *E. andrei* (Abbasi & Ramasamy, 2001), *Drawida willsi* (Gajalakshmi & Abbasi, 2004), *P. sansibaricus* (Abbasi & Ramasamy, 2001), and *Lumbricus rubellus* (Abbasi & Abbasi, 2019).

But all the studies described so far in Chapters 10 and 11 have been confined to *E. fetida,* so that the earthworm species remains a common constant while we can witness the effect of other factors. Nonetheless, it is necessary to check whether the qualities of VCs generated from the same substrate but by different species of earthworms are substantially similar and, if not, to what extent they differ.

Accordingly, four species not used in the studies described so far in Chapters 9 and 10 – *E. andrei, P. sansibaricus, L. rubellus,* and *D. willsi* – were deployed. Of these, *D. willsi* is an anaecic, while the other three species are epigeics. And the average adults of all four species differ from each other in terms of both mass and body length.

The weeds selected were ipomoea and salvinia. These two weeds were chosen over lantana and parthenium because ipomoea is the most toxic of the four, while salvinia, despite having allelopathy, does not possess animal toxicity. Hence, the two weeds occupy two ends of the harmful effect spectrum and are likely to be representative of the weeds that lie between these two ends. The studies led to the key findings described below.

11.4.1 C:N Ratio

The harvests of salvinia and ipomoea used in this study had C:N ratios of 42.2 and 28.2, respectively, Figures 11.8 and 11.9. In comparison, their vermicasts had C:N ratios in the ranges 14-14.8 and 15.2–16.7, respectively (Figures 11.8 and 11.9). While the reduction in the C:N ratios upon vermicomposting of salvinia achieved by the four earthworm species followed the order *E. andrei* (15.2) > *L. rubellus*

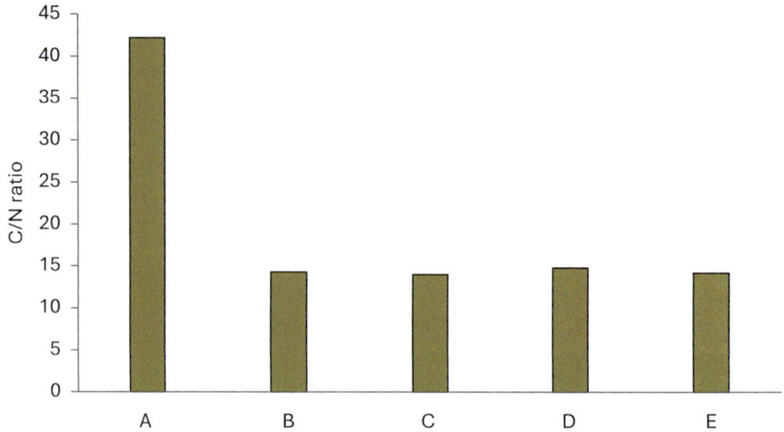

FIGURE 11.8 C N ratio of (a) salvinia plant and VC derived from (b) *E. andrei*, (c) *P. sansibaricus*, (d) *L. rubellus*, and (e) *D. willsi*

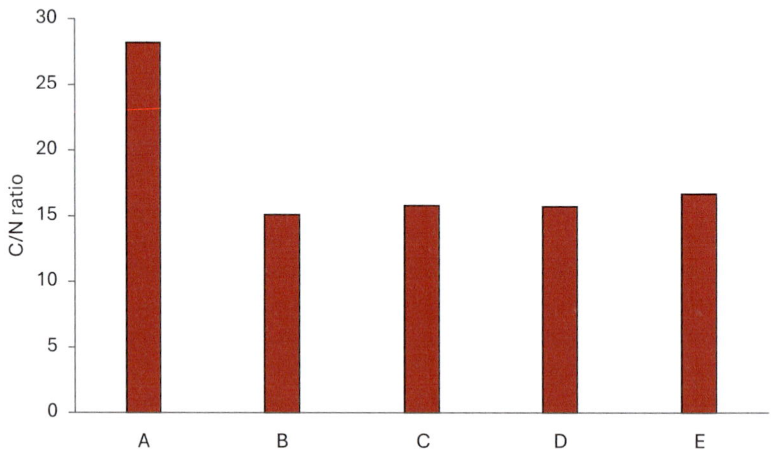

FIGURE 11.9 C N ratio of (a) ipomoea leaves and its VC of (b) *E. andrei*; (c *P. sansibaricus*; (d) *L. rubellus*; and (e) *D. willsi*.

(15.7) > *P. sansibaricus* (15.8) > *D. willsi* (16.7), the order in the case of ipomoea was *P. sansibaricus* (14) > *D. willsi* (14.2) > *E. andrei* (14.3) > *L. rubellus* (14.8). The differences across the species, as also across the two weeds, are only marginal, and the absence of a consistent order across the earthworm species shows that the C:N ratio of the VCs does not seem to be significantly influenced by the earthworm species.

11.4.2 The FTIR Spectra

The VCs of both the weeds, generated by all four species of earthworms – as illustrated with the examples of the salvinia VC generated by *L. rubellus* (Figures 11.10 and 11.11) and the ipomoea VC generated by *P. sansibaricus* (Figures 11.12 and 11.13) – show

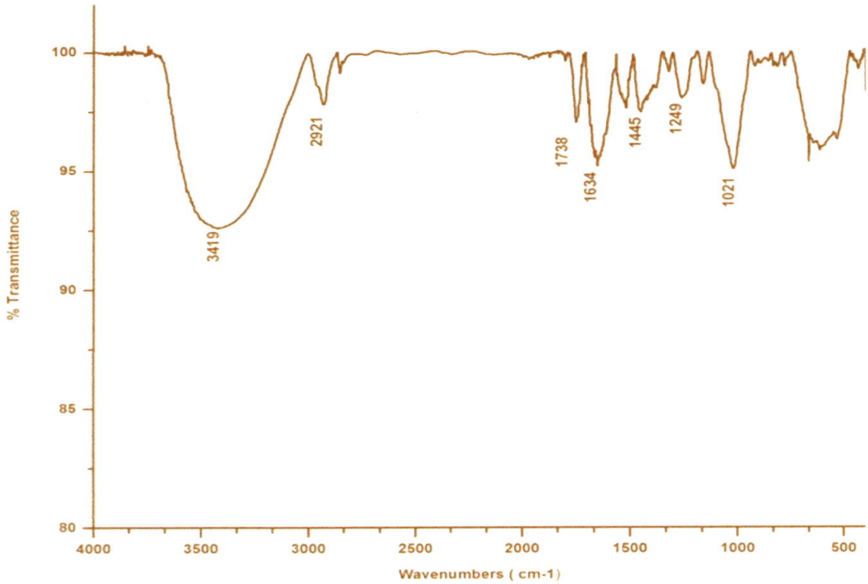

FIGURE 11.10 FTIR spectra of salvinia leaves.

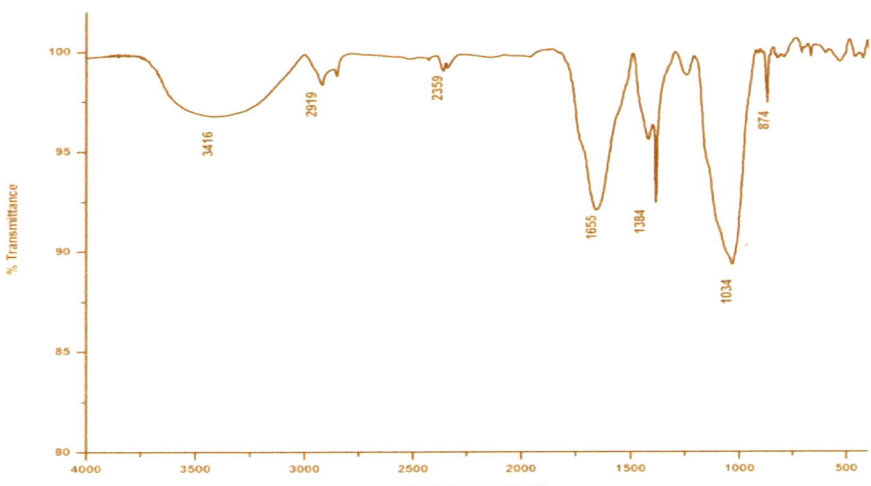

FIGURE 11.11 FTIR spectra of VC generated by *L. rubellus*.

that each of the earthworm species has significantly degraded the chemicals in the two weeds which are known to be responsible for the weeds' antagonistic nature. The extent of this effect was similar in some respects and differed slightly in some other, but, overall, there was little to choose between the performance of the four earthworm species.

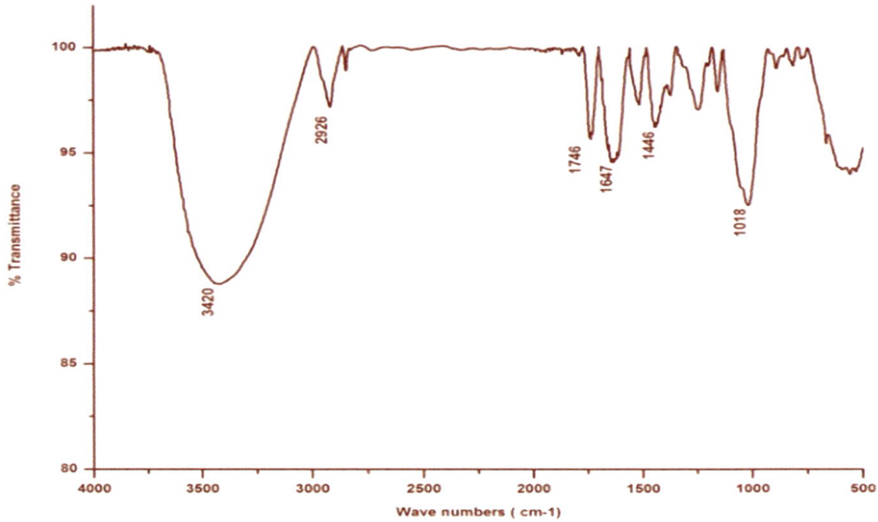

FIGURE 11.12 FTIR spectra of ipomoea leaves.

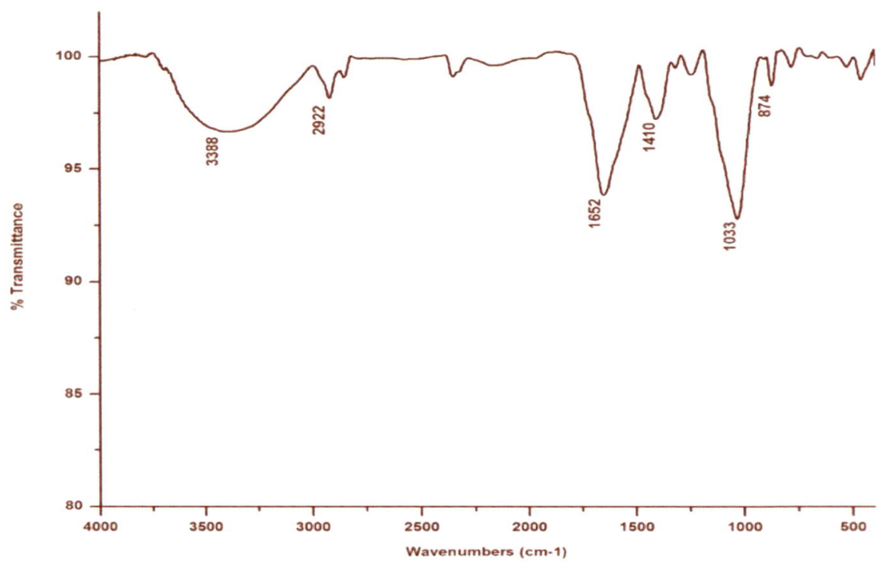

FIGURE 11.13 FTIR spectra of *P. sansibaricus* vermicast.

11.4.3 TG Analysis and DC Scans

TG curves reveal that whereas salvinia suffered an overall mass loss of 93% in the course of its heating up to 800°C (Figure 11.14), the SVCs generated by the four earthworm species had mass loss of only 70%–72%. Figure 11.15, pertaining to

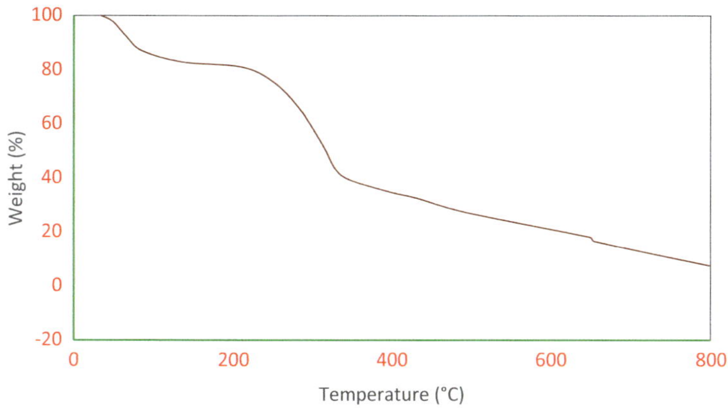

FIGURE 11.14 TG curve of the salvinia plant.

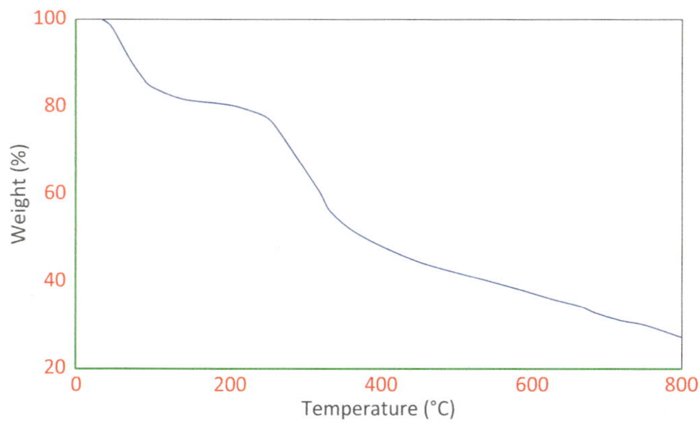

FIGURE 11.15 TG curve of VC by *D. willsi*.

D. willsi, is typical of the TG pattern seen with all four earthworm spaces. As for ipomoea, it suffered a mass loss of 90% while the IVCs produced by the four species had much lesser mass losses: 55%–79%. The range is much wider in IVCs than in SVCs, but leads to a similar conclusion when we see that the upper limit is 72% for SVCs and 79% for IVCs. In other words, all VCs had a much higher proportion of mineralized than the respective parent weeds, due to which the VCs had less volatiles to lose.

The DSC curves of ipomoea, and of *L. rubellus,* as representatives of the DSC curves of the other three species, are shown as Figures 11.16 and 11.17. The ipomoea curve (Figure 11.16) has two exothermic peaks. The first of these occurs in the 300°C–350°C range and the second one at 450°C–500°C. The first peak can be attributed to the presence of carbohydrate and cellulose, while the second peak is likely due to lignin and phenols.

VCs of all four species, too, have two exothermic peaks in each of their DSC curves, but the peaks are much more shallow, as illustrated with the DSC curve of the

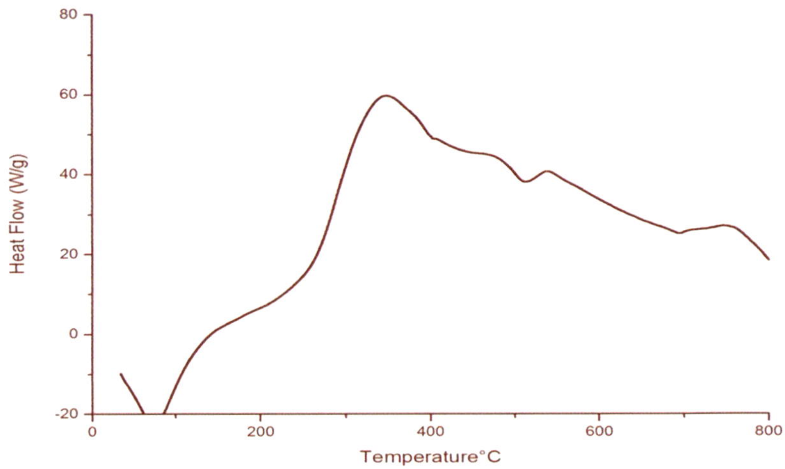

FIGURE 11.16 DSC curve of ipomoea plant.

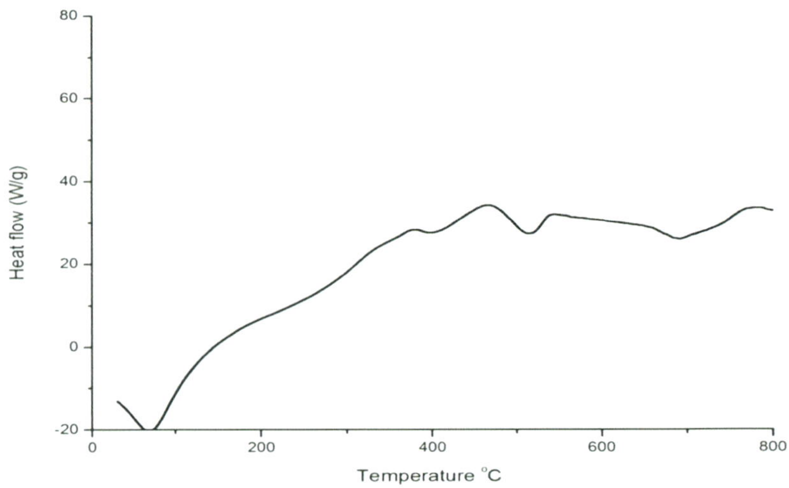

FIGURE 11.17 DSC curve of *L. rubellus* vermicast.

L. rubellus vermicast (Figure 11.17). The extent of reduction in the exothermic peak follows the trend *P. sansibaricus*> *L. rubellus*> *E. andrei*>*D. willsi*. The results indicate that in the course of ipomoea's vermicomposting, there is a breakdown of simpler aliphatic compounds as well as lignin and phenols present in it.

11.4.4 Scanning Electron Microscopy

The SEM of ipomoea leaves (Figure 11.18) reflects the weed's robust and relatively contiguous structure – seemingly bound to lignin-containing fibres. In contrast, the

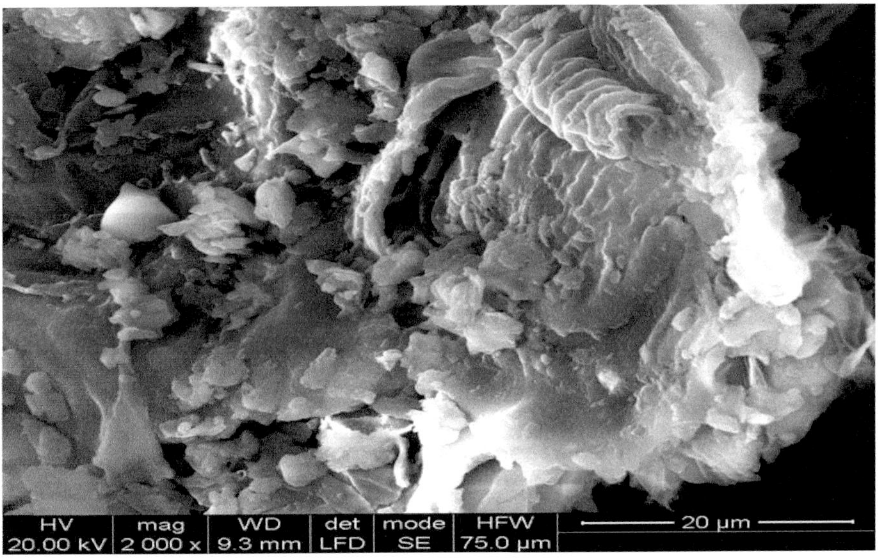

FIGURE 11.18 SEM images of ipomoea leaves.

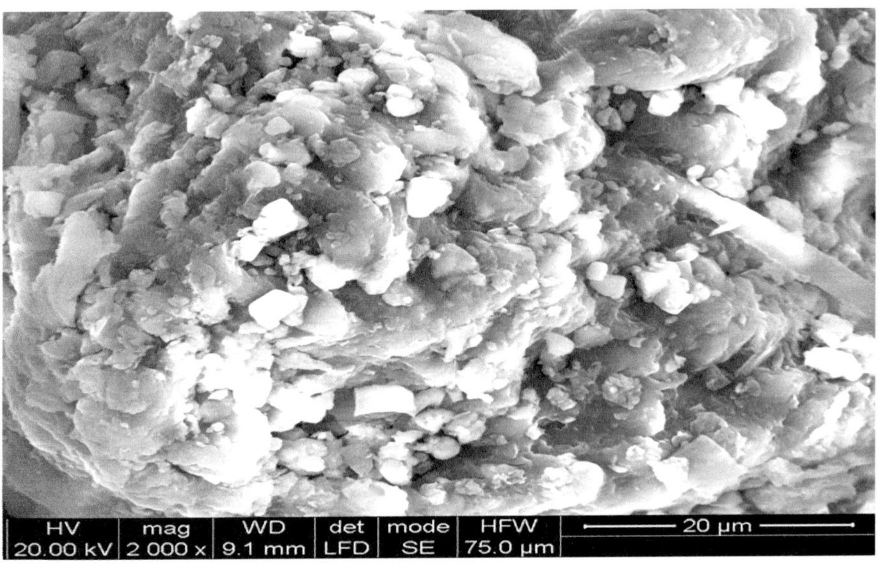

FIGURE 11.19 SEM images of *E. andrei* vermicast.

micrographs of VCs derived from all four species of earthworms reflect disaggregation and withering, as illustrated with the SEM of *E. andrei* (Figure 11.19).

When earthworms feed on the ipomoea biomass, they grind the latter with their gizzard (Ali et al., 2015); there is further disaggregation as the microbial fauna acts on the substrate during the course of its passage through the earthworm gut

(Atiyeh et al., 2000; Edward et al., 2011; Hussain, 2016). This facilitates progressive degradation of ipomoea which is reflected in the SEM micrographs.

11.5 SUMMARY

The material presented above carries significant depth and breadth of information as it encompasses four different weeds – of which three possess animal toxicity while the fourth is allelopathic – and four different species of earthworms. Based on elaborate studies, it shows that irrespective of the weed or the earthworm used to vermicompost that weed, certain transformations occur which convert even toxic and/or allelopathic phytomass into benign organic fertilizers.

REFERENCES

Abbasi, S.A, and Abbasi, T., 2019. *Current Concerns in Environmental Engineering*. Nova Science Publishers, New York, pages xvi+309, ISBN: 978-1-53613-920-4.

Abbasi, S.A., and Ramasamy, E.V., 2001. *Solid Waste Management with Vermitechnology*. Discovery Publishing House, New Delhi, xiv + 178 pages.

Albanell, E., Plaixats, J. and Cabrero, T., 1988. Chemical changes during vermicomposting (Eisenia fetida) of sheep manure mixed with cotton industrial wastes. *Biology and Fertility of Soils*, *6*, pp.266–269.

Ali, U., Sajid, N., Khalid, A., Riaz, L., Rabbani, M.M., Syed, J.H. and Malik, R.N., 2015. A review on vermicomposting of organic wastes. *Environmental Progress & Sustainable Energy*, *34*(4), pp.1050–1062.

Aparna, V., Dileep, K.V., Mandal, P.K., Karthe, P., Sadasivan, C. and Haridas, M., 2012. Anti-inflammatory property of n-hexadecanoic acid: Structural evidence and kinetic assessment. *Chemical Biology & Drug Design*, *80*(3), pp.434–439.

Arancon, N.Q., Galvis, P.A. and Edwards, C.A., 2005. Suppression of insect pest populations and damage to plants by vermicomposts. *Bioresource Technology*, *96*(10), pp.1137–1142.

Atiyeh, R.M., Arancon, N., Edwards, C.A. and Metzger, J.D., 2000. Influence of earthworm-processed pig manure on the growth and yield of greenhouse tomatoes. *Bioresource Technology*, *75*(3), pp.175–180.

Baddi, G.A., Alburquerque, J.A., Gonzálvez, J., Cegarra, J. and Hafidi, M., 2004. Chemical and spectroscopic analyses of organic matter transformations during composting of olive mill wastes. *International Biodeterioration & Biodegradation*, *54*(1), pp.39–44.

Bykov, I., 2008. Characterization of natural and technical lignins using FTIR spectroscopy. Master Thesis, Lulea University of Technology. Retrieved from https://urn.kb.se/resolve?urn=urn:nbn:se:ltu:diva-42881

Critter, S.A. and Airoldi, C., 2006. Thermal analysis of Brazilian tropical soils originating from different sources. *Journal of the Brazilian Chemical Society*, *17*, pp.1250–1258.

Deka, H., Deka, S., Baruah, C.K., Das, J., Hoque, S. and Sarma, N.S., 2011. Vermicomposting of distillation waste of citronella plant (Cymbopogon winterianus Jowitt.) employing Eudrilus eugeniae. *Bioresource Technology*, *102*(13), pp.6944–6950.

Dell'Abate, M.T., Benedetti, A. and Sequi, P., 2000. Thermal methods of organic matter maturation monitoring during a composting process. *Journal of Thermal Analysis and Calorimetry*, *61*(2), pp.389–396.

Edwards, C.A., 1998. The use of earthworms in the breakdown and management of organic wastes. In: Edwards, C.A. (Ed.). *Earthworm Ecology*. CRC Press, Boca Raton, FL, USA, pp. 327–354.

Edwards, C.A., Norman, Q.A., and Sherman, R., 2011. *Vermiculture Technology, Earthworms, Organic Waste and Environmental Management*. CRC Press, Boca Raton pp.17–19.

Gajalakshmi, S. and Abbasi, S.A., 2003. High-rate vermicomposting systems for recycling paper waste. *Indian Journal of Biotechnology*, 2, 613–615

Gajalakshmi, S. and Abbasi, S.A., 2004. Vermiconversion of paper waste by earthworm born and grown in the waste-fed reactors compared to the pioneers raised to adulthood on cowdung feed. *Bioresource Technology*, *94*(1), pp.53–56.

Ganesh, P.S., Gajalakshmi, S. and Abbasi, S.A., 2009. Vermicomposting of the leaf litter of acacia (Acacia auriculiformis): Possible roles of reactor geometry, polyphenols, and lignin. *Bioresource Technology*, *100*(5), pp.1819–1827.

Godson, S.G.A., Gajalakshmi, S., and Abbasi, S.A., 2015. Sustainable vermicomposting of *Mimosa pudica* in two different forms by employing three epigeic earthworm species. *International Journal of Environmental Science and Engineering Research*, *6*(1), pp.1–6.

Huang, G.F., Wu, Q.T., Wong, J.W.C. and Nagar, B.B., 2006. Transformation of organic matter during co-composting of pig manure with sawdust. *Bioresource Technology*, *97*(15), pp.1834–1842.

Hussain, N., 2016. Gainful utilization of toxic and allelopathic weeds for the generation of biofertilizer through vermicomposting: An assessment of the fertilizer value of the resultant vermicompost, PhD Thesis, Pondicherry University, p. 175.

Hussain, N. and Abbasi, S.A., 2018. Efficacy of the vermicomposts of different organic wastes as "clean" fertilizers: State-of-the-art, *Sustainability*, *10*(4), p.1205.

Hussain, N., Abbasi, T., and Abbasi, S.A., 2018. Generation of highly potent organic fertilizer from pernicious aquatic weed *Salvinia molesta*. *Environmental Science and Pollution Research* (Springer), *25*, pp.4989–5002.

Hussain, N., Abbasi, T., and Abbasi, S.A., 2015. Vermicomposting eliminates the toxicity of lantana (*Lantana camara*) and turns it into a plant friendly organic fertilizer. *Journal of Hazardous Materials* (Elsevier), *298*, pp.46–57.

Hussain, N., Abbasi, T., and Abbasi, S.A., 2016a. Vermiremediation of an invasive and pernicious weed salvinia (*Salvinia molesta*). *Ecological Engineering* (Elsevier), *91*, pp.432–440.

Hussain, N., Abbasi, T., and Abbasi, S.A., 2016b. Transformation of toxic and allelopathic lantana into a benign organic fertilizer through vermicomposting. *Spectrochimica Acta Part A: Molecular and Biomelecular Spectroscopy* (Elsevier), *163*, pp.162–169.

Hussain, N., Abbasi, T., and Abbasi, S.A., 2016c. Transformation of the pernicious and toxic weed parthenium into an eco-friendly organic fertilizer by vermicomposting. *International Journal of Environmental Studies*, *73*, pp.731–745.

Hussain, N., Abbasi, T., and Abbasi, S.A., 2016e. Vermicomposting transforms allelopathic parthenium into a benign organic fertilizer. *Journal of Environmental Management* (Elsevier), *180c*, pp.180–189.

Hussain, N., Abbasi, T., and Abbasi, S.A., 2016e. Vermicomposting transforms allelopathic parthenium into a benign organic fertilizer. *Journal of Environmental Management* (Elsevier), *180c*, pp.180–189.

Hussain, N., Abbasi, T., and Abbasi, S.A., 2016d. Vermicomposting-mediated conversion of the toxic and allelopathic weed ipomoea into a potent fertilizer. *Process Safety and Environmental Protection*, *13*, pp.97–106.

Hussain, N., Abbasi, T., and Abbasi, S.A., 2017. Toxic and allelopathic ipomoea yields plant-friendly organic fertilizer. *Journal of Cleaner Production* (Elsevier), *148*, pp.826–835.

Iranshahi, M., Emami, S.A. and Mahmoud, S.M., 2007. Detection of sesquiterpene lactones in ten Artemisia species population of Khorasan provinces. *Iranian Journal of Basic Medical Sciences*, *10*(3), pp.183-188. https://doi.org/10.22038/ijbms.2007.5293.

Islam M.T., Ali, E.S., Uddin, S.J., Shaw, S., Islam, M.A., Ahmed, M.I., Chandra Shill, M., Karmakar, U.K., Yarla, N.S., Khan, I.N., Billah, M.M., Pieczynska, M.D., Zengin, G., Malainer, C., Nicoletti, F., Gulei, D., Berindan-Neagoe, I., Apostolov, A., Banach, M., Yeung, A.W.K., El-Demerdash, A., Xiao, J., Dey, P., Yele, S., Jóźwik, A., Strzałkowska, N., Marchewka, J., Rengasamy, K.R.R., Horbańczuk, J., Kamal, M.A., Mubarak, M.S., Mishra, S.K., Shilpi, J.A., and Atanasov, A.G. 2018. Phytol: A review of biomedical activities. *Food and Chemical Toxicology*, *121*, pp.82–94. doi: 10.1016/j.fct.2018.08.032.

Kumar, D.S., Kumar, P.S., Rajendran, N.M. and Anbuganapathi, G., 2013. Compost maturity assessment using physicochemical, solid-state spectroscopy, and plant bioassay analysis. *Journal of agricultural and Food Chemistry*, *61*(47), pp.11326–11331.

Masciandaro, G., Macci, C., Doni, S. and Ceccanti, B., 2010. Use of earthworms (Eisenia fetida) to reduce phytotoxicity and promote humification of pre-composted olive oil mill wastewater. *Journal of the Science of Food and Agriculture*, *90*(11), pp.1879–1885.

Mochochoko, T., Oluwafemi, O.S., Jumbam, D.N. and Songca, S.P., 2013. Green synthesis of silver nanoparticles using cellulose extracted from an aquatic weed; water hyacinth. *Carbohydrate Polymers*, *98*(1), pp.290–294.

Morais, F. and Queda, C., 2003. Study of storage influence on evolution of stability and maturity properties of MSW composts advances for a sustainable society part II: Proceedings of the Fourth International Conference of ORBIT Association on Biological Processing of Organics Perth. In *Proceedings of the Fourth International Conference of ORBIT Association on Biological Processing of Organics*, Perth, Australia.

Pavia, D.L., Lampman, G.M., Kriz, G.S. and Vyvyan, J.A., 2001. *Introduction to Spectroscopy*. Thomson Learning. Inc., USA, pp.24–29.

Pietro, M. and Paola, C., 2004. Thermal analysis for the evaluation of the organic matter evolution during municipal solid waste aerobic composting process. *Thermochimica Acta*, *413*(1–2), pp.209–214.

Putnam, A.R., 1988. Allelochemicals from plants as herbicides. *Weed Technology*, *2*(4), pp.510–518.

Ravindran, B., Sravani, R., Mandal, A.B., Contreras-Ramos, S.M. and Sekaran, G., 2013. Instrumental evidence for biodegradation of tannery waste during vermicomposting process using *Eudrilus eugeniae*. *Journal of Thermal Analysis and Calorimetry*, *111*, pp.1675–1684.

Rizvi, S.J.H. and Rizvi, V., 1992. *Allelopathy: Basic and Applied Aspects*. Chapman & Hall, New York, p. 480.

Rontani, J.F. and Volkman, J.K., 2003. Phytol degradation products as biogeochemical tracers in aquatic environments. *Organic Geochemistry*, *34*(1), pp.1–35.

Smidt, E. and Meissl, K., 2007. The applicability of Fourier transform infrared (FT-IR) spectroscopy in waste management. *Waste Management*, *27*(2), pp.268–276.

Smith, J.L. and Paul, E.A., 2017. The significance of soil microbial biomass estimations. In: Bollag, J.-M. (Ed), *Soil Biochemistry*. Routledge, New York,, pp. 357–398.

Teh, C.Y., Ta Yeong W., and Joon Ching J., 2014 Potential use of rice starch in coagulation–flocculation process of agro-industrial wastewater: Treatment performance and flocs characterization. *Ecological Engineering*, *71*, pp.509–519.

Wang, Y., Schuchardt, F., Sheng, F., Zhang, R., and Cao, Z., 2004. Assessment of maturity of vineyard pruning compost by Fourier transform infrared spectroscopy, biological and chemical analyses. *Landbauforschung Volkenrode*, *54*, 163–169.

Xu, R., Ye, Y., and Zhao, W. eds., 2012. *Introduction to Natural Products Chemistry*. CRC Press, Boca Raton, pp. 101–118. https://doi.org/10.1201/b11017

Zbytniewski, R. and Buszewski, B., 2005. Characterization of natural organic matter (NOM) derived from sewage sludge compost. Part 2: Multivariate techniques in the study of compost maturation. *Bioresource Technology*, *96*(4), pp.479–484.

12 Long-Term Impact on the Growth and Fecundity of Successive Generations of Earthworms Fed Exclusively on One or the Other Species of Weeds

12.1 INTRODUCTION

The information that has been covered in the previous chapter puts it beyond reasonable doubt that in the course of feeding upon a toxic weed and depositing its vermicompost, earthworms are able to neutralize the toxicity and transform the parent substrate into not only a harmless but also a very agriculture-friendly product. But if earthworms are made to feed upon the same toxic weed, generation after generation – as would happen when the weed's vermicomposting will have to be carried out continuously for utilizing its abundant phytomass – will it not adversely affect the growth and the reproductive ability of successive generations of earthworms? The ones that will be born and will be growing in those vermireactors? Will it not lead to any difficulty in running such reactors uninterruptedly and at the top level of efficiency for several months/years? Answers to these questions are not available in the prior art. They are also not expected to be available because, as detailed in the previous two chapters, reports on the direct vermicomposting of phytomass – without any pre-composting and/or manure supplementation – are themselves largely missing in the pre-existing literature. The only exception appears to be a report from this author's group, by Gajalakshmi and Abbasi (2004), which describes the extent of vermicomposting of paper waste achieved by earthworms born and raised in the waste-fed reactors compared to the pioneers raised to adulthood on cow dung feed. But the paper waste used was spiked with cow dung, albeit by a much lower concentration of cow dung than has been used by other authors. Hence, it did not deal with unadulterated phytomass. Also, the study was confined to assessing a single generation change. The studies described in this chapter cover three generations of four species of earthworms, while vermicomposting two different weeds.

DOI: 10.1201/9780429331886-14

12.2 DEFINING ATTRIBUTES OF THE METHODOLOGY

The objective of the investigations has been to run the vermireactors uninterruptedly for several months so that at least three generations of earthworms can complete their life span in the respective reactors while we assess vermicast production per the pre-set number of adult earthworms of each generation. At the same time, it was necessary to dampen as much as possible the effect of other factors such as decomposition of feed occurring independently of earthworms, and the rising rate of feed uptake due to rising populations of earthworms. To address those concerns, we invoked the Pseudo-Discretized Continuous reactor OPeration (PDCOP) protocol, conceived and developed by S.A. Abbasi and co-workers (Gajalakshmi & Abbasi, 2003, 2004; Nayeem-Shah et al., 2015; Abbasi et al., 2020; Bhat et al., 2023; Tabassum-Abbasi et al., 2023). The protocol, to quote from Abbasi et al. (2020), enables reactor operation which is not actually continuous but approximates continuity; hence the term 'pseudo-discretized continuous.'

> In PDCOP, the vermireactors are initiated with a pre-set quantity of the substrate and a certain fixed number of adult earthworms. After allowing the earthworms to effect vermicomposting for a set number of days, say 20 or 25, the reactor contents are transferred to another container for determining the extent of conversion of the substrate to vermicast as also assessing the fecundity by counting the offspring in terms of the numbers of juveniles and cocoons produced by the earthworms. Soon after removing the reactor contents, the reactors are restarted with fresh ipomoea feed but with the same adult animals that were deployed initially, while excluding the juveniles and cocoons. This makes it possible to measure the rate of vermicast production per adult animal and per unit time.
>
> Since the unused substrate—which, if not removed, would have biodegraded even without the action of the earthworms—is removed every 20–25 days, the effect of factors other than ingestion of the feed by the earthworms is minimized.
>
> PDCOP thus ensures that the earthworms graze only upon totally fresh, or almost fresh, feed as they would be doing in the 'high-rate' vermireactor operation based on low solid retention times of just 20–25 days. Further, since the juveniles and the cocoons that are generated in the vermireactors are separated before they could grow to the stage where they begin consuming significant quantities of the feed, their influence, too, on the reactor performance is sharply dampened.

All of the above enable assessment of the quality of vermicomposting garnered as a function of the number of earthworms and time, thereby providing avenues of process control and monitoring. Other procedural details were as reported by Abbasi et al. (2020) and Tabassum-Abbasi et al. (2023).

Salvinia and ipomoea were chosen for the study on the basis of the same logic as explained in Section 11.4 of Chapter 11. *Eisenia andrei*, *Perionyx sansibaricus*, *L. rubellus*, and *D. willsi* were the four species chosen as representatives of species suitable for vermicomposting because they differ from each other in average body size, body mass, and other attributes, even as all are known to be efficient vermicomposters.

12.3 RELATIVE PERFORMANCE OF THREE EARTHWORM GENERATIONS

Figures 12.1–12.3 summarize all findings pertaining to salvinia, while Figures 12.4–12.6 reflect the gist of all the observations recorded on ipomoea. While the first generation of *E. andrei*, which had been nurtured on animal manure and the adults

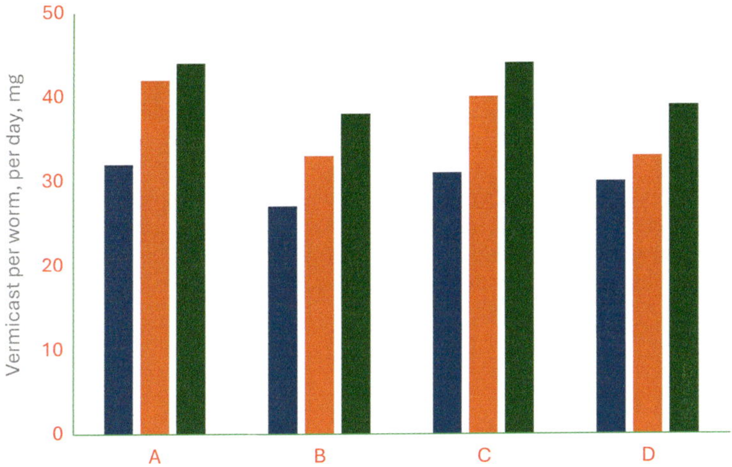

FIGURE 12.1 Comparative performance of three successive generations of earthworms in the vermicomposting of salvinia. A: *E. andrei*; B: *P. sansibaricus*; C: *L. rubellus*; D: *D. willsi*. ■ First generation; ■ Second generation; ■ Third generation.

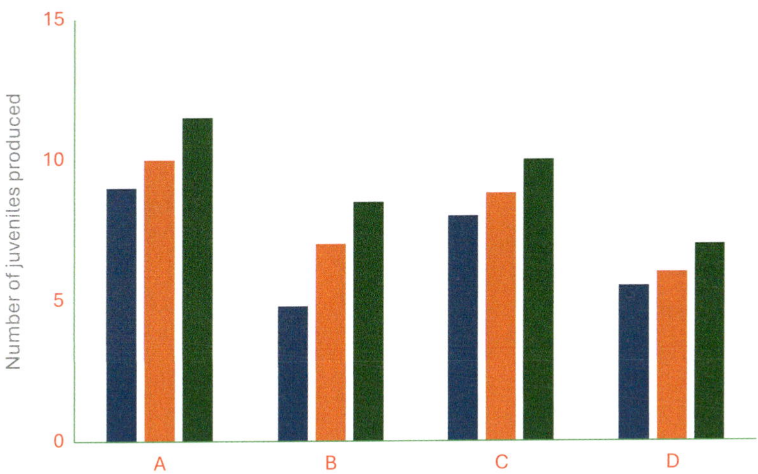

FIGURE 12.2 Comparative performance in the production of juveniles of two successive generations of earthworms born and grown in salvinia-fed vermireactors with the fecundity of manure-reared pioneers. A: *E. andrei*; B: *P. sansibaricus*; C: *L. rubellus*; D: *D. willsi*. ■ First generation; ■ Second generation; ■ Third generation.

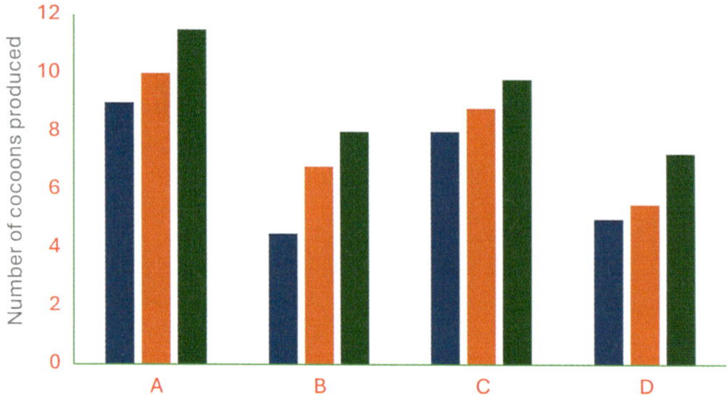

FIGURE 12.3 Relative capability in terms of cocoon production of the second and third generation of earthworms born and grown in salvinia-fed vermireactors and the yield achieved by manure-reared pioneers. A: *E. andrei*; B: *P. sansibaricus*; C: *L. rubellus*; D: *D. willsi*. ■ First generation; ■ Second generation; ■ Third generation.

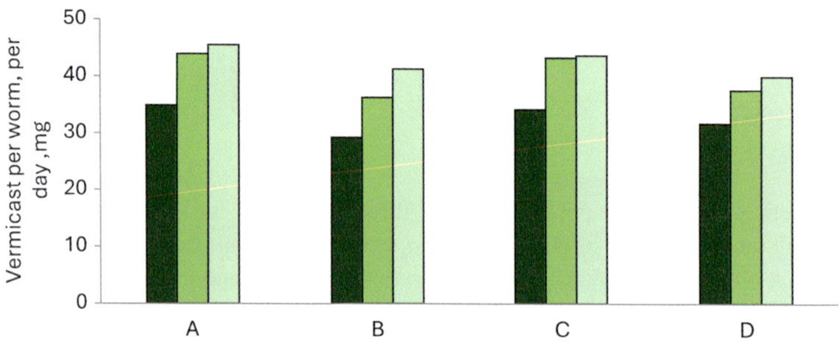

FIGURE 12.4 Relative efficiency of three generations of earthworms in vermicomposting ipomoea. A: *E. andrei*; B: *P. sansibaricus*; C: *L. rubellus*; D: *D. willsi*. First generation ■; Second generation □; Third generation □.

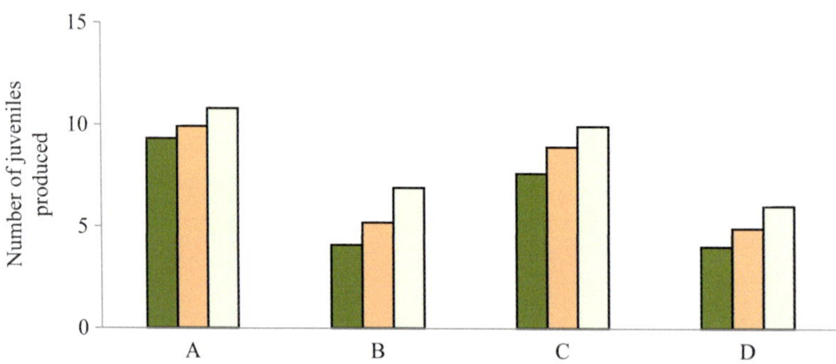

FIGURE 12.5 Juveniles produced by the second and third generation of earthworms born and grown in ipomoea-fed vermireactors in comparison to the output of manure-reared pioneers. A: *E. andrei*; B: *P. sansibaricus*; C: *L. rubellus*; D: *D. willsi*. First generation ■; Second generation □; Third generation □.

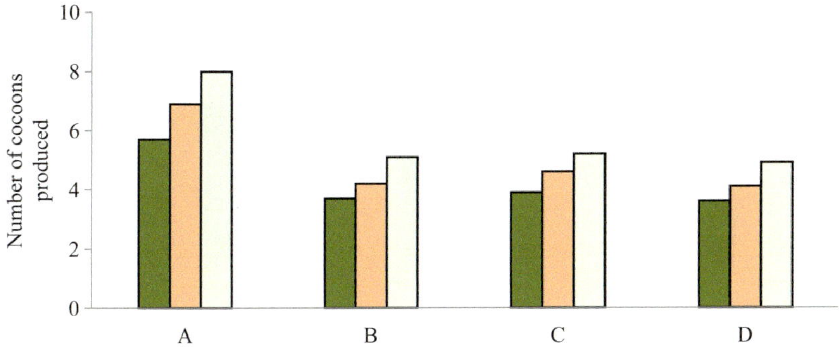

FIGURE 12.6 Cocoons produced by the second and third generation of earthworms born and grown in ipomoea-fed vermireactors in comparison to the output of manure-reared pioneers. A: *E. andrei*; B: *P. sansibaricus*; C: *L. rubellus*; D: *D. willsi*. First generation ■; Second generation □; Third generation □.

of which were used to vermicompost salvinia, could produce 31.6 ± 4 mg of vermicompost (VC) per worm, per day, the second generation did significantly better, to record a VC output of 41.6 ± 4.7 mg/worm.day. The third generation was the most prolific – producing VC to the extent of 44.7 ± 2.7 mg/worm.day.

Coming to ipomoea, *E. andrei* appeared to relish it more than it did salvinia – leading to VC production per worm per day of 34.9 ± 3.9 by the first generation, 43.9 ± 4.5 by the second generation, rising further to 45.5 ± 5.6 in the third generation. *P. sansibaricus* was again slow in adapting to the weed feed, producing only 29.3 ± 5.1 mg of VC/earthworm.day. But, as it had done with salvinia, it picked up very well, and the second and third generations were highly productive, reaching close to the output of *E. andrei*. Once again, *P. sansibaricus* was the lane laggard; *L. rubellus* and *D. willsi* had again much better adaptive responses similar to that of *E. andrei*.

The differences in the vermicast output across the three generations were statistically significant, at confidence levels of 95% or higher, in most cases.

It is noteworthy that in all cases the vermicast output (hence rate of feeding) per earthworm per day was markedly lower during the first 20 days of start of the experiments than the eventual average in the first earthworm generation: 19.4 against 31.6 ± 4 for *E. andrei*, 18.7 against 27.3 ± 3.2 for *P. sansibaricus*, and 25.7 against 30.2 ± 3.4 for *D. willsi*. These are the statistics pertaining to salvinia. The pattern was similar to ipomoea. This indicates that the first generation, which consisted of earthworms that had been raised to adulthood on cow dung as the sole feed, could not immediately adapt to salvinia or ipomoea as their sole feeds. But they did better by the 40th day and by the 60th day had reached the levels of the eventual average output. In all cases, the figures of vermicast output as a function of the number of days of reactor operation showed a sharply rising trend in the case of the respective first earthworm generation. This is illustrated in Figure 12.7. The trend was less sharp in the second generation (Figure 12.8) and became still more muted in the third generation. This indicates that by the time the third generation of earthworms appeared in vermireactors fed solely with either salvinia or ipomoea, they had more or less been fully adapted. And this was seen with all four species of earthworms.

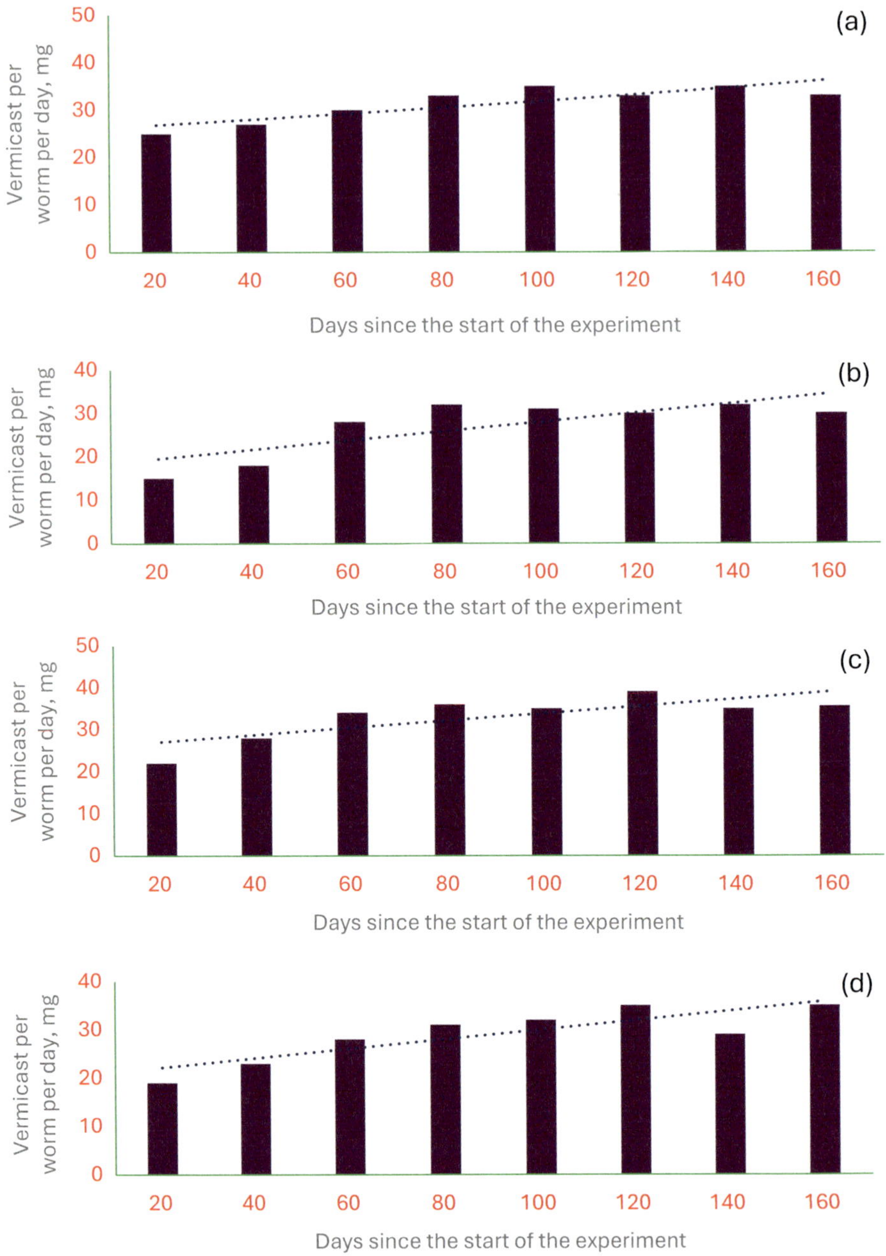

FIGURE 12.7 Production of vermicompost in pulse-fed, semi-continuous reactors operated with (a) *E. andrei*, (b) *P. sansibaricus*, (c) *L. rubellus*, and (d) *D. willsi* by earthworms cultured to adulthood on cow dung and fed with fresh ipomoea. Trend lines are also shown.

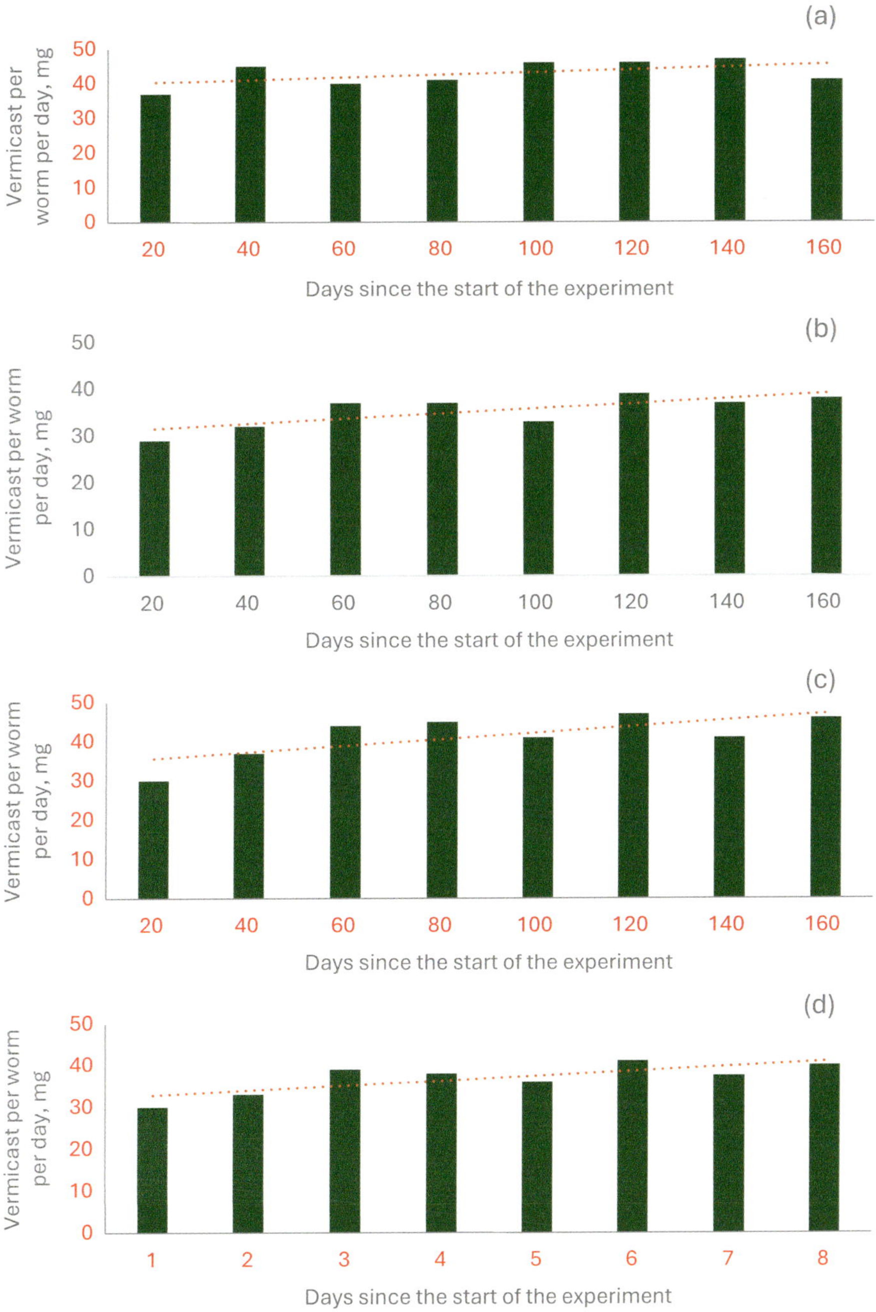

FIGURE 12.8 Production of vermicompost by the second generation of earthworms born and grown in ipomoea-fed vermireactors. (a) *E. andrei*, (b) *P. sansibaricus*, (c) *L. rubellus*, and (d) *D. willsi*.

The reproductive ability, as reflected in the number of juveniles and cocoons produced by successive generations, is seen in Figures 12.2, 12.3, 12.5, and 12.6. In all cases, the second generation had higher fecundity than the first generation, and the third generation had it better than the second.

12.4 ASPECT TO PONDER

Interestingly, while all earthworm species and their generations produced more VC/earthworm.day from ipomoea than they did from salvinia, the trend was reversed in the matter of juvenile and cocoon production. Indeed, the reproductive thrust seen in salvinia-fed reactors was significantly greater than witnessed in ipomoea.

These findings are important in the context of long-term reactor operation and need further exploration. Since successive generations have shown increasing fecundity, it is possible that after the fourth, fifth, or higher generation, the difference in fecundity with respect to the two weeds may vanish. If it doesn't happen, it will mean that some ingredients in ipomoea suppress earthworm fecundity. The practical implication of it will be that in long-term continuous operation, salvinia-fed vermireactors will score over ipomoea-fed vermireactors because the former will generate a larger number of animals per unit time to process the feed (Tabassum-Abbasi et al., 2018).

REFERENCES

Abbasi, S.A., Patnaik, P., Tabassum-Abbasi, Khamrang, C. and Abbasi, T., 2020. A comparative study of the reproductive and the vermicomposting ability of three generations of epigeic and anecic earthworms when subjected to a toxic weed (Ipomoea) as the sole feed. *Compost Science & Utilization*, *28*(1), pp.28–35, doi.org/10.1080/1065657X.2019.1709918.

Bhat, M.A., Tabassum-Abbasi, Abbasi, T., and Abbasi, S.A., 2023. An inexpensive phytoremediation system for treating 50,000 L/day of sewage. *International Journal of Phytoremediation*, *25*(8), pp.1029–1041. https://doi.org/10.1080/15226514.2022.2125497.

Gajalakshmi, S. and Abbasi, S.A., 2003. High-rate vermicomposting systems for recycling paper waste. *Indian Journal of Biotechnology*, *2*, pp.613–615.

Gajalakshmi, S. and Abbasi, S.A., 2004. Vermiconversion of paper waste by earthworm born and grown in the waste-fed reactors compared to the pioneers raised to adulthood on cowdung feed. *Bioresource Technology*, *94*(1), pp.53–56.

Nayeem-Shah, M., Gajalakshmi, S. and Abbasi, S.A., 2015. Direct, rapid and sustainable vermicomposting of the leaf litter of neem (Azadirachta indica). *Applied Biochemistry and Biotechnology*, *175*, pp.792–801.

Tabassum-Abbasi, Abbasi, T. and Abbasi, S.A., 2018. Role of HAZOP in assessing risk of accidents in chemical process industries: capability and lacunae. *International Journal of Engineering, Science and Mathematics*, *7*(1), pp.482-487. https://www.ijesm.co.in/article_html.php?did=4688&issueno=0

Tabassum-Abbasi, Khamrang, C., Abbasi, T., and Abbasi, S. A. 2023. The fecundity and the vermicomposting efficiency of three generations of four earthworm species in high-rate vermireactors fed with the pernicious aquatic weed salvinia. *Environmental Science and Pollution Research*, *30*(10), pp.27382–27393. doi:10.1007/s11356-022-24085-3.

13 First-of-Its Kind Success in Rapidly Vermicomposting a Xerophyte and Utilizing the Product as an Organic Fertilizer

13.1 INTRODUCTION

Till a recent report by these authors (Patnaik et al., 2022), no study has existed on the direct vermicomposting of any xerophyte. Our group (Ganesh et al., 2009) had earlier vermicomposted a xerophyte, acacia (*Acacia auriculiformis*), but only after pre-composting it. There is no report even on any failed attempt made to vermicompost a xerophyte without augmenting it with pre-composting, enzyme fortification, or blending it with animal manure. This is not surprising in view of the fact that by virtue of their adaptation to the highly demanding, water-scarce, and harsh desert environments, most species of xerophytes are exceptionally stout and hardy. This makes them much more capable of resisting biodegradation than other macrophytes are. In the course of the adaptation, some xerophytes have acquired a waxy coating over their stomata to minimize loss of water and ensure even temperature. This has made those species even more resistant to biodegradation.

Prosopis (*Prosopis juliflora*) is a xerophyte that has become one of the world's most widely distributed, invasive, and colonizing plants (Patnaik et al., 2024). Due to its rapid and unbridled growth, prosopis generates millions of tonnes of phytomass per year. Its woody part is utilizable as fuel, and for some other purposes like fashioning inexpensive furniture, but its leaves have no known utility value which could be significant enough to enable the leaves' large-scale utilization. Moreover, prosopis leaves are stout, fibrous, and wax-coated and are very hard to biodegrade. Due to this, when prosopis leaves fall on the ground, they remain in their original shape and form for several months. They keep leaching allelochemicals and other toxic exudates into the underlying soil, harming its productivity and biodiversity (Tabassum-Abbasi et al., 2018). Additionally, they pose a great hazard to animals, including humans, due to the big, sturdy, and disease-causing thorns they contain (Patnaik et al., 2024).

There have been very intense efforts to utilize prosopis leaves as a source of medicinals, cosmetics, nanoparticles, activated carbon, fibres, compost, etc,

DOI: 10.1201/9780429331886-15

(Tabassum-Abbasi et al., 2019; Patnaik et al., 2024) but all these applications put together can utilize only an insignificant proportion of the enormous quantities of prosopis leaves that are generated every year.

After seeing it repeatedly – as detailed in Chapter 11 – that the process of vermicomposting transforms even toxic weeds like ipomoea and parthenium into useful and harmless fertilizers, we decided to explore the possibility of vermicomposting prosopis leaves and checking whether they, too, get converted to benign fertilizers. We also decided to explore direct vermicomposting of the prosopis leaves, instead of trying pre-composting, with or without further supplementing the leaves with animal manure, for similar reasons as described in the earlier chapters. The reasons, to recapitulate, are:

Those steps make the process more time consuming and material consuming, hence economically unviable.

It puts a severe limitation on the quantities of leaves that can be vermicomposted because the quantities of animal manure that are available are very small compared to the quantities of prosopis leaves required to be processed.

13.2 REACTION OF EARTHWORMS TO PROSOPIS LEAVES

Given that it was the first-ever attempt to directly vermicompost a xerophyte, we were apprehensive. The fact that the xerophyte was prosopis, which is not only extraordinarily resistant to biodegradation but also contains highly allelopathic and toxic constituents, made us even more apprehensive. Hence, we decided to explore seven species of earthworms – the epigeics *Eisenia fetida*, *E. andrei*, *Eudrilus eugeniae*, *Lumbricus rubellus*, *Perionyx excavatus*, and *P. sansibaricus*, and the *anecic Drawida willsi* – in the hope that by this wide choice, the prospects of identifying suitable species may get better.

The apprehensions we had were quickly justified because, upon being released in vermireactors fed with prosopis leaves, all species of earthworms promptly died. An attempt was then made to detoxify the prosopis leaves by soaking them for 24 hours in water so that the toxic constituents leach out. But when earthworms were released in the soaked prosopis leaves, they soon died. We then pre-soaked the leaves for 4 days, then 6, 8, and more days. We also started adding fresh cow dung (CD) at 20 g/L in the water used for soaking, equivalent to 4.7 g of dry weight when oven dried to constant weight at 105°C. This was done with the expectation that the CD smell and contents may cushion the shock that the earthworms may be experiencing when put in reactors fed with prosopis leaves. This ploy did not stop earthworms from drying, but it did begin prolonging the onset of death with increasing duration of soaking. Eventually, at the soaking duration of 15 days, there was no earthworm mortality in any species except *D. willsi*. With it, the vermicomposting of prosopis became, *prima facie*, feasible. It was also seen that the water in which prosopis leaves were soaked had extracted significant concentrations of volatile fatty acids (VFAs) from the prosopis leaves. When fed to an anaerobic digester, the VFAs were fermented to yield a clean gaseous fuel in the form of biogas. This made the pre-soaking step remunerative, as the energy gain added to the benefit of utilizing prosopis biomass.

13.3 DEMONSTRATION OF UNINTERRUPTED AND SUSTAINED VERMICOMPOSTING OF PROSOPIS LEAVES FOR CLOSE TO TWO YEARS

Using the pseudo discretized continuous reactor operation protocol (PDCOP), described in the previous chapter, the systematic vermicomposting of prosopis was started, with 30 days as the solids retention time (SRT). The findings are summarized in Table 13.1. As may be seen during the first 30-day operation, all the species fed voraciously on prosopis leaves which had been pre-soaked for 15 days in water containing 20 g/L of fresh CD. But when the reactors were disbanded and

TABLE 13.1
Vermicast Yield, mg per Earthworm per day by Six Species of Earthworms Feeding on Prosopis Leaves Pre-Soaked in Water for 15 days

Duration of Reactor Operation, Days	Vermicast Produced, mg per Earthworm per day by					
	E. eugeniae	P. sansibaricus	P. excavatus	L. rubellus	E. fetida	E. andrei
0–30[a]	64.1[a]	51.0[a]	51.7[a]	46.6[a]	41.6[a]	52.2[a]
31–60[a]	31.3[a]	39.4[a]	27.8[a]	18.5[a]	30.2[a]	20.1[a]
61–90[a]	21.7[a]	46.7[a]	20.0[a]	17.0[a]	21.3[a]	15.3[a]
91–120	Insignificant	Insignificant	Insignificant	Insignificant	Insignificant	Insignificant
121–150	7.4	9.5	5.8	8.1	11.1	6.8
151–180	14.5	6.4	14.5	7.8	5.6	6.1
181–210	10.7	14.8	5.4	12.6	5.7	9.5
211–240	22.0	15.4	15.1	12.6	14.0	15.6
241–270	36.6	19.3	18.4	10.6	7.0	9.3
271–300	24.7	29.0	21.6	15.3	10.7	6.9
301–330	6.9	11.1	13.8	10.0	13.1	10.0
331–360	5.8	10.8	10.9	12.7	5.2	9.6
361–390	7.3	9.5	6.0	8.1	11.4	6.9
391–420	14.5	6.3	14.4	7.8	5.4	6.0
421–450	10.8	14.5	5.1	12.6	5.6	9.7
451–510	11.7	8.7	10.9	10.7	11.6	7.1
511–570	25.8	19.9	23.2	22.7	23.9	17.1
571–630	27.0	23.4	26.4	27.1	28.0	19.1
Average ± SD (of the data of 91–630 days)	15 ± 10.6	13.2 ± 7.4	12.8 ± 7.2	11.9 ± 6.4	10.5 ± 7.8	9.3 ± 4.4

[a] Before the start of these reactors, prosopis leaves were soaked for 15 days in water containing 4.7 mg/L CD. Once the reactors started, they were fed with prosopis leaves, which had been pre-soaked in water for 15 days but without any CD. Yet the effect of initial pre-treatment with CD seemed to have lasted for 90 days. Hence, the first 90 days' data have been omitted from consideration.

immediately restarted as per the PDCOP, we fed them with prosopis leaves soaked for 15 days in water without having added CD to the water. This was done with an aim to rid prosopis vermicomposting of dependence on CD – or any other form of animal manure. The reactors responded with a precipitous drop in the vermicast production, which became poorer during 61–90 days and fell to negligible levels during the next pulse of 91–120 days. But there was no earthworm mortality, and the drop in the vermicompost (VC) output seemed to have occurred as the earthworms struggled to adapt to prosopis-only feed. During 121–150 days' pulse, measurable quantities of VC were produced, and the gain continued. After 450 days, the SRT was increased to 60 days. It caused a spike in VC production (Table 13.1). When the VC output of each species of earthworm was plotted against time for the 91–630 day duration, the corresponding statistical trend lines showed a sharp upward slope, as illustrated with the examples of *E. eugeniae* (Figure 13.1) and *E. andrei* (Figure 13.2). It shows that even though the earthworms were taking much longer to adapt to prosopis leaves as a sole feed than they did for other weeds (Chapters 7 and 8), the adaptation was sure and steady. The production of juveniles and cocoons also displayed sharply rising trends, as illustrated in Figures 13.3 and 13.4, pertaining to *P. sansibaricus*.

By all indications, it appeared clear that vermireactors fed solely on prosopis leaves will take a much longer time to reach the peak level of productivity than taken by other species of plants. But eventually they will. And once the adaptation/gestation/incubation phase is over, the prosopis leaves will get vermicomposted as rapidly as other phytomass. This has been established by the follow-up work of

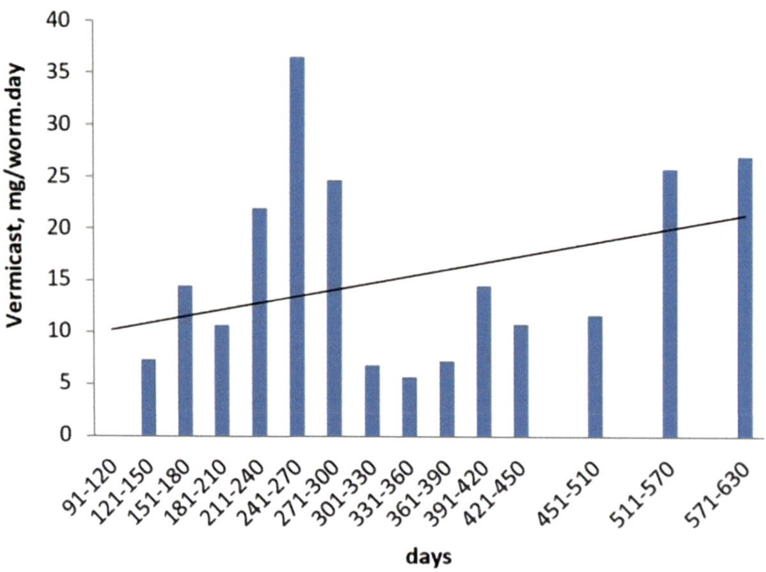

FIGURE 13.1 Vermicast generated by *E. eugeniae* in reactors containing 15-day pre-soaked (in water) prosopis leaves. The statistical trend line is also seen.

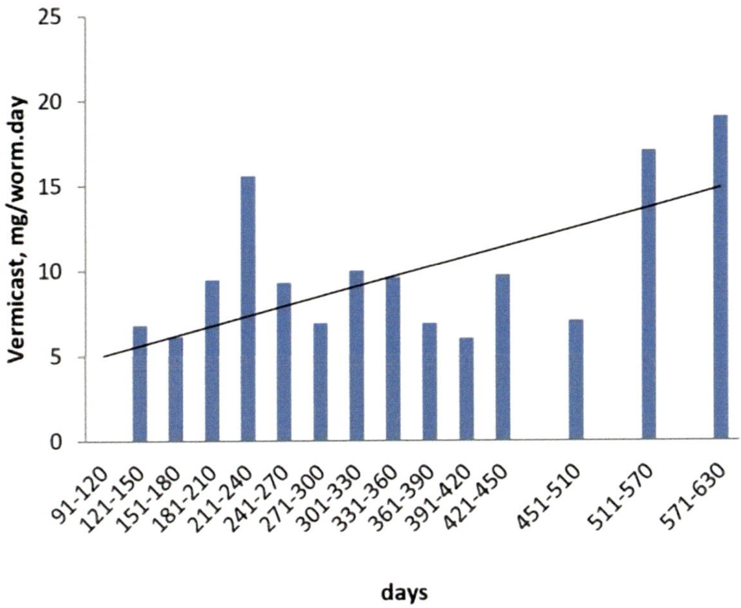

FIGURE 13.2 Vermicast generated by *E. andrei* in reactors containing 15-day pre-soaked (in water) prosopis leaves. The statistical trend line is also seen.

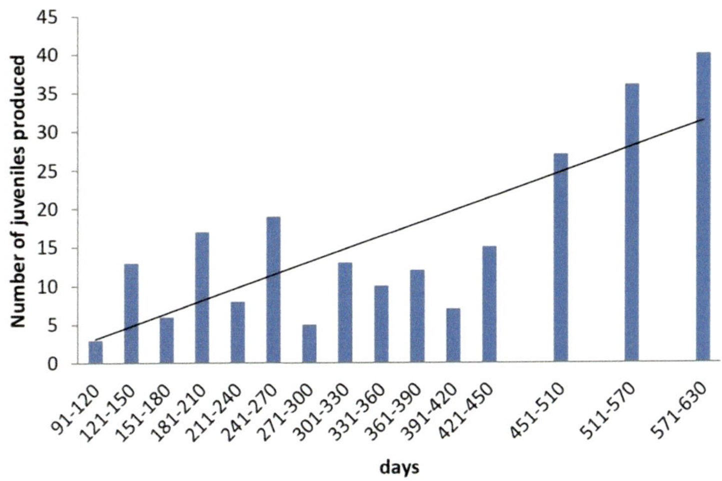

FIGURE 13.3 Juveniles produced by *P. sansibaricus* in reactors containing 15-day pre-soaked (in water) prosopis leaves. The statistical trend line is also seen.

these authors, whose vermireactors are by now converting prosopis leaves into VC within 25 days.

It is also most likely that successive generations of earthworms that have been brought up on prosopis-only feed will get increasingly adapted and attuned to such a feed. The adaptation will also enable them to reach the peak of their fecundity.

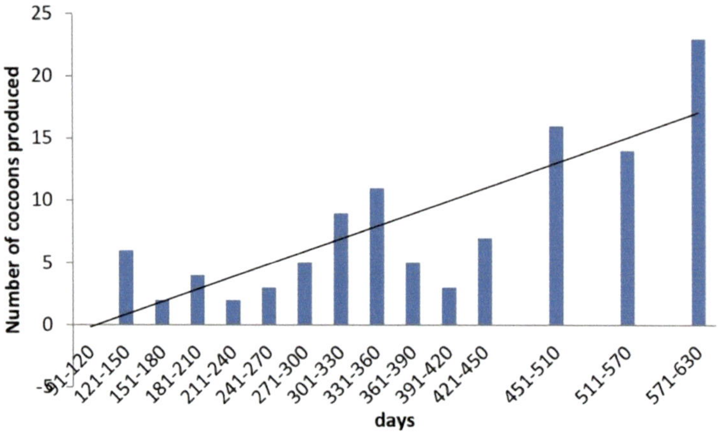

FIGURE 13.4 Cocoons produced by *P. sansibaricus* in reactors containing 15-day pre-soaked (in water) prosopis leaves. The statistical trend line is also seen.

13.4 PROSOPIS VERMICOMPOST (PRVC) AS A FERTILIZER

The effect of applying PRVC, which had been obtained after the earthworms had come out of the influence of the spiked CD, as fertilizer was explored as the next logical step. For it, PRVC was applied at the rates of 0%, 0.75%, and 1.5% by weight to the soil on which the generation and early growth of five common food plants were assessed: ladies finger (*Abelmoschus esculentus*), cucumber (*Cucumis sativus*), tomato (*Lycopersicon esculentum*), brinjal (*Solanum melongena*), and chilli (*Capsicum annuum*). Subsequently, tomato and brinjal were grown till they fructified to learn how PRVC affected their yields and nutrient content.

The effect of PRVC on the germination of the seeds of the five plants is displayed in Figure 13.5. While prosopis leaves are known to strongly suppress the germination of seeds of other species, PRVC enhanced germination in all seeds except the seeds of ladies finger. PRVC also enhanced shoot length, root length, shoot dry weight, and root dry weight – as typified with the findings on brinjal (Figure 13.6). Likewise, as illustrated with the case of brinjal, PRVC boosted chlorophyll levels (Figure 13.7) and enhanced the soil quality (Figure 13.8) by reducing the bulk and the particle densities while enhancing total porosity and water holding capacity. In most cases, the benefits followed the order $0\% < 0.75\% < 1.5\%$, and more often than not, the differences were statistically significant at $\geq 95\%$ confidence level.

When the studies were extended to the effect of PRVC on full growth and fructification of brinjal and tomato, the beneficial effects were witnessed throughout. As illustrated with the example of tomato, PRVC enhanced stem length, root length, and number of branches per plant (Figure 13.9). It induced earlier flowering (Figure 13.10) and had a positive effect on fruit weight, its size, its length, and its number per plant (Figure 13.11).

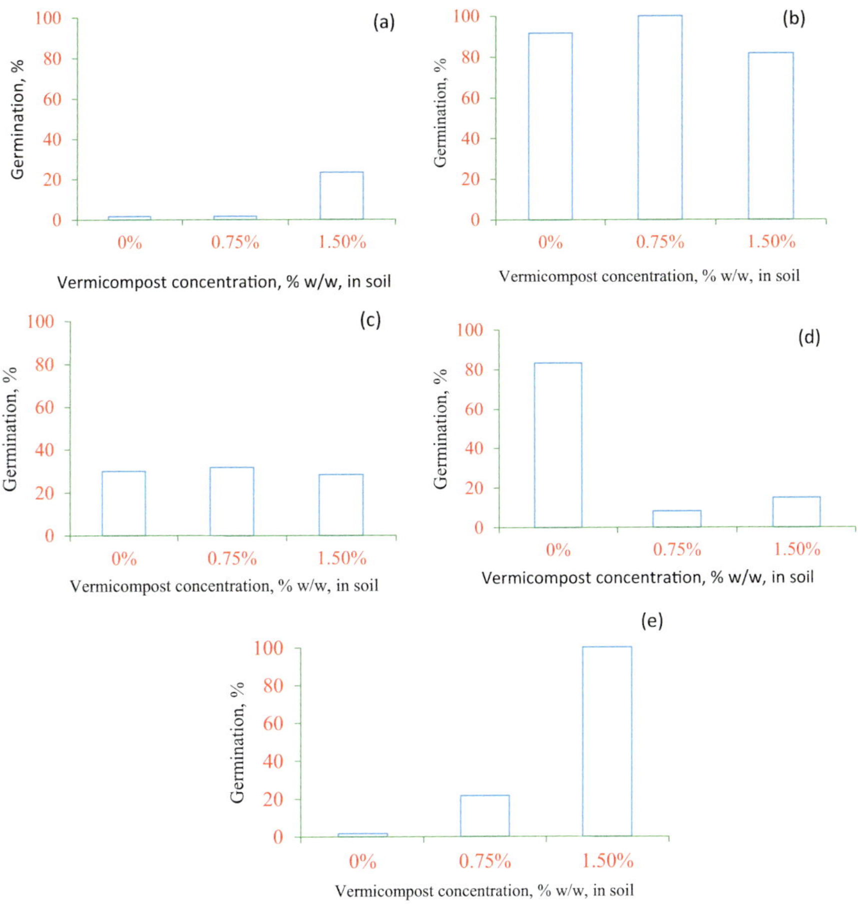

FIGURE 13.5 Effect on germination of the seeds of (a) brinjal, (b) chilli, (c) cucumber, (d) ladies finger, and (e) tomato – as reflected in percent germinated till 8 days after sowing in soils with or without PRVC.

Since no past reports were available on the range of concentrations of nutrients recommended or expected in the brinjal and tomato fruits, we used the fruits obtained from the market to see the contents in comparison to the contents of the PRVC-fertilized brinjal and tomato. It was seen that PRVC-fertilized brinjal had significantly more carbohydrates, calcium, potassium, and magnesium but significantly less proteins, sodium, and phosphorus. The PRVC-fertilized tomatoes had higher levels of carbohydrates, potassium, magnesium, and phosphorus, but lower proteins, calcium, and sodium, than the fruits purchased from the market. Hence, there was no clear pattern of superiority or inferiority, even as the findings do endorse that PRVC is at least as effective as fertilizer – if not better – as the synthetic chemical fertilizers

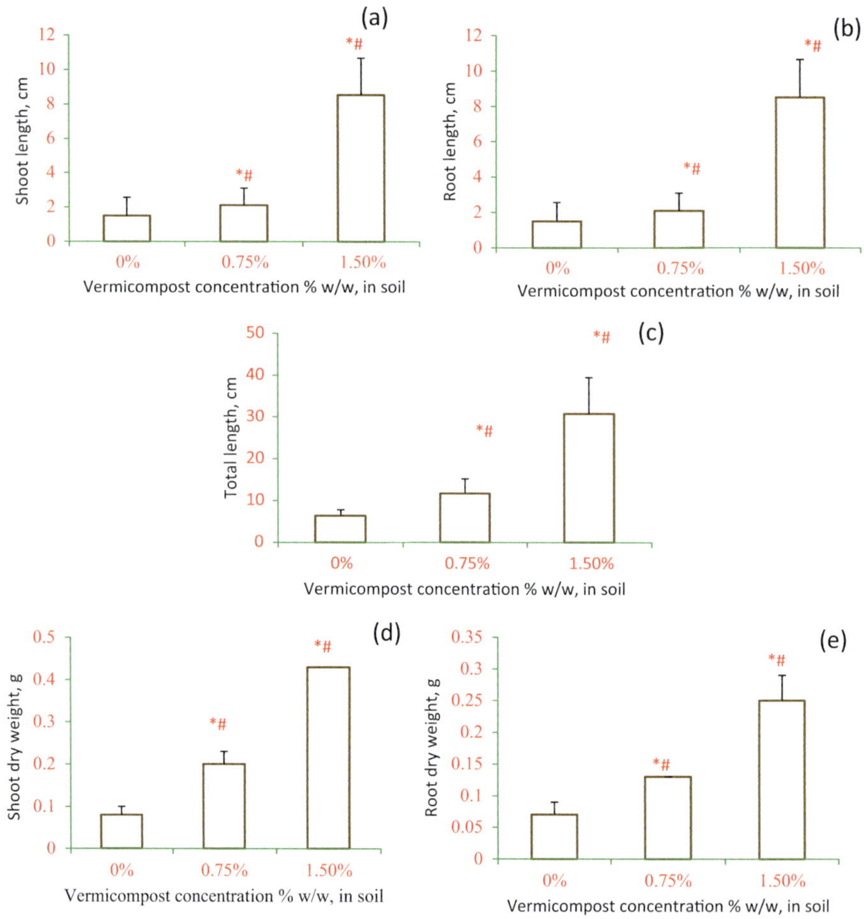

FIGURE 13.6 Influence of PRVC (a) shoot length, (b) root length, (c) total length, (d) shoot dry weight, (e) and root dry weight, of brinjal. A star (*) above a bar indicates that there is a significant difference between PRVC fortified soil and control soil. A hash (#) mark represents a significant difference between one PRVC concentration and the other (≤95% confidence level).

on which the marketed vegetables had been grown. Between the two PRVC treatments, also, there was no clear pattern. While PRVC boosted germination, prosopis leaves suppressed it (Figures 13.12 and 13.13).

13.5 PRVC AS CONTROLLER OF PLANT PATHOGENS

PRVC appeared to be repelling pests and inducing pathogen resistance in the plants fertilized by it. Perhaps it was due to this that none of the fruits caught any infection. Only in the early growth stage, the seedlings were infested by leaf miner (Figure 13.14). But as the growth increased, the number of incidents decreased in proportion.

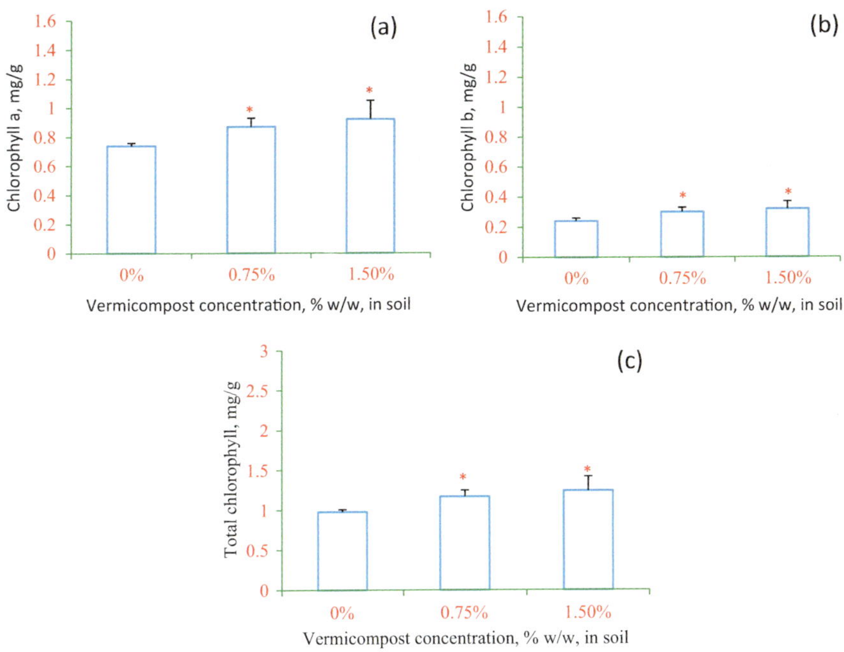

FIGURE 13.7 Influence of PRVC on (a) chlorophyll "a", (b) chlorophyll "b", and (c) total chlorophyll levels in 40-day-old seedlings of brinjal. A star (*) above a bar indicates that there is a significant difference between PRVC fortified soil and control soil. A hash (#) mark represents a significant difference between one PRVC concentration and the other (≤95% confidence level).

13.6 HOW DOES PRVC COMPARE WITH CD AND COW DUNG VC (CDVC)?

The VC that is uniformly accepted as an excellent organic fertilizer-cum-pest repellent is the one derived from CD – the CDVC. CD itself is a good fertilizer, while the parent substrate of PRVC is highly anti-soil and allelopathic. So much so that in all the experiments in which prosopis leaves were added to the soil instead of PRVC, no germination or growth of any other plant species took place.

Given this backdrop, how does PRVC compare with CD and CDVC?

Detailed analysis by us, summarized in Figure 13.15, reveals that of the three fertilizers, PRVC provides the most extractable potassium, NH_3–N, iron, and copper. It also possesses more NO_3–N, total-N, and zinc than CD. Both CD and CDVC are ahead of PRVC only with respect to extractable P.

13.7 FACTORS WHICH TRANSFORM HIGHLY TOXIC AND INSIDIOUS PROSOPIS LEAVES INTO A HIGHLY BENIGN AND BENEFICIAL FERTILIZER-CUM-PEST-REPELLENT?

The manner in which the constituents of the prosopis leaves, responsible for its toxicity, allelopathy, and intransigence, get progressively changed into plant-friendly and soil-friendly entities was explored. The same sequence of investigations, which was

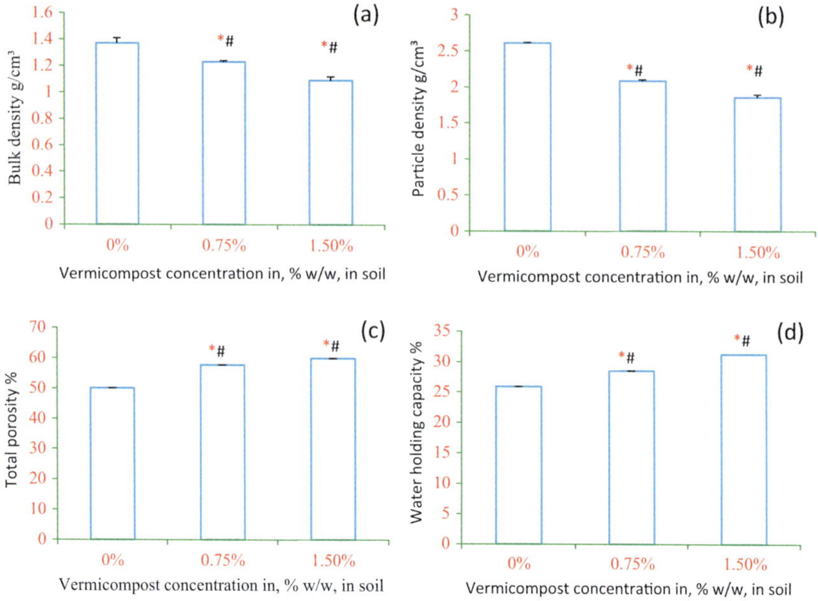

FIGURE 13.8 The values of (a) bulk density, (b) particle density, (c) total porosity, and (d) water holding capacity of the soil in which brinjal seeds had been germinated. A star (*) above a bar indicates that there is a significant difference between PRVC fortified soil and control soil. A hash (#) mark represents a significant difference between one PRVC concentration and the other (≤95% confidence level).

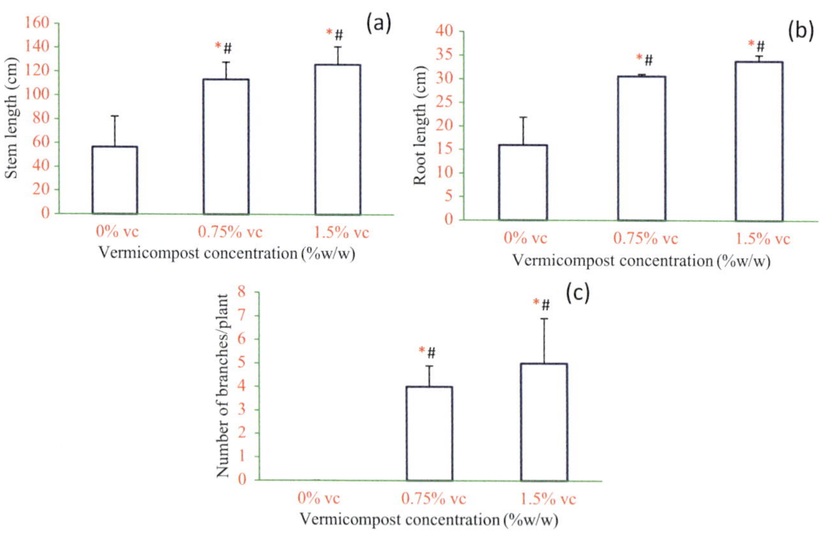

FIGURE 13.9 Influence of PRVC on (a) stem length; (b) root length; and (c) number of branches per plant of tomato. A star (*) above a bar indicates that there is a significant difference between PRVC fortified soil and control soil. A hash (#) mark represents a significant difference between one PRVC concentration and the other (≤95% confidence level).

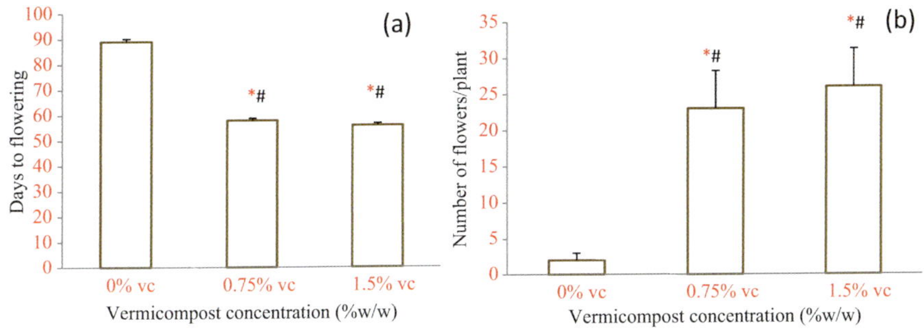

FIGURE 13.10 Influence of PRVC on (a) early flowering and (b) number of flowers per plant of tomato. A star (*) above a bar indicates that there is a significant difference between PRVC fortified soil and control soil. A hash (#) mark represents a significant difference between one PRVC concentration and the other (≤95% confidence level).

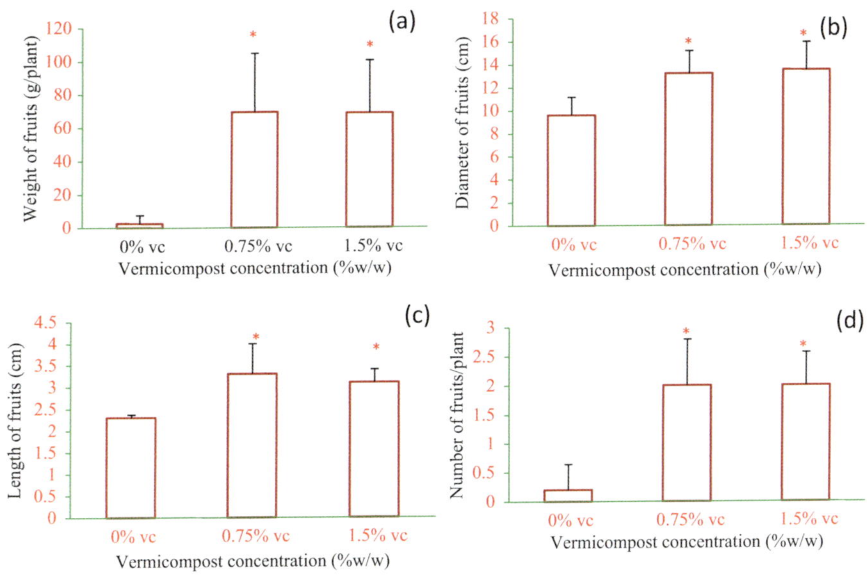

FIGURE 13.11 Influence of PRVC on (a) weight of fruits per plant; (b) diameter of fruits; (c) length of fruits; and (d) number of fruits per plant of tomato. A star (*) above a bar indicates that there is a significant difference between PRVC fortified soil and control soil. A hash (#) mark represents a significant difference between one PRVC concentration and the other (≤95% confidence level).

used for four weeds and four species of earthworms, as described earlier in Chapter 11, was followed for prosopis leaves and PRVC:

a. change in the C:N ratio;
b. destruction of allelochemicals and other toxic constituents as discerned using Fourier transform infrared (FTIR) spectrometry;

FIGURE 13.12 Plants (a) cucumber, (b) ladies finger, (c) tomato, (d) chilli, and (e) brinjal germinated and grown on prosopis vermicast amended soil.

c. extent of mineralization as revealed by thermogravimetric analysis (TGA) and differential scanning calorimetry (DSC);
d. molecular rearrangements indicating the general breakdown of complex organics into more diverse and simpler structures; and
e. changes in surface characteristics as reflected by scanning electron microscopic visuals.

FIGURE 13.13 Soil amended with (a) fresh prosopis leaves and (b) pre-soaked prosopis leaves shows no sign of germination after 90 days.

FIGURE 13.14 Leaves of brinjal infested with leaf miner.

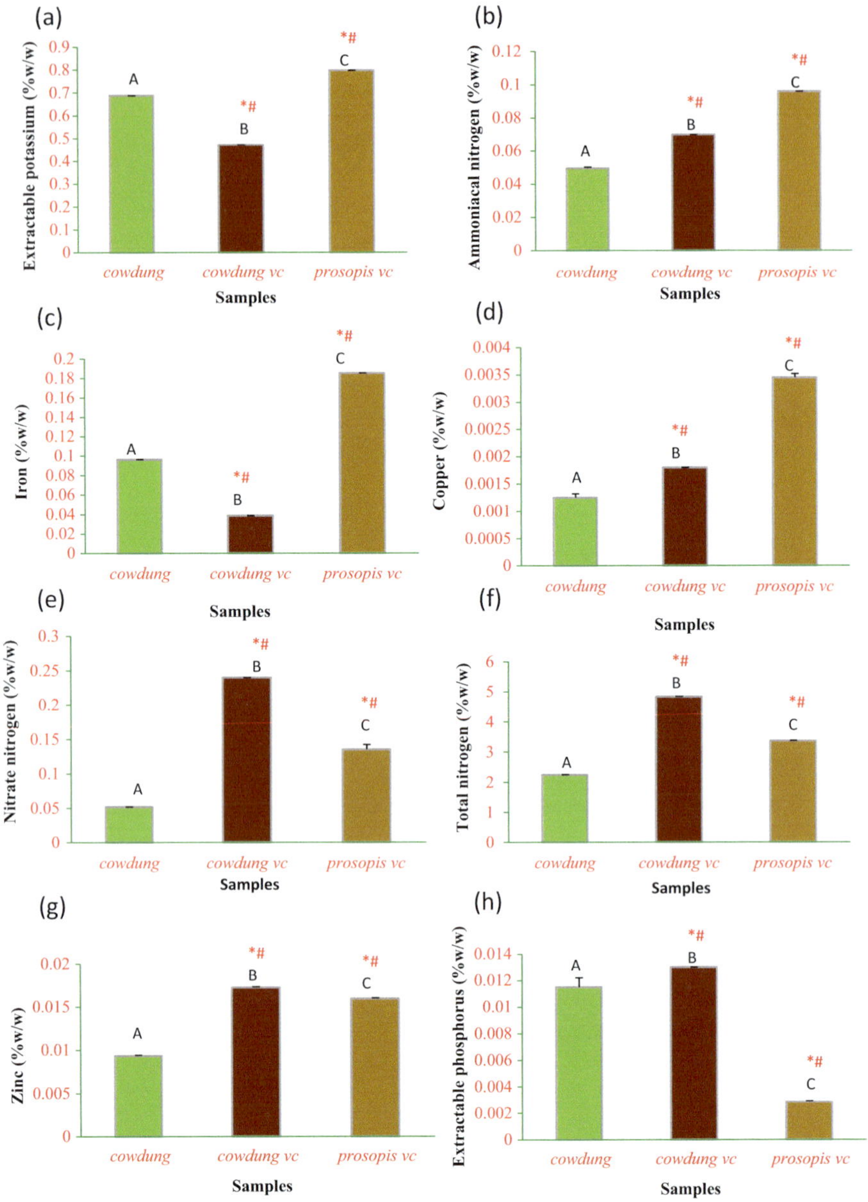

FIGURE 13.15 Relative nutrient content in CD, CDVC, and PRVC.

The findings are also similar. Vermicomposting of 15-day pre-soaked prosopis leaves causes a fall in the C:N ratio from 22.3 to less than 15 for all earthworm species (Figure 13.16). This brings PRVC to the category of fertilizers, which are "highly desirable for agronomic use" (Patnaik et al., 2022, 2023). Fourier transform infrared spectroscopy reveals destruction of polyphenols and other biochemicals responsible for much of the allelopathy and toxicity associated with prosopis. Thermogravimetric analysis and differential scanning calorimetry confirm that

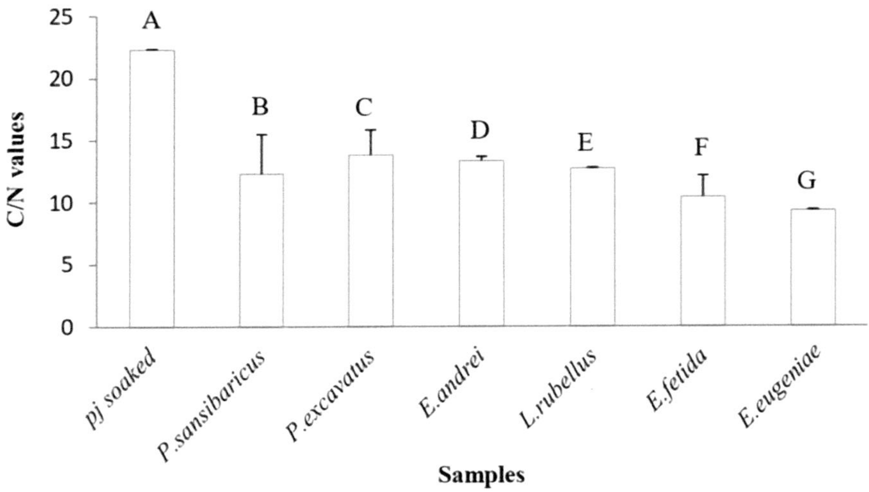

FIGURE 13.16 Change in the C:N ratio of 15 days pre-soaked (in water) prosopis leaves (A) in comparison to the vermicompost generated from it by earthworms *P. sansibaricus* (B); *P. excavatus* (C); *E. andrei* (D); *L. rubellus* (E); *E. fetida* (F); and *E. eugeniae* (G).

PRVC of all six earthworm species studied had attained significant mineralization, thereby unlocking the nutrients from the leaves' ligninous matrices and making them bioavailable. GC-MS showed strong rearrangement of the constituent chemicals: The concentration of a few chemicals present in several times higher mass fractions than the average gave way to a larger number of chemicals, which had more even presence and simpler structures. Lastly, SEM of the PRVC of all the earthworm species showed that vermicomposting breaks the waxy coating on the prosopis leaves and erodes the smoothness of their surface, thereby enhancing their biodegradability.

REFERENCES

Ganesh, P.S., Gajalakshmi, S., and Abbasi, S.A., 2009. Vermicomposting of the leaf litter of acacia (*Acacia auriculiformis*): possible roles of reactor geometry, polyphenols, and lignin. *Bioresource Technology, (Elsevier) 100*, 1819–1827.

Patnaik, P., Abbasi, T. and Abbasi, S.A., 2023. Fate of chemical constituents of prosopis leaves during the conversion of the leaves into an organic fertiliser. *International Journal of Environmental Analytical Chemistry, 103*(5), pp.1123–1139.

Patnaik, P., Abbasi, T. and Abbasi, S.A., 2022. Production of Vermicompost by Utilizing the Leaves of the Hazardous Xerophyte Prosopis (*Prosopis juliflora*). *Waste and Biomass Valorization, 13*(2), 1243–1254.

Patnaik, P., Abbasi, T. and Abbasi, S.A., 2024. *Prosopis Juliflora: Attributes, Impact, Utilization.* 1st Edition, Academic Press, Elsevier, London, 430pp. ISBN: 9780128198254.

Tabassum-Abbasi, Abbasi, T. and Abbasi, S.A., 2018. Role of HAZOP in assessing risk of accidents in chemical process industries: capability and lacunae. *International Journal of Engineering, Science and Mathematics, 7*(1), pp.482–487. https://www.ijesm.co.in/article_html.php?did=4688&issueno=0

Tabassum-Abbasi, Abbasi, T. and Abbasi, S.A., 2019. Microbial fuel cells as source of clean energy – potential and pitfalls. *Nature Environment and Pollution Technology, 18*(3), pp.789–797. https://neptjournal.com/upload-images/NL-69-12-(10)-B-3554.pdf

14 Vermicomposting as an Ideal Foil to Upgrade Macrophyte/Algae-Based Sewage Treatment Systems into Close-Loop Biorefineries

14.1 INTRODUCTION: PROBLEMS ASSOCIATED WITH SEWAGE AND ITS TREATMENT

Household sewage is the most voluminous and ubiquitous of all wastewaters. It is generated wherever human beings dwell, and its global production is estimated to reach volumes of the order of 8 billion kiloliters per year. In its urban regions alone, India has been estimated to generate 2.65 billion litres of sewage per year (CPCB, 2024) – with similarly large quantities generated in the rural and peri-urban parts of India. Due to the absence of affordable technology, as much as 80% of all the sewage generated in India is discharged untreated (Bhat et al., 2023; CPCB, 2024). So much of wastewater, laden with enteric parasites and other disease-causing microflora, when let loose in lakes, ponds, and rivers, plays havoc with the receiving water courses, stressing the ecosystems beyond redemption. It is also the prime cause of debilitating diseases (Benarde et al., 1973; Esrey et al., 1991; Abbasi, 1999, 2018; Arnone & Walling 2007; Abbasi & Abbasi 2018; Sengupta 2018). It is believed that the situation in other developing countries is similar, if not worse (Abbasi et al., 2019).

14.2 THE INADEQUACY OF CONVENTIONAL TECHNOLOGIES AND THE PHYTOREMEDIATION SYSTEMS

The conventional sewage treatment technologies, which are essentially based on the activated sludge process and its variants (Spellman, 2009; Abbasi & Abbasi, 2018, 2019), are not only expensive to commission but also to maintain because of their high energy requirements, supervisory manpower, and machine depreciation. This also enhances their carbon footprint. To get around these hurdles and make sewage treatment affordable – as well as simpler and cleaner – a number of technologies

DOI: 10.1201/9780429331886-16

have been developed, which are based on the use of macrophytes or algae in different forms of reactors for treating sewage. The resulting systems are essentially of three kinds:

1. The "constructed" or "artificial" wetland systems, which aim to mimic the way wastewater gets treated when discharged into natural wetlands. (Subject to the condition that the quantity of sewage disposed of is within the capacity of the wetland to treat). The constructed wetlands use plants like common reed (*Phragmites australis*), typha (*Typha latifolia*), colocasia (*Colocasia esculenta*), etc, rooted in either soil or in gravel of different sizes. These systems have different hydrologic regimes (horizontal flow, vertical flow, surface/sub-surface flow, etc), and the support media play a key role in the system performance along with the macrophytes deployed.
2. The tank/channel systems in which free-floating aquatic weeds like water hyacinth (*Eichhornia crassipes*), salvinia (*Salvinia molesta*), and pistia (*Pistia stratiotes*) are made to proliferate on the water surface and bring about the treatment of sewage passing underneath. The plants remain free-floating, needing no anchorage.
3. Algae-based systems in which one or other species of algae is mixed with wastewater and subsequently removed based on a pre-determined hydraulic retention time (HRT) deemed adequate to effect the desired level of treatment by the added algae.

Based on the habits of different species of macrophytes, different types of reactor hydrology, different reactor geometrics, different modes of operation, and different needs for integration with other systems, etc, a large number of macrophyte-based sewage treatment systems have been developed (Tang et al., 2020a,b; Zhang et al., 2021; Hazra & Durso, 2022). In like manner, different designs of algae-based wastewater treatment systems have been introduced.

In recent years, a new category of macrophyte-based sewage treatment systems has also been attempted, which are called "floating wetlands" (Rome et al., 2023; Vo et al., 2023). These comprise one or more species of vascular plants which are entrapped within circular bounds, often of about 5–10 m^2. The disk-shaped contraptions are allowed to float over lakes/ponds polluted with sewage, causing *in situ* treatment and obviating the need for land to set up the treatment system. But these systems are far too slow and mild to treat raw sewage within a reasonable time (Bhat et al., 2023).

But even as macrophyte/algae-based sewage treatment systems are much less machine-intensive, energy-intensive, and expensive than conventional activated-sludge-based technologies, they suffer from slow kinetics. Most systems require a HRT of several days. Even after persistent work on improving the design and operation of these systems ever since they were introduced to the world by the German hydrobiologist Kathe Sidel in the 1950s (Brix, 1994), the rate of sewage treatment in the macrophyte/algae-based systems is three times and more slower than achievable by the conventional technologies. This necessitates three times or larger space requirements for installing macrophyte/algae-based systems in contrast to the conventional ones. But land is at a premium everywhere, especially in the population-dense developing world. Due to this, macrophyte/algae-based sewage treatment systems are in much less widespread use than conventional systems. Contrary to popular perception, algae have a much slower rate of growth than many macrophytes. For instance (Putri & Huang, 2020), duckweed (*Lantana minor*) is nearly twice as fast-growing as the algae chlorella (*Chlorella vulgaris*). Additionally, algae are much more difficult to manage and harvest than macrophytes.

14.3 THE SHEFROL TECHNOLOGY

In order to remove the biggest constraint being faced by macrophyte/algae-based sewage treatment systems, S. A. Abbasi and coworkers have developed the SHEFROL technology, which matches, even exceeds, the rate at which conventional technologies achieve sewage treatment. How the invention was conceived, developed, tested, and validated at bench-scale (Figure 14.1) has been documented in the prior art (Abbasi & Tauseef, 2018a,b,2019; Abbasi et al., 2013, 2018a,b, 2019). Subsequently, several pilot plants of different capacities were set up in the open and operated continuously for several years with sewage coming in real time from human dwellings of different sizes (Abbasi et al., 2024; Bhat et al., 2023, 2021, 2020; Tabassum-Abbasi et al., 2022). Some of these plants are seen in Figures 14.2–14.5. These studies have established not only the efficacy, efficiency, and ease of operation but also the robustness of the SHEFROL® technology because it was able to efficiently handle not only diurnal variations in the sewage flows but also shock loads. At present, the technology

FIGURE 14.1 Set up for bench-scale studies on SHEFROL under controlled conditions. Artificial lights have been used to simulate sunshine.

FIGURE 14.2 A 50,000 L/day SHEFROL pilot plant with duckweed as the main bioagent.

FIGURE 14.3 A 35,000 L/day SHEFROL pilot plant based on salvinia as the main bioagent.

FIGURE 14.4 A 1,10,000 L/day SHEFROL pilot plant based on water hyacinth.

FIGURE 14.5 A 30,000 L/day pilot plant using water hyacinth in the sedimentation tank and *Alternanthera sessilis* in the channels.

FIGURE 14.6 A SHEFROL plant treating all the sewage generated by the town of Hariyawan, UP. The SHEFROL (bounded rectangular formation) is at left, and the pond in which the treated sewage is disposed of is at right.

is open to commercialization, and full-scale plants are being installed at different locations (Figure 14.6).

The acronym SHEFROL® stands for a bioreactor with "sheet-flow, root level" hydrology in which short-statured terrestrial, aquatic, or amphibious plants are used in a densely packed fashion in narrow channels through which sewage is made to flow as a sheet of thickness adequate to cover the plant roots. The dimensions of the channels are optimized to ensure maximum effluent-root contact and minimum short-circuiting. To quote from Bhat et al. (2023),

> A plug-flow operation is approximated, which makes the reactor highly space-efficient. The various design features which enable much more efficient yet inexpensive treatment of sewage in SHEFROL® reactors in comparison to the pre-existing macrophyte-based sewage treatment systems have been detailed earlier (Abbasi et al., 2019). Of the several attributes, the two most significant are:
>
> a. Since the phytoremediator plants are packed to capacity in the SHEFROL® channels, it enables the plants to support each other in remaining erect, obviating the need for any anchor or scaffold for support.
> b. The combination of the sheet-flow-root-level (SHEFROL®) hydrology, the low depth-to-surface-area ratio of the water column, and the turbulence caused when wastewater flows through a dense assemblage of plant roots enables very brisk and efficient aeration. This overcomes one of the most basic shortcomings associated with macrophyte-based wastewater treatment systems as mentioned earlier. Moreover, this advantage is achieved without the need to invest in auxiliary aeration devices.

Indeed, the ability of SHEFROL® to function efficiently without the use of pumps, aerators, or scaffolds reduces the system's ecological footprint to almost negligible. As stated in Bhat et al. (2023), to quote: The system primarily runs on solar energy as captured and utilized by the plants while the dependence of the plants on sewage as their only source of nutrients drives the treatment of the sewage. As a further indication of the novelty of the SHEFROL® invention, a patent has been recently granted to it (Abbasi et al., 2018a). The process has also been trademarked.

> A series of bench-scale studies, with one or other of a wide variety of macrophytes deployed as the phytoremediator, have confirmed the efficacy of SHEFROL® in treating sewage. These studies have shown that SHEFROL® can be run with not only aquatic and amphibious plants as bioagents but also terrestrials. The mechanism by which macrophytes cause wastewater treatment in SHEFROL®, and how the SHEFROL® design maximizes the gains, has been described earlier (Abbasi et al., 2019). It was seen time and again that the SHEFROL® systems could be operated interminably without the need to invest in energy for pumping or recycling. Whatever plant debris was generated could be easily vermicomposted (Hussain et al., 2017, 2018).

A very special attribute of SHEFROL is that, despite being essentially a single-pot process, it enables not only near-total secondary treatment but substantial primary and tertiary treatment as well. All these factors reflect the potential of SHEFROL® as not only a clean-green, simple, and inexpensive option, but also a very quick and versatile option for sewage treatment.

14.4 UPGRADING MACROPHYTE/ALGAE-BASED SEWAGE TREATMENT SYSTEMS INTO BIOREFINERIES

Biorefinery is a term that has gained currency in recent years to cover all aspects of biomass utilization, in which biomass, especially phytowaste, is converted to energy and other beneficial products. As per the International Energy Agency (IEA, 2009), to quote, *biorefining is the sustainable processing of biomass into a spectrum of bio-based products (food, feed, chemicals, materials) and bioenergy (biofuels, power, and/or heat).*

According to the US Department of Energy, to quote, *a biorefinery is a processing plant concept for the extraction of a wide variety of products from biomass feedstock* (Chandra et al., 2018).

Cheruhini (2010) is credited for having introduced the biorefinery concept, basing it on the features of the traditional oil refineries as benchmarks to lessen global oil dependence and mitigate climate change. Later, a classification of biorefineries was proposed by Ubando et al. (2019) based on the type of biomass used as raw material: lignocellulosic, algal, and waste. In reality, there is little novel about "biorefineries" because attempts to use biomass – especially phytomass and phytowaste – for producing liquid, solid, and gaseous fuels, as well as innumerable other byproducts such as chemicals, cosmetics, household furnishing, etc, have been going on intensively for a very long time. From late 1970 onwards, especially after the 1973 and 1979 oil shocks, these efforts have been very intense (Abbasi & Abbasi, 2010a,b; Abbasi et al., 2012).

The belief has been that if we can find commercially viable ways to utilize biomass/biowaste, then we can go on cultivating biomass and using it, *ad infinitum*. But this belief does not make allowance for the fact that biomass can be carbon neutral (having sequestered the carbon in the course of its growth and so not generating any surplus greenhouse gases when it is degraded or burnt), but it is far from being "nutrient-neutral" (Abbasi & Abbasi, 2010a). With every round of biomass harvested from a piece of land, that many nutrients are taken from the corresponding soil.

Moreover, by the mid-1980s, a concept more inclusive and value-adding than the concept of biorefinery had gained prevalence. That was of integrating wastewater treatment with biomass production so that the nutrients present in wastewater can be phytoextracted even as the system treats sewage (and other wastewater) and converts the wastewater to reusable water. It was then aimed to integrate the wastewater treatment step with a downstream step in which the biomass generated from the wastewater is converted to energy and/or other useful products. In a way, the system aimed to treat wastewater while transforming the treatment system into a net profit-making venture rather than one that costs hard-to-afford money. Hence, the world has already been exploring options, of which the more recently coined term *biorefinery* is but a component.

Among the many publications of those times is a voluminous book on the subject (Smith & Reddy, 1987), which describes a myriad of studies addressing this concept. It is based on an international conference held at Orlando, Florida, in 1986, in which one of the authors (SAA) had also presented his research (Abbasi, 1987).

These and other aspects notwithstanding, the coinage of the term biorefinery has given a new spin and attraction to an age-old concept.

Indeed till, as described in Chapters 6–13, these authors have, in recent years, developed the paradigm of high-rate vermicomposting, which has now enabled all kinds of phytomass – even toxic weeds and degradation-resistant xerophytes – to be vermicomposted into benign, plant-friendly, and soil-enriching organic fertilizers, no option was available with which the billions of tonnes of phytomass – whether generated on wastewater or in other ways – could be utilized in a commercially viable manner. The very many options to utilize biomass – as a source of feed, food and solid/liquid/gaseous feeds, drugs, cosmetics, other chemicals, fibres, household bags/trophies/furniture, etc (Abbasi & Abbasi 2010a,b; Abbasi et al., 1986; Tabassum-Abbasi et al., 2019; Smith & Reddy, 1987) – that were fervently explored since long, especially since the 1970s, either did not yield products which were commercially viable or had viable products such as some Ayurvedic, Siddha, homeopathic, or other forms of traditional medicines, and a few allelopathic drugs but in such quantities of demand that only an insignificant portion of the available biomass could be utilized through their manufacture. Only since 2015 (Abbasi et al., 2015) has the situation begun to change due to the copious evidence gathered since then on:

a. the large-scale viability of vermicomposting various species of aquatic, amphibious, and terrestrial phytomass in a direct and rapid manner and
b. on the excellent fertilizer value of the resulting vermicompost, independent of whether the parent substrate was benign or had toxicity/allelopathy.

We have also added a step of extracting volatile fatty acids (VFAs) from the phytomass before it is put for vermicomposting. This step involves soaking the phytomass with cow dung-inoculated water in sequential batch aerobic reactors for a solid retention time of about 5 days. The extracted VFAs serve as energy precursors and can be easily, quantitatively, and rapidly converted to energy in the form of methane-rich biogas in any available anaerobic digester (Abbasi et al., 2012a,b; Tauseef et al., 2013). VFAs comprise chemicals that are entirely made up of carbon, oxygen, and hydrogen. Due to this, no medium or minor nutrients are lost from the phytomass in the step of VFA extraction. On the other hand, soaking in aqueous medium for 5–6 days makes the phytomass soft and withered, making it much more amenable to vermicomposting than unsoaked phytomass.

Given that all the phytomass coming out of the VFA extraction step is quantitatively converted to organic fertilizer, leaving no waste to dispose, which is then added to the soil, resulting in the sequestration of its carbon content, the overall sewage treatment-energy generation–fertilizer production system is not merely carbon neutral but is also carbon negative.

In summary, the success of the high-rate vermicomposting technology, as described in Chapters 6–13, promises to make the biorefinery concept also a likely major success.

REFERENCES

Abbasi, S.A., 1987. A new method for the microdetermination of molybdenum (VI) in steel, biological materials and waters. *Analytical Letters*, *20*(12), pp.2021–2034.

Abbasi, S.A., 1999. Water quality indices – state of the art. *Journal of the Institution of Public Health Engineers*, (1), pp.13–24. https://www.researchgate.net/publication/317824095_Water_quality_indices_-_state_of_the_art

Abbasi, S.A., 2018. The myth and the reality of energy recovery from municipal solid waste. *Energy, Sustainability and Society*, 8(1), 1–15, doi: 10.1186/s13705-018-0175-y.

Abbasi, S.A. and Abbasi, T., 2019. *Current Concerns in Environmental Engineering*, Nova Science Publishers, New York, pages xvi+309, ISBN: 978-1-53613-920-4.

Abbasi, S.A. and Nipaney, P.C., 1986. Infestation by aquatic weeds of the fern genus Salvinia: its status and control. *Environmental Conservation*, *13*(3), pp.235–241.

Abbasi, S.A. and Tauseef, S.M., 2018a. Use of the terrestrial weed Alternanthera ficoidea in treating greywater in soil-less SHEFROL® bioreactors. *Water Science and Technology*, *77*(8), pp.2005–2013.

Abbasi, S.A. and Tauseef, S.M., 2018b. A system for rapid and inexpensive treatment of sewage using the weed Eclipta prostrata in SHEFROL® bioreactor. *Water and Environment Journal*, *32*(4), pp.573–584.

Abbasi, S.A. and Tauseef, S.M., 2019. Rapid treatment of greywater (household sewage) by terrestrial weed Achyranthes aspera in SHEFROL® reactors. *Environmental Progress & Sustainable Energy*, *38*(2), pp.467–476.

Abbasi, S.A., Nayeem-Shah, M. and Abbasi, T., 2015. Vermicomposting of phytomass: limitations of the past approaches and the emerging directions. *Journal of Cleaner Production*, *93*, pp.103–114.

Abbasi, S.A., Ponni, G. and Tauseef, S.M., 2018a. Marsilea quadrifolia: a new bioagent for treating wastewater. *Water, Air, & Soil Pollution*, *229*, pp.1–8.

Abbasi, S.A., Ponni, G. and Taussef, S.M., 2018b. Proficiency of Brahmi (Indian Pennywort) Hydrocotyle asiatica in One-pot Secondary and Tertiary Treatment of Sewage in SHEFROL® System. *Nature Environment and Pollution Technology*, *17*(2), pp.603–609.

Abbasi, S.A., Rahi, R., Abbasi, T., Patnaik, P. and Abbasi, T., 2024. A pilot-scale assessment of five common weeds in the sustainable treatment of sewage utilizing SHEFROL®, with prospects of a closed-loop biorefinery. *International Journal of Phytoremediation*, pp.1–15.

Abbasi, S.A., Tabassum-Abbasi, Ponni, G. and Tauseef, S.M. 2019. Potential of joyweed *alternanthera sessilis* for rapid treatment of domestic sewage in SHEFROL® bioreactor. *International Journal of Phytoremediation*, *21*(2), 160–169. doi:10.1080/15226514.2018.1488814.

Abbasi, T. and Abbasi, S.A., 2010a. Biomass energy and the environmental impacts associated with its production and utilization. *Renewable and Sustainable Energy Reviews*, *14*(3), pp.919–937.

Abbasi, T. and Abbasi, S.A., 2010b. Enhancement in the efficiency of existing oxidation ponds by using aquatic weeds at little or no extra cost—the macrophyte-upgraded oxidation pond (MUOP). *Bioremediation Journal*, *14*(2), pp.67–80.

Abbasi, T., and Abbasi, S.A., 2018. *Perspectives in Pollution Control and Sustainable Development,* Discovery Publishing House, New Delhi; xi+569 pages; ISBN: 978–9–3505–6889-7.

Abbasi, T., Poornima, P., Kannadasan, T. and Abbasi, S.A., 2013. Acidrain: past, present, and future. *International Journal of Environmental Engineering (Inderscience)*, *5*, pp.229–272. https://doi.org/10.1504/IJEE.2013.054703.

Abbasi, T., Tauseef, S.M. and Abbasi, S.A. 2012b. *Biogas Energy*, Springer, New Yark & Londan, xiv+169 Pages; ISBN 978-1-4614-1039-3.

Abbasi, T., Tauseef, S.M. and Abbasi, S.A., 2012a. Anaerobic digestion for global warming control and energy generation—An overview. *Renewable and Sustainable Energy Reviews*, *16*(5), pp.3228–3242.

Arnone, R.D. and Perdek Walling, J., 2007. Waterborne pathogens in urban watersheds. *Journal of Water and Health*, *5*(1), pp.149–162.

Benarde, M.A., 1973. Land disposal and sewage effluent: appraisal of health effects of pathogenic organisms. *Journal-American Water Works Association*, *65*(6), pp.432–440.

Bhat, M.A., Abbasi, T. and Abbasi, S.A., 2020. A household-scale on-site embodiment of the novel bioreactor SHEFROL® for treating greywater. *Taiwan Water Conservancy*, *68*(2), pp.26–35.

Bhat, M.A., Abbasi, T. and Abbasi, S.A., 2021. A pilot plant of SHEFROL® phytoremediation Technology for treating greywater of a typical Indian village. *International Journal of Environmental Pollution and Remediation*, *41*, pp.668–672.

Bhat, M.A., Tabassum-Abbasi, Abbasi, T. and Abbasi, S.A., 2023. An inexpensive phytoremediation system for treating 50,000 L/day of sewage. *International Journal of Phytoremediation*, *25*(8), pp.1029–1041.

Brix, H., 1994. Use of constructed wetlands in water pollution control: historical development, present status, and future perspectives. *Water Science and Technology*, *30*(8), pp.209–223.

Central Pollution Control Board: CPCB, 2024. *Guidelines for reuse of treated sewage in reference to item of circular economy.* Ministry of Environment, Forest & Climate Change, Govt. of India. pp 30. https://cpcb.nic.in/openpdffile.php?id=TmV3c0ZpbGVzLzEwNV8xNzA3Mzc4MTYwX21lZGlhcGhvdG8yMDUzNC5wZGY=. Last accessed December 2024.

Chandra, R., Castillo-Zacarias, C., Delgado, P. and Parra-Saldívar, R., 2018. A biorefinery approach for dairy wastewater treatment and product recovery towards establishing a biorefinery complexity index. *Journal of Cleaner Production*, *183*, pp.1184–1196.

Cherubini, F. and Jungmeier, G., 2010. LCA of a biorefinery concept producing bioethanol, bioenergy, and chemicals from switchgrass. *The International Journal of Life Cycle Assessment*, *15*, pp.53–66.

Esrey, S.A., Potash, J.B., Roberts, L. and Shiff, C., 1991. Effects of improved water supply and sanitation on ascariasis, diarrhoea, dracunculiasis, hookworm infection, schistosomiasis, and trachoma. *Bulletin of the World Health organization*, *69*(5), p.609.

Hazra, M. and Durso, L.M., 2022. Performance efficiency of conventional treatment plants and constructed wetlands towards reduction of antibiotic resistance. *Antibiotics*, *11*(1), p.114.

Hussain, N., Abbasi, T. and Abbasi, S.A., 2017. Enhancement in the productivity of ladies finger (Abelmoschus esculentus) with concomitant pest control by the vermicompost of the weed salvinia (Salvinia molesta, Mitchell). *International Journal of Recycling of Organic Waste in Agriculture*, *6*, pp.335–343.

Hussain, N., Abbasi, T. and Abbasi, S.A., 2018. Generation of highly potent organic fertilizer from pernicious aquatic weed Salvinia molesta. *Environmental Science and Pollution Research*, *25*, pp.4989–5002.

IEA, 2009, Bio-based chemicals, value added products from biorefineries, *IEA Bioenergy-Report*, 36 pages. https://www.ieabioenergy.com/blog/publications/bio-based-chemicals-value-added-products-from-biorefineries/ (Last accessed July 2025)

Putri, F.E. and Hung, T.C., 2020. Comparison of nutrient removal and biomass production between macrophytes and microalgae for treating artificial citrus nursery wastewater. *Journal of Environmental Management*, *264*, p.110303.

Rome, M., Happel, A., Dahlenburg, C., Nicodemus, P., Schott, E., Mueller, S., Lovell, K. and Beighley, R.E., 2023. Application of floating wetlands for the improvement of degraded urban waters: Findings from three multi-year pilot-scale installations. *Science of the Total Environment*, *877*, p.162669.

Sengupta, A., 2018 Water. In: Narain S, (ed) *State of India's Environment*, Centre for Science and Environment, New Delhi.

Smith, W.H. and Reddy, K.R. 1987. *Aquatic Macrophytes for Water Treatment and Resource Recovery*. Magnolia Publishing, Orlando (FL). ISBN 09-414-63001.

Spellman, F.R. 2009. *Water and Wastewater Treatment Plant Operation*. CRC Press, Boca Raton. p. xxxix + 960.

Tabassum-Abbasi, P., and Abbasi, S.A., 2022. A Closed loop wastewater biorefinery based on the recently introduced SHEFROL reactor and two freely available macrophytes. *Clean–Soil, Air, Water, 51*(3), p.2100287. https://doi.org/10.1002/clen.202100287.

Tabassum-Abbasi, Abbasi, T. and Abbasi, S.A., 2019. Microbial fuel cells as source of clean energy – potential and pitfalls. *Nature Environment and Pollution Technology*, *18*(3), pp.789–797. https://neptjournal.com/upload-images/NL-69-12-(10)-B-3554.pdf.

Tang, S., Liao, Y., Xu, Y., Dang, Z., Zhu, X. and Ji, G., 2020a. Microbial coupling mechanisms of nitrogen removal in constructed wetlands: A review. *Bioresource Technology*, *314*, p.123759.

Tang, X., Wen, Y., He, Y., Jiang, H., Dai, X., Bi, X., Wagner, M. and Chen, H., 2020b. Full-scale semi-centralized wastewater treatment facilities for resource recovery: Operation, problems and resolutions. *Water Science and Technology*, *82*(2), pp.303–314.

Tauseef, S.M., Abbasi, T. and Abbasi, S.A., 2013. Energy recovery from wastewaters with high-rate anaerobic digesters. *Renewable and Sustainable Energy Reviews*, *19*, pp.704–741.

Ubando, A.T., Rivera, D.R.T., Chen, W.H. and Culaba, A.B., 2019. A comprehensive review of life cycle assessment (LCA) of microalgal and lignocellulosic bioenergy products from thermochemical processes. *Bioresource Technology*, *291*, p.121837.

Vo, T.K., Vo, T.D., Ntagia, E., Amulya, K., Le, S.L., Le, L.T., Tran, C.S., Ha, T.L., Bui, X.T. and Lens, P.N., 2023. Pilot and full scale applications of floating treatment wetlands for treating diffuse pollution. *Science of the Total Environment*, *899*, p.165595.

Zhang, H., Tang, W., Wang, W., Yin, W., Liu, H., Ma, X., Zhou, Y., Lei, P., Wei, D., Zhang, L. and Liu, C., 2021. A review on China's constructed wetlands in recent three decades: Application and practice. *Journal of Environmental Sciences*, *104*, pp.53–68.

Part III

Use of Black Soldier Flies in Zoocomposting

15 The Unique Role of Black Soldier Flies in the Zoocomposting of Food Waste

15.1 INTRODUCTION

As has been projected in the preceding chapters of this book, the high-rate vermicomposting paradigm and the associated technology developed by these authors have enabled direct vermicomposting of all types of plant biomass (phytomass), besides animal manure. It has made it possible to rapidly vermicompost even toxic/allelopathic weeds like lantana (*Lantana camara*), ipomoea (*Ipomoea crassipes*), parthenium (*Parthenium hysterophorus*), and xerophytes like prosopis (*Prosopis juliflora*), while transforming them into highly benign, plant-friendly, and soil-friendly organic fertilizers. Their benignness and efficacy as fertilizers have been extensively checked by carrying out detailed chemical analysis of the vermicomposts relative to the parent substrates. It has also been tested in terms of the relative effect on the germination, growth, and yield of different species of plants at bench-scale as well as field-scale.

In Chapters 17 and 18, we have described our novel studies which show that termites can be harnessed to make them decompose – or "termicompost/termigrade" – different types of ligninous and "hard" biowaste which neither earthworms can process nor processes like anaerobic digestion aimed at generating liquid or gaseous fuels can handle. Examples of such biowastes are thermocol, coir pith, and press mud.

But neither earthworms nor termites are capable of feeding upon cooked food or abattoir waste (Figures 15.1–15.4) to stabilize it. Nor are such waste streams amenable to other processes like anaerobic digestion, with which animal manure, sewage, and several other types of biowastes can be stabilized. The main reasons are the presence of high concentrations of proteinous substances and fat in the abattoir waste, and of oil, grease, and spices in the cooked food (plus animal protein if the food is non-vegetarian). None of these are palatable to earthworms or termites, and all of them not only resist decomposition in aerobic or anaerobic digesters, but tend to make the digesters malfunction.

As a result of this, large portions of waste cooked food and abattoir waste are disposed of untreated in drains meant to carry only sewage, or on land, forming one of the major sources of pollution of the global environment (Figures 15.5 and 15.6).

DOI: 10.1201/9780429331886-18

FIGURE 15.1 Typical pile of cooked–uncooked food waste. (https://commons.wikimedia.org/wiki/File:Food_waste_with_moldy_bread.jpg.)

FIGURE 15.2 Piles of food waste awaiting disposal. (https://commons.wikimedia.org/wiki/File:Food_waste_for_sewage_oil.jpg.)

FIGURE 15.3 Miscellaneous constituents of abattoir waste. (https://upload.wikimedia.org/wikipedia/commons/a/a8/Solid_Wastes_from_Slaughterhouse.jpg.)

15.2 FOOD WASTE AND ABATTOIR WASTE

Food waste has been defined as "food that is of good quality and fit for human consumption but that does not get consumed because it is discarded either before or after it spoils" (Figures 15.7 and 15.8). Food waste typically, but not exclusively, occurs at the retail and consumption stages in the food value chain and is the result of negligence or a conscious decision to throw food away (Brian et al., 2013). Globally, about one-third of the food produced ends up as waste and constitutes a major fraction of the municipal solid waste (Varelas, 2019). In developing countries like India, these highly putrefiable wastes pose a major threat to environmental health (Thi et al., 2015). It has been reported that citizens of developed and developing nations generate 107 kg/year and 56 kg/year of food waste per capita, respectively. Next to China, India is the biggest contributor of food waste in the world (Figure 15.9). About 51% of municipal solid waste (MSW) produced in India is comprised of food waste (Thi et al., 2015). The global food waste generation has been estimated to increase by 44% between 2005 and 2025 (Antonis, 2013). In India, food waste is mostly dumped in the landfills (Figure 15.10), which leads to the release of methane, carbon dioxide, and other toxic and/or global-warming gases. It also contributes to the proliferation of pests and diseases. Only a small fraction (5%–6%) of the food waste is bioprocessed either by animal feeding, composting, or anaerobic digestion. But all three of these options fall short in processing various kinds of food waste, and also take an impractically long time to accomplish.

FIGURE 15.4 Examples of very difficult-to-manage constituents of abattoir waste. (https://commons.wikimedia.org/wiki/File:307_Meats_Slaughterhouse_-_53259185977.jpg.)

As for abattoir waste, it comprises animal faeces, blood, bone, fat, animal trimmings, paunch content, and urine from operations or areas like lairage, stunning or bleeding, carcass processing, and by-product processing (Figures 15.11 and 15.12). These constituents pose a serious threat to the environment, public health, and animal health, thereby contaminating natural resources and stressing the economy of the country. The known management methods for these wastes are burial, composting, incineration, anaerobic digestion, and blood processing (Bandaw & Herago, 2017). But most of these methods are too expensive to use. Worse, most lead to newer problems of pollution, proving remedies worse than the disease.

FIGURE 15.5 Insanitary conditions created by food waste piled on the roadside. (https://commons.wikimedia.org/wiki/File:Food_waste_onto_small_truck.jpg.)

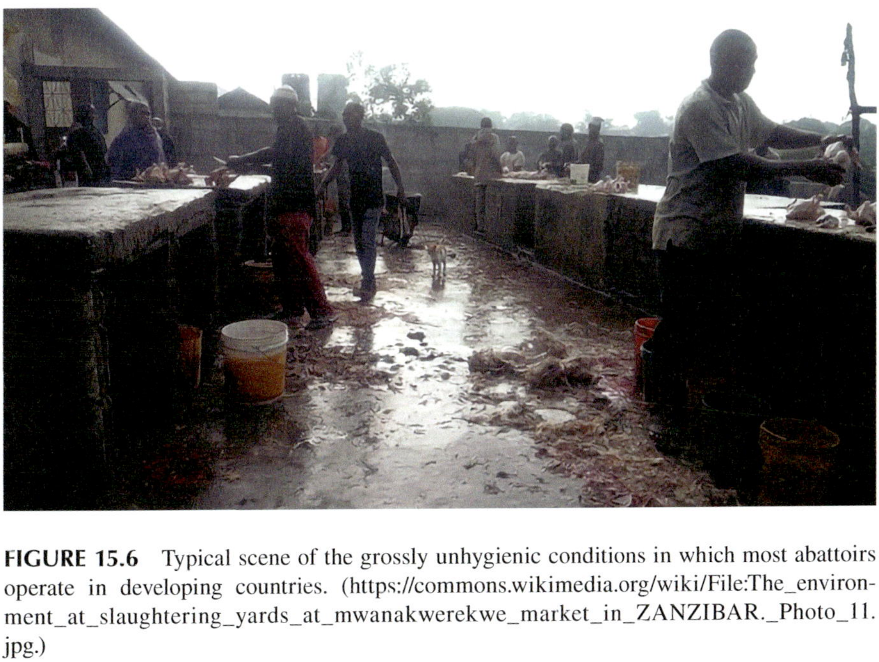

FIGURE 15.6 Typical scene of the grossly unhygienic conditions in which most abattoirs operate in developing countries. (https://commons.wikimedia.org/wiki/File:The_environment_at_slaughtering_yards_at_mwanakwerekwe_market_in_ZANZIBAR._Photo_11.jpg.)

FIGURE 15.7 A discarded pile of food waste with several constituents still fit for human consumption. (https://commons.wikimedia.org/wiki/File:Trashed_vegetables_in_Luxembourg.jpeg.)

15.3 POTENTIAL OF INSECT-BASED SYSTEMS FOR THE MANAGEMENT OF FOOD WASTE AND ABATTOIR WASTE

As was explained in Chapter 1, many insects carry out composting in nature by feeding on decaying organic wastes. This provides nutrients to the insects while reducing the amount of waste material and stabilizing most of it into easily bioavailable nutrients. Coprophagous and carrion-feeding flies play a particularly important role in the recycling of organic matter in nature. The increasing realization of this potential of using insects to consume organic waste and convert it into food, feed, biofuels, fertilizer, and other valuable secondary products is gaining momentum both as a research curiosity and as a business opportunity. Insect rearing has a lower environmental impact than rearing of conventional livestock, with much lower greenhouse gas emissions, much greater water and space use efficiency, and much higher feed conversion rates (Premalatha et al., 2011; Tabassum-Abbasi et al., 2016; Varelas, 2019).

Due to their traditional role as insects that feed upon, and breed in, carrion, several species of flies, including house fly (*Musca domestica*), black soldier fly (BSF, *Hermetia illucens*), green bottle fly (*Lucilia sericata*), face fly (*Musca autumnalis*), common flesh fly (*Sarcophaga carnaria*), and stable fly (*Stomoxys calcitrans*)

FIGURE 15.8 An illustrative example of how massive quantities of food items are allowed to go to waste. (https://commons.wikimedia.org/wiki/File:US_Navy_100626-N-9634R-134_Seaman_Fridah_N._Kelley_wades_through_a_dumpster_of_expired_bread_that_will_be_given_to_local_farmers_to_use_as_compost_during_a_community_service_project_at_the_Feed_Nova_Scotia_food_bank.jpg.)

(Figures 15.12–15.18), have been explored as food waste composters. But except for BSF, all other species are pests in one form or another. Some are ectoparasites; some tend to spread diseases while they bite, suck blood, or mechanically spread disease vectors around (Cickova et al., 2015; Brits et al., 2016; Dehghani et al., 2018). For instance, the stable fly (*S. calcitrans*) – also called barn fly, biting fly, and dog fly (Figure 15.18) – can infest cattle so heavily as to make them anaemic. It bites humans as well and carries with it trypanosomatid parasites. It is a vector for several ailments in humans and other animals. Likewise, *L. sericata*, the green bottle fly (Figure 15.15), is a scourge of animals, especially sheep. It can colonize human body

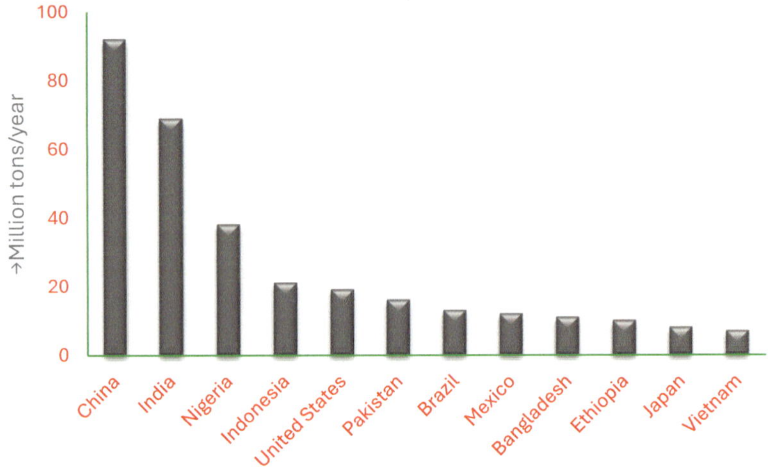

FIGURE 15.9 Manual household food waste by some countries.

FIGURE 15.10 Landfills receive a lot of food waste, which is dumped mixed with other forms of municipal solid waste. (https://commons.wikimedia.org/wiki/File:Basura_en_el_relleno_sanitario_de_onia.jpg.)

FIGURE 15.11 Various constituents of abattoir waste (https://commons.wikimedia.org/wiki/File:Exe_works.jpg.)

FIGURE 15.12 Complexity of abattoir waste. (https://commons.wikimedia.org/wiki/File:Abattoir-veau-mort.jpg.)

FIGURE 15.13 The house fly (*Musca domestica*). (https://commons.wikimedia.org/wiki/File:Musca.domestica.-.lindsey.jpg.)

FIGURE 15.14 The BSF (*Hermetia illucens*). (https://commons.wikimedia.org/wiki/File:Hermetia_illucens_Black_soldier_fly_edit1.jpg.)

FIGURE 15.15 The green bottle fly (*Lucilia sericata*). (https://commons.wikimedia.org/wiki/File:Lucilia_sericata_8663.jpg.)

FIGURE 15.16 The face fly (*Musca autumnalis*). (https://commons.wikimedia.org/wiki/File:Musca.autumnalis.female.jpg.)

FIGURE 15.17 The common flesh fly (*Sarcophaga carnaria*). (https://commons.wikimedia.org/wiki/File:Graue_Fleischfliege.JPG.)

FIGURE 15.18 The stable fly (*Stomoxys calcitrans*). (https://commons.wikimedia.org/wiki/File:Stomoxys_calcitrans_02.JPG.)

parts (Bank, 1964; Cetinkaya et al., 2008). Because of these disadvantages, these species have not found large-scale deployment as processors of food and abattoir wastes.

Only BSF among the family of flies has the ability to clearly and efficiently process various types of food wastes, abattoir wastes, fish offal, raw fruit and vegetable waste, cadavers, manure, faeces, plant residues etc – which cannot be processed by other methods – and without posing hazards to the environment and the health of humans and other animals. The uniqueness of the BSF as zoocomposters has been described in Section 15.6, but before it, we present an explanation of why the word "soldier" has been associated with the common name of *H. illucens.*

15.4 WHY IS BSF SO CALLED?

As we will see by-and-by, BSF is the most harmless, laid-back, trouble-free, accommodating, and almost sluggish member of the fly family. Nor does it come barging into our houses, nor has any war-like temperament been witnessed in this species. Why, then, has BSF been given the name "soldier"?

Before exploring the answer, let us get an idea of what a BSF looks like. BSF (*Hermetia illucens*) belongs to the family stratiomyidae of the order Diptera. Its adults (Figure 15.14) are about 1.5 cm long, making it a medium-sized fly species. Its body is predominantly black, giving reflections ranging from blue to green, which create a metallic ambiance. Its head is wide, carrying antennae about twice as long. The legs are black with whitish tarsi, and the wings are membrane-like. When in flight, the BSF wings generate a loud, buzzing noise, which can be mistaken for a wasp in flight. But that mimicry is for self-preservation, to scare away predators. Otherwise, the BSF flight causes no harm. Nor does a BSF bite or sting others. Then why the "soldier" sobriquet?

L.C. Maill, in the 1912 book, *The Natural History of Aquatic Insects*, writes that in the case of BSF, "from the back of the thorax two spines stand out towards the abdomen [this is a characteristic of adults in the genus *Stratiomys*], and hence the name (*Stratiomys*), which means the 'armed fly'". Earlier entomologists like Jan Swammerdam (1637–1680) had made similar conjectures. Indeed, in Europe, the soldier sobriquet so fascinated the entomologists that it was extended to give a military-like hierarchy to BSF with names like the *Silver Colonel*, the *Barred Green Colonel*, and the *Ornate Brigadier.* Apparently, the stripes present on the BSF body also reminded people of "gaudy military uniforms".

15.5 THE UNIQUENESS OF THE BSF

There are several attributes of BSF which make it very special for the humankind (Wang & Shelomi, 2017; Singh & Kumari, 2019; Bassa et al., 2020; Tepper et al., 2024):

- To start with, due to their larger size, BSF larvae outcompete the larvae of much less human-friendly houseflies and blowflies from colonizing decaying organic matter (Figure 15.19). This is achieved by the ability of the BSF

FIGURE 15.19 Larvae monopolize food waste, to the elimination of the larvae of other species. (https://commons.wikimedia.org/wiki/File:Hermetia_illucens,_Poznan.jpg.)

larvae to consume larvae of other fly species. A major advantage of it is that it drastically reduces the stench that is associated with the consumption of decaying organic matter by houseflies and blowflies.

- Most species of housefly are disease vectors, while BSF is not. It is because BSF has a much smaller sponging mouthpart than possessed by the housefly species, enabling it to only consume liquids such as nectar of flowers (Figure 15.20). It can't eat, nor regurgitate food, as a housefly does. This prevents BSF from spreading diseases. Unlike houseflies and other pests, BSF do not encroach human dwellings in search of food. This prevents them from being an intruder and being considered a nuisance to humankind, which many other insects become in human-dominated areas. Being detritivores and coprophagous, the egg-bearing BSF females are only attracted by rotting food or degrading manure.
- BSF is sedentary and sluggish in its habits, not taken to flying around as houseflies and other species of insects do. This is because BSF can have only a limited food intake due to its inability to have anything except a liquid diet. Such a diet provides it with much less expandable energy than is available to other species of winged insects. BSF is also docile, making it easy to catch and relocate it because it does not resist being picked up, nor bite or sting when touched.
- It is the larvae of BSF (Figure 15.20) that carry out the zoocomposting of the organic waste. While consuming the waste, BSF larvae kill *E. coli,*

FIGURE 15.20 A BSF feeding on nectar. (https://commons.wikimedia.org/wiki/File:Hermetia_illucens_MHNT_Fronton.jpg.)

Salmonella enterica, and other pathogens and also biodegrade harmful organic compounds present in the waste into benign products. BSF larvae are among the most efficient waste processors among the animals, feeding voraciously as they metamorphose into pupae through six instars. Upon reaching the pre-pupal stage (Figure 15.21), they stop foraging, clear their gut of food, and tend to migrate towards cool, dark, and dry substrates to pupate. This migration instinct, similar to the one that induces earthworms to migrate to darker, adequately moist, and richer food sources, is harnessed to accomplish self-harvesting of the BSF pupae. Holes provided with ramps and/or holes on the upper sides of the waste processing chambers allow the BSF pupae to climb out of the composter and drop into collection bins.

- BSF-driven zoocomposting accompanies rapid reduction of the volume and the mass of the waste. The larvae break apart the biowaste and churn it, leading to accelerated decomposition and concomitant generation of heat. This causes rapid water loss via evaporation and rapid conversion of organic carbon into CO_2. The bulk of the waste is soon reduced by half.
- While nutrition-rich pupae form the main product of the BSF-driven zoocomposting of waste (Gardner, 2023; Amrul et al., 2022; Siddique et al., 2022), it also yields a granulated and odourless residue, called "frass" (Figure 15.22). The BSF frass is predominantly larval excreta, with some larval exuviae, and unconsumed feed (Lomonaco et al., 2024). It is an excellent organic fertilizer (Rehman et al., 2023; Lopes et al., 2022),

FIGURE 15.21 The "pre-pupa" which is the sixth instar larva of the BSF. (https://commons.wikimedia.org/wiki/File:Hermetia_Illucens_(Larve_sixieme_stade)_(sixth_instar_larva).jpg.)

FIGURE 15.22 Freshly generated BSF frass. (https://commons.wikimedia.org/wiki/File:Wet_black_soldier_fly_frass.jpg.)

capable of supplying easily bioavailable nutrients and organic matter to the soil, modifying soil micoflora, and positively impacting plant behaviour (Schmitt and de Vrines, 2020). Additionally, BSF pupae possess chitin, which can be extracted and used as an anti-fouling agent, a water purifier, and a raw material for several drugs, cosmetics (Suryati et al.,

2023), and other chemicals (Rana et al., 2015; Khayrva et al., 2019; Xiu et al., 2021; Kaczor et al., 2023). They are also being explored for possible use in making degradable plastics (Wooley and Tibbetts, 2023) and biodiesel (Jung et al., 2022; Tettamanti and Bruno, 2024). It is no wonder that the number of research papers per year on BSF has risen 350 times during the last 11 years (Kaczor et al., 2023), while the number of patents granted per year on BSF use has grown similarly large, especially in China.

15.6 NATIVITY AND DISTRIBUTION OF BSF

As is currently agreed (Niass et al., 2021), BSF has its origins in the Americas, more specifically in the South of the USA. But, like very many other species of plants and animals, BSF is now widely distributed across most of the world, thriving as well in temperate regions as it does in the tropics. Indeed, the ranges of environmental conditions – especially temperature, light, humidity, and weather shifts – that BSF can tolerate are very broad (Kuppusamy et al., 2020). Due to this adaptability, BSF is now present, as seen in Figure 15.23, on five continents (Oliveira et al., 2015).

When any plant or animal species attains such widespread distribution, there is always a possibility of it becoming overly invasive, colonizing, and harmful in at least some of the regions. Such fears have been expressed for BSF also (Demetriou et al., 2022), but as of now, no evidence exists of BSF becoming a nuisance in any region it has moved into. What has been documented is the positive role BSF plays in suppressing the reproduction of the otherwise very abundant and harmful housefly *M. domestica*, thereby functioning as an agent for the biological control of *M. domestica* (Furman et al., 1959).

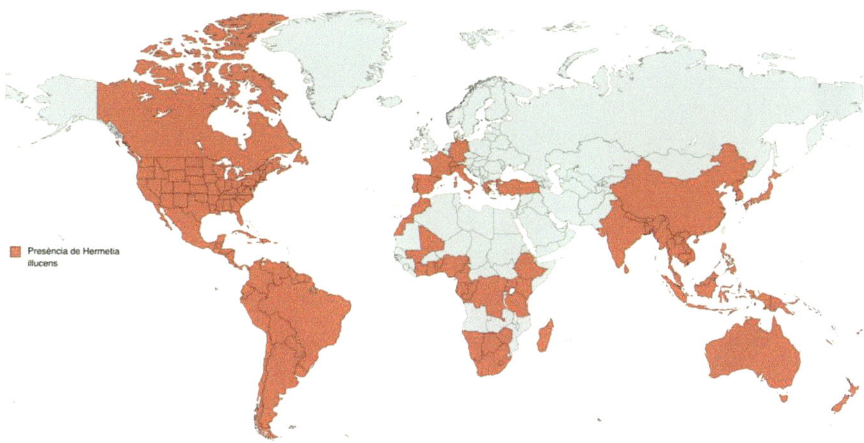

FIGURE 15.23 BSF has spread to almost two-thirds of the world's land area. (https://zh.wikipedia.org/zh-cn/File:Distribucio_hermetia_illucens.png.)

15.7 THE BSF LIFE CYCLE

In ideal conditions of the environment and the waste-to-process, the BSF life cycle (Figure 15.24) spans 41 ± 4 days. The active zoocomposting phase is carried out in about half of this duration. In other words, if deployed in near-ideal conditions, BSF achieves conversion of waste to pupal zoomass and frass – which are the two main products of the process and are a raw material for various other products – in about 3 weeks.

The BSF life cycle (Smets et al., 2020; Tran et al., 2022) begins with the pre-mating selection and then the mating of adult males and females. Each female lays 500–1000 eggs close to, though not upon, the rotting biowaste. BSF, being an eurygamous insect that mates during flight, needs broad open areas to be able to mate before the female begins laying eggs in dry crevices near the food source (Surendar et al., 2020). The eggs incubate for 3–4 days and hatch into the BSF

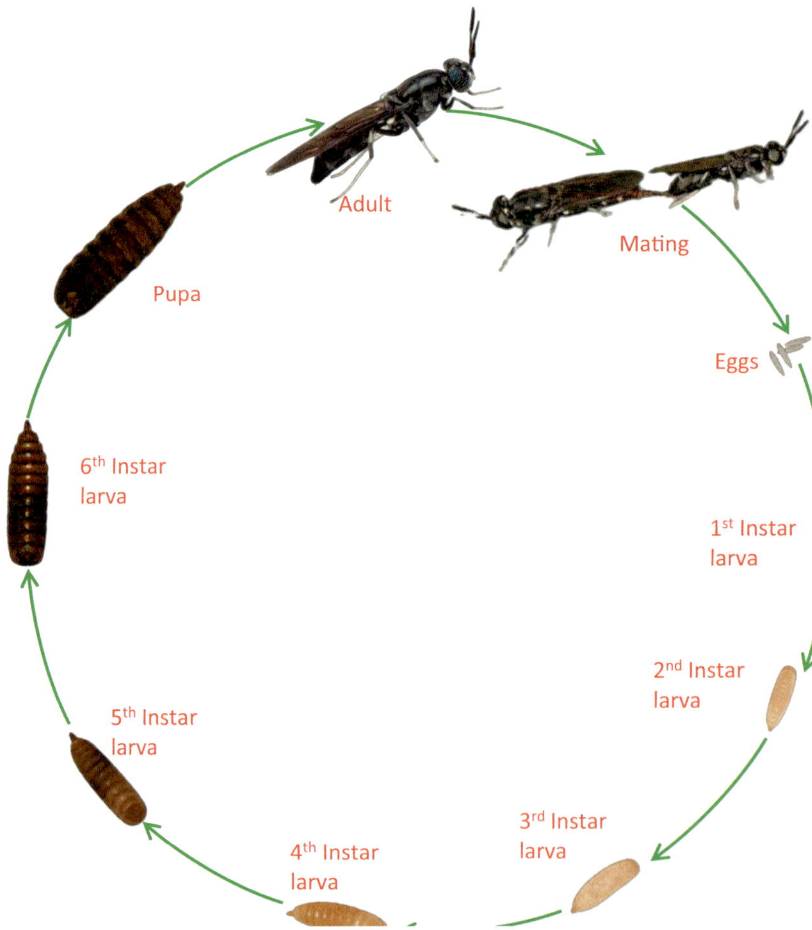

FIGURE 15.24 The BSF life cycle.

larvae. This is the first instar, 5–19 mm long, with lipid and protein content of the order of 31% and 41%, respectively. These larvae begin to feed upon the waste in right earnest, gaining weight through five instars, rapidly growing and metamorphing even as they voraciously consume the biowaste and transform it into carbon dioxide, larval zoomass, and frass. In the course of this series of transitions, every new instar is bigger and more intensely coloured than the previous one. The fifth instar is the final stage of larval metamorphosis. At this stage, it may reach 27 and 6 mm in length and width, respectively, and can weigh up to 225 mg (Tomberlin et al., 2023; Makkaer et al., 2014). It then moves to the pre-pupal stage, which lasts for 7–10 days. In this time span, the pre-pupa stops feeding. This means that the stage during which the biowaste was consumed began and ended with the larval stage of the BSF. This notwithstanding, the valorization of the biowaste continues, because as the pre-pupae grow to pupae, it accompanies the formation of chitin and melanin, even as the progressively darkening colour of the instars and the pre-pupae now acquire a distinctly dark hue. Eventually, the pupae shed their exoskeletons upon completing metamorphosis into BSF. It is these exoskeletons that are rich in chitin and melanin, providing raw material for obtaining these chemicals. They are particularly rich sources of chitin, which makes up about 35% of the BSF exoskeleton mass (Xiong et al., 2023).

It must be emphasized that the 5–6 week of time span shown above to cover the life cycle of BSF corresponds to ideal ambient conditions of warmth and humidity and the luxury availability of the waste to feed upon. Otherwise, the life cycle may take longer, going up to 16 weeks or more (Furman, 1959; Diener et al., 2011). If the larvae have to brave harsh winters, they hibernate, slowing down their growth to the extent of taking 9 months to complete their metamorphosis into the pre-pupa stage.

15.8 DESIGNING BSF-BASED BIOWASTE PROCESSING FACILITIES AT DIFFERENT SCALES: GENERAL CONSIDERATIONS

From the study of the life cycle of BSF, the following pointers emerge for designing BSF-based food processing facilities at different scales of operation.

1. A ploy is needed to attract BSF towards the zoocomposting facility.
2. There is an imperative need to control ambient temperature, humidity, light, moisture content, and pH to near-ideal levels, as it is the most essential requirement to ensure the completion of each BSF life cycle in the shortest-possible duration (Aval et al., 2022; Salam et al., 2022; Siddiqui et al., 2023). A waste stream of reasonably consistent ingredients – for example, mixed food waste from different eateries – also adds to the zoocomposting efficiency. These controls would, in turn, ensure that the zoocomposting phase, which begins with the hatching of BSF eggs and ends with the larvae reaching the pre-pupal stage, occurs at its highest rate. The machinery needed to install such controls and maintain them may not be viable at household scales, but it can be accommodated with increasing ease at progressively larger scales.

3. It must be ensured that the speed with which waste is provided for processing matches the speed of the BSF life cycle. If in any given situation the supply of the waste is faster than the speed with which larvae can consume it in a given container, it may lead to overcrowding of larvae – with associated stunting – as well as waste pile up. On the other hand, if the supply is too slow, the reactors may be running sub-optimally.
4. The use of the migratory instinct of the BSF pre-pupae should be maximized to facilitate their harvesting. The emphasis is to design the exit routes of the pre-pupae in the zoocomposting vessels so that the pre-pupae easily move out and get dropped into designated containers.

   In pursuit of these goals, a huge variety of designs have been proposed for BSF-based zoocomposting facilities, as discussed below.

15.9 AN OVERVIEW OF THE DESIGN APPROACHES

Presented below is an overview of the plethora of different design approaches and their combinations – with or without added innovations – that are available in the prior art on BSF-based zoocomposting systems. Some authors, for instance Eriksson (2021), have described in great detail, with elaborate examples of the relevant calculations, the finesse that can go into the design of a BSF-zoocomposting facility. The CAD models of their final designs of the key components of their BSF facility are reproduced in Figures 15.25–15.29.

At the Swedish University of Agricultural Sciences (SLUS), Dr Cesila Lalander and her team have designed and are operating a research facility in which 200–250 kg of waste is processed by BSF per day. The group conducts BSF rearing trials on different biowaste streams (canteen waste, brewery spent grains, abattoir waste, vegetable cuttings, etc) and assesses the process efficiency in terms of waste-to-biomass conversion efficiency, material reduction, larval survival, final larval weight, ventilation requirement, etc. In addition, they assess the hygienic aspects of the process. They also conduct research on the generated frass to assess its fertilizer potential and other impacts on plant health. The larvae are evaluated for their feed quality for primarily fish and poultry, in both live and post-processed (dried, as a protein meal) forms. Other ways of utilizing the larvae are also explored, or different post-treatment methods of the larvae are assessed. Some of the facility's key units are depicted in Figures 15.30–15.33. The facility uses the separator designed by Eriksson (2021), as seen in Figures 15.28 and 15.29, as do some other BSF-based Swedish companies like Ragn-Sells. Glimpses of one of the Ragn-Sells separators can be had from Figures 15.34 and 15.35.

Most large-scale systems rely on artificial rearing of BSF, but the ones using natural fly populations also exist. They are mostly used to treat different kinds of manure. Such systems are usually simple, made up of concrete pits or lagoons, with a ramp. While these systems are relatively inexpensive, they have little provision for process control and can become easily infected by undesirable insects and microflora (Sheppard et al., 1994; Cickova et al., 2015).

Artificial rearing and mechanized waste processing systems developed so far have mostly targeted the treatment of fermented agricultural by-products, raw fruit waste,

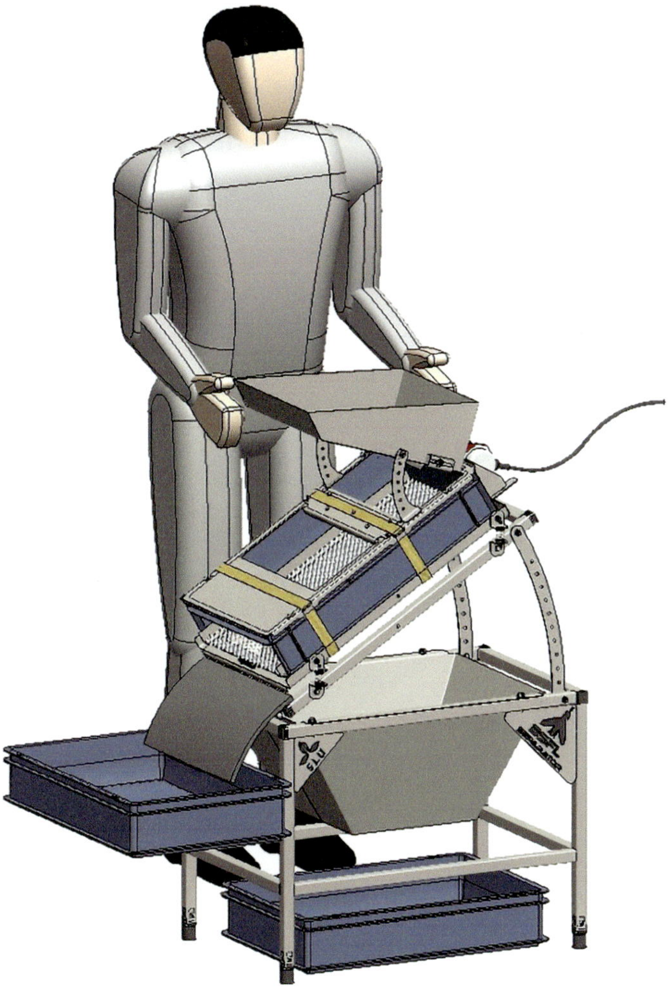

FIGURE 15.25 CAD model of the new separator with a 183 cm tall human CAD model for size and scale comparison. (Image credit: Ismail Eriksson, 2021.)

vegetable waste, and other types of food waste. Most of these systems have mating cum egg laying chambers, egg-hatching chambers, larval rearing cum waste processing systems, and provisions for the pre-pupae migration aided by artificial lights. The feed is pre-processed and then fed to the reactors. The systems are extensively automated, mechanized, climate-controlled, and aerated (Cickova et al., 2015; Cheng et al., 2017; Kumar et al., 2018; Lalander et al., 2019).

Newton et al. (2005) suggest that pit-based manure waste treatment systems can be provided with conveyor belts for collection, a ramp for migration, an efficient drainage system, and an additional greenhouse for adult flies. A medium-scale biodegradation plant in Costa Rica set up by Diener et al. (2011) maintains BSF in a netted greenhouse, and waste processing is done in zinc-coated steel trays with

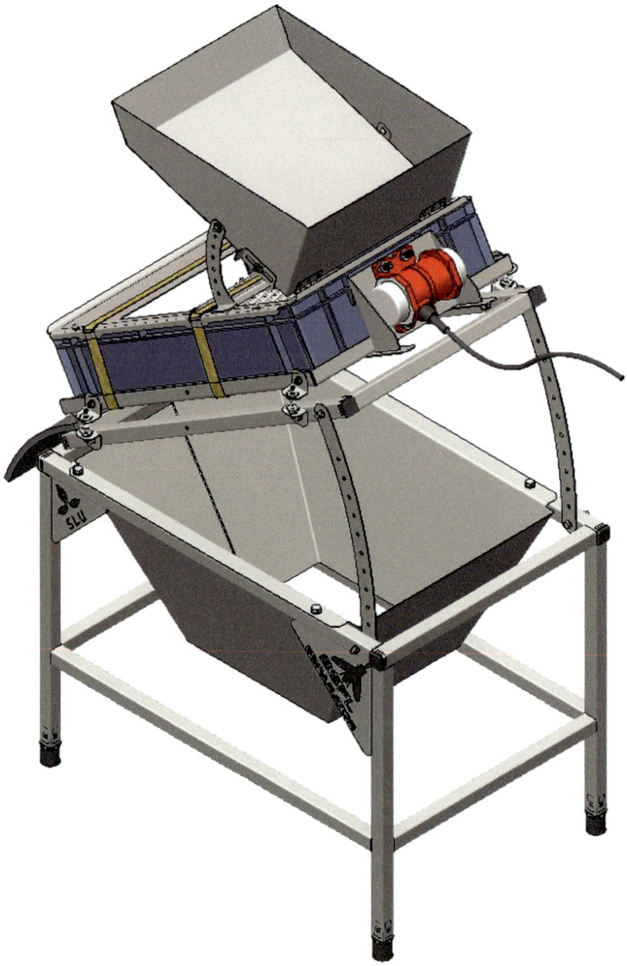

FIGURE 15.26 CAD model of the new vibrating sieve set at 30.75° operation angle. (Image credit: Ismail Eriksson, 2021.)

two ramps. Lee (2011) has developed a BSF breeding farm made of net, consisting of breeding induction apparatus having multiple grooves, induction medium like foam and wood, air conditioners, illuminators, etc. Similar breeding apparatuses have been developed by other authors having containers or space/house covered by mesh (KR101123249B1, 2010; CN203913000U, 2014; CN206043152U, 2016; KR101846282B1, 2017; CN207911828, 2018; CN208390638U, 2018).

Eby and Morgan (1977) have described a mechanized system consisting of a rotating larva culture drum reactor linked to a perforated screen belt for harvesting the larvae using heat lamps. Roller-type reactors with an air inlet and screen mesh have also been developed (CN102351394A, 2011; CN20674233U, 2017; CN207135997U, 2017). Perednia (2015) reports a semi-enclosed rotating barrel system for feeding,

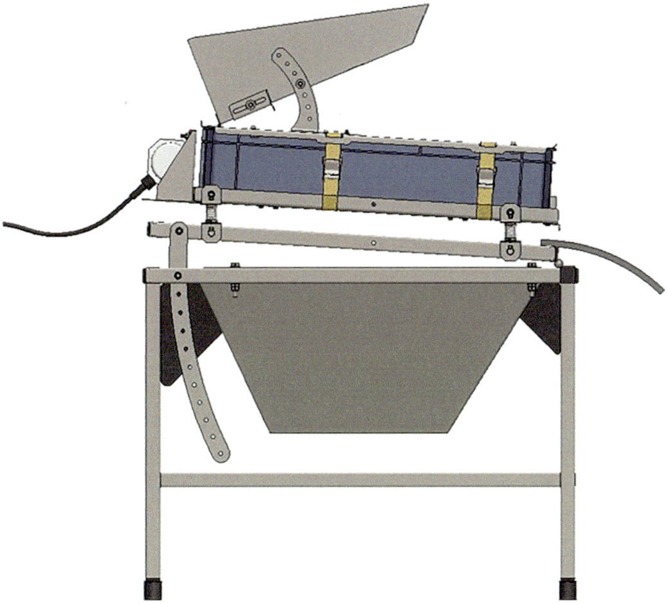

FIGURE 15.27 CAD model of the new vibrating table set at 3.5° operation angle. (Image credit: Ismail Eriksson, 2021.)

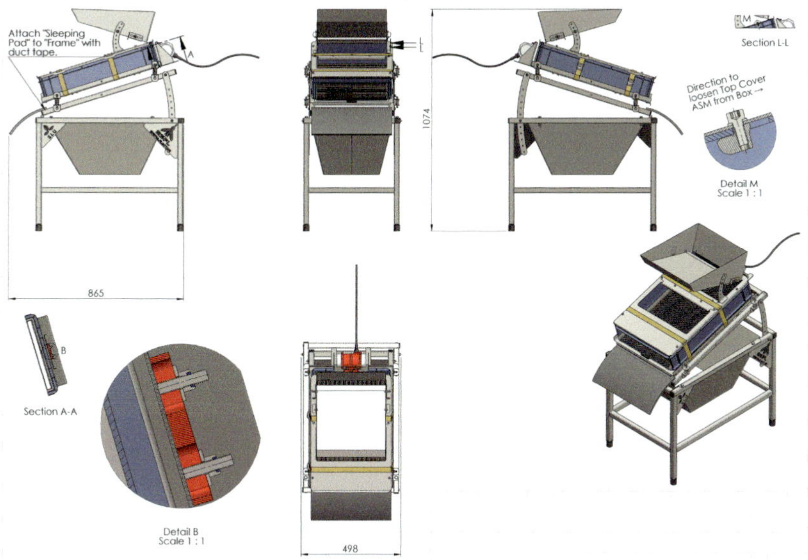

FIGURE 15.28 Sheet 1 of Separator ASM technical assembly drawing. (Image credit: Ismail Eriksson, 2021.)

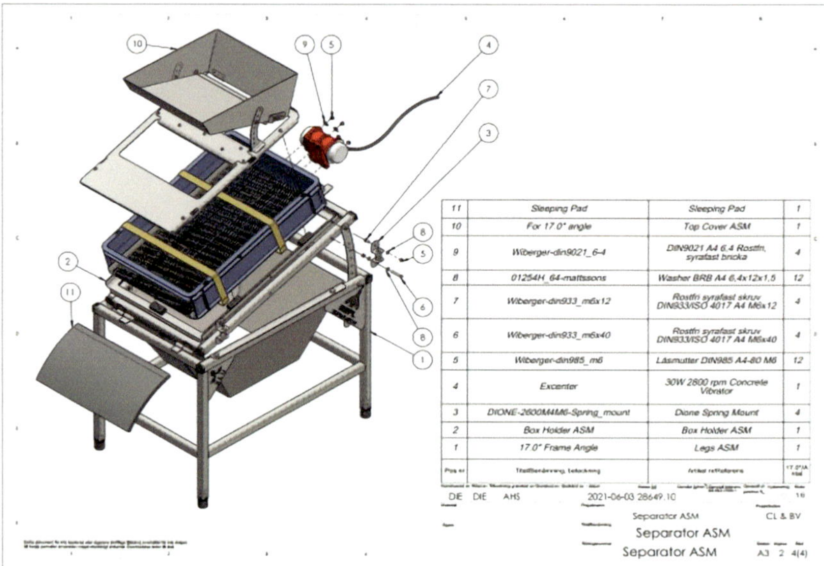

FIGURE 15.29 Sheet 4 of Separator ASM technical assembly drawing. (Image credit: Ismail Eriksson, 2021.)

FIGURE 15.30 Exterior of the BSF-based zoocomposting facility at the Swedish University of Agricultural Sciences (SUAL). (Image Credit: Dr C. Lalander, SUAL.)

FIGURE 15.31 BSF colonies at the SUAL facility. (Image Credit: Dr C. Lalander, SUAL.)

culturing, and harvesting BSF larvae. The barrel has removable conical ends for larval migration and an automated feeder. A similar apparatus with a conical cover has been developed for processing organic wastes (WO214137179A1, 2014). Olivier (1998, 1999) has described a mechanized and automated system having a conveyor belt equipped with a device for adding and distributing ground waste to the belt. It is followed by a means of depositing fly eggs onto the waste and a means of removing the larvae from the waste with the help of infrared lights, and off the belt. Additionally, the system has been provided with a means of removing the treated waste from the belt. On similar lines, systems for rearing and breeding of BSF larvae have been developed by Urbanski et al. (2019) and have been patented (KR101339105B1, 2013; CN206051041U, 2016; CN107497833A, 2017; CN207271797U, 2017; CN208159863U, 2018). All consist of an autonomous conveyor belt made of plastic with inwardly bending lateral side walls on both sides. The trough profile also comprises a feeding supply system.

The system developed by Olivier (2002, 2003) consists of a disposal track (with bins) where waste is processed by larvae at the surface, and treated material is removed from the bottom using subsurface scrapers, so that treated material is removed without disturbing the larvae. The frass is further treated using composting or vermicomposting. Olivier (2004) has designed another apparatus for converting putrescent wastes by utilizing round culture containers without moving parts, with ramps for collecting migrating BSF larvae.

FIGURE 15.32 Zoocomposting in progress at the SUAL facility. (Image Credit: Dr C. Lalander, SUAL.)

There is another class of systems that are based on stackable trays for incubation, growth of insect larvae, and waste processing in a controlled environment. These systems consist of automatic waste shredding, grinding, pasteurizing, and delivery units. Mechanized collection of frass and its post-processing using dryers is done. Some systems also include a mating and oviposition chamber, a pre-pupae chamber, and artificial lights to facilitate larval migration to the larval collection container (Kappelt & Levenhagen, 1998; KR101762186B1, 2016; Newton & Sheppard, 2017; Vickerson et al., 2017; Marchant et al., 2018; Lolley, 2020; Fotiadis, 2020). Popa and Nealson (2018) describe a system for mass propagation of BSF larvae. It consists of

FIGURE 15.33 Waste being delivered to the SUAL facility. (Image Credit: Dr C. Lalander, SUAL.)

larval incubation containers in the form of trays having a larva bedding covered by a net. The trays can be stacked in racks and are provided with spreaders, aerators, pre-pupae collection boxes, and introduced funnels. The feed is ground and sieved before being fed to the reactors. The apparatus by El-Massri (2019) consists of a compartmentalized tray having a pivot along one side and a sloping wall with a lip along the opposite side. It also consists of a feeding system and a conveyor belt for removing frass.

Reactors comprising tanks/boxes with removable covers, ramps or slopes for larval migration, heat lamps to induce migration, automatic waste distributing and frass/ larvae collection systems, air conditioning, and temperature control systems, as well as leachate collection systems have also been designed to derive maximum benefits from BSF-mediated zoocomposting (KR20090121779A, 2008; Popa and Green, 2012; CN203327766U, 2013; CN203648978U, 2013; WO2017173849A1, 2016; Caprio et al., 2016; Villamar, 2017; CN207322428U, 2017; CN207444048U, 2017; CN207678686U, 2017; CN108739684A, 2018; CN208159865U, 2018; CN208390636U, 2018; Popa & Nealson, 2019).

Coming to highly mechanized systems with automatically functioning parts, Alrayya (2019) has developed a highly sophisticated, automated system for breeding and rearing BSF on a large scale. It consists of various chambers which are placed one above the other and are interconnected. The system includes a large hollow treatment chamber that receives ground feed and larvae, and is connected to a smaller

FIGURE 15.34 The pupae separator, designed by Eriksson (2021), operating the Ragn-Sells facility at Orsa, Sweden. (Image Credit: Audinisa Fadhila, Ragn-Sells.)

dark pupation chamber placed above it. The dark chamber has an opening to the meshed fly chamber, having a light source and vibrator, for mating and egg laying in the bio-balls. The bio-balls can be operatively moved to the hatching chamber. All the chambers are provided with air circulation and environmental control systems. Similarly, Whitaker and Hillman (2020) have developed a modular system for breeding and harvesting flies consisting of an egg-growth chamber, larval chamber, pupation chamber, release box, and breeding chamber. The organic waste treatment system invented by Bakhsh (2019) consists of a tower having a first and second section. Both sections have many floors, wherein the first floor and its section receive

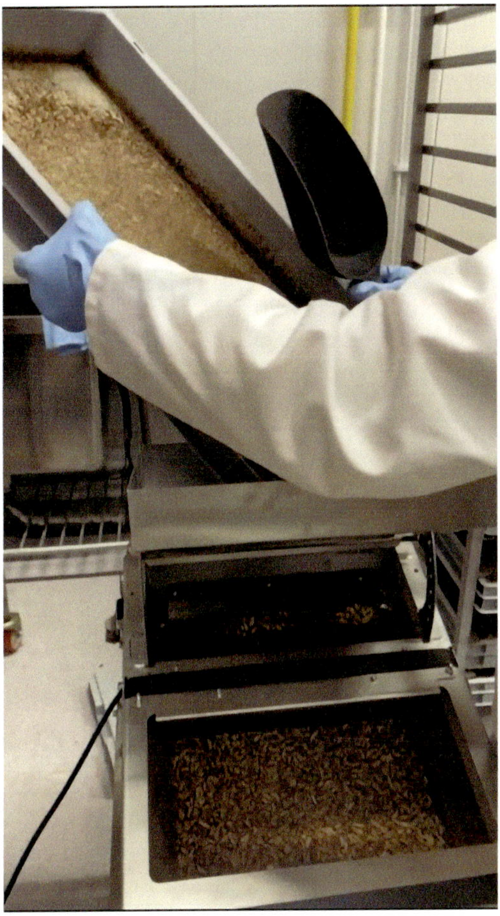

FIGURE 15.35 Another view of the separator. (Image Credit: Audinisa Fadhila, Ragn-Sells.)

and process waste. Harvesting is carried out on the second floor and in its section. The wastes are pre-processed, chemically treated, and ground before being treated with 5-day-old larvae. The system also has an air circulation system, a collection tray, conveyors, and separators. A similar container having upper and lower chambers for continuous feeding and self-harvesting of larvae has been designed by Font et al. (2015). Ruben (2015) has developed a highly computerized and mechanized system for converting food waste into animal feed. After the food waste is pre-treated with steam, radiation, chemicals, etc. to remove bio-hazards, it is processed, and data are maintained for quality monitoring and tracking. The food waste is spread on planar surfaces for larval feeding under a controlled environment, and movement and harvesting are done by robots and conveyors.

Auxiliary systems have also been developed to facilitate artificial breeding of BSFs. As light plays a pivotal role in the breeding behaviour of BSF, several types of artificial lighting systems have been developed, including ultraviolet (UV) and light emitting diodes (LED) (Aldana et al., 2016; Drew et al., 2018; Józefiak et al., 2018). Systems have also been designed with a focus on harvesting and separation of larvae from the digested materials – such as a rotary screen with an air-heater and sieve (CN206491189U, 2017; CN207978745U, 2018).

Several papers and patents also report contraptions developed to pre-process, sterilize, and mix wastes, and the use of additives like saw dust and microbial consortia (CN101889629B, 2010; CN102329155A, 2011; CN102329157A, 2011; Lee, 2011; KR101339105B1, 2013; CN203648978U, 2013; CN203913000U, 2014; Ruben, 2015; CN206043152U, 2016; CN206051041U, 2016; WO2017173849A1, 2016; CN106810301A, 2017; CN107549126A, 2017; CN108094692A, 2017; CN206747233U, 2017; CN207271797U, 2017; CN207322428U, 2017; CN207444048U, 2017; CN207544109U, 2017; Tinder et al., 2017; CN207911828U, 2018; CN207978745U, 2018; CN208159863U, 2018; CN208159865U, 2018; CN208390636U, 2018; CN208390638U, 2018; CN208829512U, 2018; CN109006699A, 2018; CN109007284A, 2018; CN109090043A, 2018; WO2019056762A1, 2018; CN109757440A, 2019; CN110201972A, 2019; Barroso et al., 2019; Isibika et al., 2019; Cai et al., 2019; Lim et al., 2019; Ganda et al., 2019; Lalander et al., 2019; Beesigamukama et al., 2020; da Silva et al., 2020; Fowles and Nansen, 2020; Liu et al., 2020; Wang et al., 2021).

Many companies, to name a few, Agriprotein, Nextalim, Alapre, Entofood, Insectum, Nextprotein, Hermetia, F4F, Innovafeed, and Enterrafeed, have been set up in various countries – including South Africa, France, Colombia, Malaysia, Lithuania, Germany, Chile, and Canada. These companies focus on producing larval meal, fertilizer, and bio-oil by utilizing the BSF larvae to treat agricultural wastes, manure, and fruit/vegetable wastes (Fowles and Nansen, 2020). Some additional information on the prior art is summarized in Table 15.1.

The assessment leads to the broad conclusion that most BSF-based zoocomposting systems are geared to process fresh fruit and vegetables, manure, and fermented agricultural wastes (Fowles and Nansen, 2020). Processing and management of cooked food wastes, abattoir wastes, and mixed wastes (cooked and uncooked) have been much less explored. Most of the reported work addressing those difficult wastes is confined to lab-scale studies.

Yang (2019) describes a facility at Baotou, a city of 3 million people, which processes 100 tonnes of eatery waste per day, employing BSF. After collecting the restaurant waste, it is divided into four categories: solids, water, oil, and contaminants like broken glasses or plastic bags. Only the waste of human food grade is collected, and municipal solid waste is excluded.

The shredded food waste is cooked with steam in order to eliminate extra oil and sanitize the remaining waste. The watery slurry and the oil are sent for biogas production.

The BSF-mediated zoocomposting has now commenced. The feed is transformed into larval zoomass and frass in about eight days. The frass becomes dark brown in colour and loses all of its texture at harvest.

The larvae are separated from the frass; both can be used to make a variety of products. The larvae are dried in a microwave oven designed like a tunnel, producing a product with the appearance and natural colour of living things.

The facility generates 1,300 kg of dried product per day when running to full capacity.

TABLE 15.1
An Illustrative List of BSF-Based Zoocomposting Systems Showing the Diversity of Available Designs

	Inventor/Patent No.	Type of Organic Waste Processed by the Facility	Pre-Processing of Waste	Based on Natural BSF Populations or Artificially Reared Ones	Key Features of the Bioreactor System
1.	Whitaker and Hillman (2020)	Brewer's grain, coffee grain, vegetable matter	Waste is ground	Artificial	Modular system for breeding and harvesting flies, consists of an egg laying chamber, larval chamber, pupation chamber, release box, and breeding chamber.
2.	El-Massri (2019)	Digested excrement	Waste is ground	Artificial	Compartmentalized tray having a pivot along one side and a sloping wall along the opposite side. It has a feeding system and a conveyor belt for frass removal.
3.	Alrayya (2019)	Rotten vegetables	Waste is ground	Artificial	Various chambers are placed one above the other and are interconnected. The base is a large hollow treatment chamber, connected to a smaller dark pupation chamber placed above it. The dark chamber has an opening to a meshed fly chamber having a light source, vibrator, and bio-balls for mating and egg laying. The bio-balls are automatically moved to the hatching chamber.
4.	Baksh (2019)	Industrial waste, animal and plant by-products, faeces	Waste is processed by shredding and then is chemically treated and ground	Artificial	Towers having a first and second section. Both the sections have many floors, wherein the first floor and its section process the waste, and the second section with its floors helps in harvesting.
5.	Popa and Nealson (2018)	Fermenting leachates	Waste is ground and sieved	Artificial	Stalked trays as larval incubation containers. The trays are provided with bedding and are covered by a net. They are also provided with a spreader, aerator, and pre-pupae collection box with a funnel.

(*Continued*)

TABLE 15.1 (*Continued*)
An Illustrative List of BSF-Based Zoocomposting Systems Showing the Diversity of Available Designs

	Inventor/Patent No.	Type of Organic Waste Processed by the Facility	Pre-Processing of Waste	Based on Natural BSF Populations or Artificially Reared Ones	Key Features of the Bioreactor System
6.	CN206491189U (2017) and CN207978745U (2018)	-	-	-	Systems for harvesting and separating larvae from the frass. They consist of a rotary screen with an air-heater and sieve.
7.	Aldana et al. (2016), Drew et al. (2018), abd Józefiak et al. (2018)	-	-	-	Artificial lighting systems to facilitate breeding. The lights used are UV, LED, IR, etc.
8.	Ruben (2015)	Food waste	Waste is pre-treated with steam, radiation, chemicals, etc.	Artificial	Pre-treated food waste is spread on planar surfaces for larval feeding, and subsequent harvesting is done in conveyors and by robots. It is a highly mechanized and computerized system.
9.	Font et al. (2015)	Organic waste	Waste is ground	Artificial	One reactor which contains a separating element defining an upper and lower chamber connected by a plurality of channels. This equipment is for continuous feeding and self-harvesting of larvae.
10.	Cickova et al. (2015) and Sheppard et al. (1994)	Manure (poultry, pig)	not done	Natural	Concrete pits or lagoons with ramps

(*Continued*)

TABLE 15.1 (*Continued*)
An Illustrative List of BSF-Based Zoocomposting Systems Showing the Diversity of Available Designs

	Inventor/Patent No.	Type of Organic Waste Processed by the Facility	Pre-Processing of Waste	Based on Natural BSF Populations or Artificially Reared Ones	Key Features of the Bioreactor System
11.	Perednia (2015) and WO214137179A1 (2014)	Manure, food waste	Waste is ground	Artificial	Semi-enclosed rotating barrel system for feeding, culturing, and harvesting larvae. The barrel has removable conical ends for larval migration and an automatic feeder.
12.	Courtright (2014)	Dried distiller grain, modified wet cake, spent brewer's grain	-	-	A vessel for containing substrate for rearing larvae, a cage for retaining the insects, a tubular duct connecting the vessel to the cage, a moist substrate for egg laying, and a light source to enhance mating
13.	Diener et al. (2011)	Manure (poultry), municipal organic waste	not done	Artificial	Zinc-coated steel trays with ramps; Netted greenhouse for breeding BSF
14.	Lee (2011), CN208390638U (2018), CN207911828 (2018), KR101846282B1 (2017) CN206043152U (2016), CN2039113000U (2014), and KR101123249B1 (2010)	Manure (livestock, poultry, swine), municipal organic waste	Waste is ground and is mixed with sawdust or rice husk.	Artificial	Containers for waste processing, which are basically breeding farms made of net and consisting of breeding induction apparatus, air conditioners, illuminators, etc.

(*Continued*)

TABLE 15.1 (*Continued*)
An Illustrative List of BSF-Based Zoocomposting Systems Showing the Diversity of Available Designs

Inventor/Patent No.	Type of Organic Waste Processed by the Facility	Pre-Processing of Waste	Based on Natural BSF Populations or Artificially Reared Ones	Key Features of the Bioreactor System
15. KR20090121779A (2008), Popa and Green (2012), WO217173849A1 (2016), CN203327766U (2013), CN203648978U (2013), Caprio et al. (2016), Villamar (2017), CN207322428U (2017), CN207444048U (2017) CN207678686U (2017), CN108739684A (2018), CN208159865U (2018), CN208390636U (2018), and Popa and Nealson (2019)	Organic solute from liquid, kitchen waste, putrescent wet waste, organic waste	Waste is ground, solid/liquid separation is done of the wet waste, addition of Bacillus bacteria or EM bacteria is carried out followed by mixing of auxillary material to the feed, or addition of media like coco coir into the feeding container.	Artificial	Tanks/boxes with removable cover, and ramps or slopes for larval migration. There are feeding and frass/leachate collection systems of air conditioners, and temperature controls.
16. Newton et al. (2005)	Manure (livestock, poultry, swine)	not done	Artificial	Pits with conveyor belts for collection, a ramp for migration, a drainage system, and a greenhouse for breeding flies
17. Olivier (2004)	Putrescent waste	Waste is ground	Artificial	Round culture containers with ramps.
18. Olivier (2002, 2003)	Putrescent waste, human and animal sludge, food waste	Waste is ground	Artificial	A disposal track (with bins) where waste is processed by larvae at the surface and treated material is removed from the bottom by scrapers. The frass is further treated by composting or vermicomposting

(*Continued*)

TABLE 15.1 (*Continued*)
An Illustrative List of BSF-Based Zoocomposting Systems Showing the Diversity of Available Designs

	Inventor/Patent No.	Type of Organic Waste Processed by the Facility	Pre-Processing of Waste	Based on Natural BSF Populations or Artificially Reared Ones	Key Features of the Bioreactor System
19.	Olivier (1998,1999), Urbanski et al. (2019), CN208159863U (2018), CN107497833A (2017), CN207271797U (2017), CN206051041U (2016), KR101339105B1 (2013),	Putrescent waste, fruit and vegetable waste	Waste is ground, pre-fermented, and mixed with wheat bran and yeast.	Artificial	Conveyor belt, where ground waste is added by a device along with fly eggs. Removal of larvae is done with the help of infrared light and treated waste by machines. The conveyor belts are made up of either metal or plastic
20.	Kappelt and Lolly (2020), Fotiadas (2020), Marchant et al. (2018), Newton and Sheppard (2017), Vickerson et al. (2017), KR101762186B1 (2016), and Levenhagen (1998)	Manure, brewer's grain, restaurant waste, fruit and vegetable waste, fish offal, coffee pulp, meat, grains	Waste is shredded, ground, and pasteurized	Artificial	Stackable trays for incubation, growth, and waste processing in an artificially controlled environment. Mechanized collection of frass and its post-processing is done using dryers. The system also consists of various chambers for breeding of BSF
21.	Eby and Morgan (1977), KR101762186B1 CN20674233U (2017), CN207135997U (2017), and CN102351394A (2011)	Manure (cow, chicken)	Waste is ground	Artificial	A rotating larval culture drum reactor linked to a perforated screen belt for harvesting using heat lamps

At a much lower scale, but no less significantly (UNDP, 2023), a BSF-zoocomposting facility at the town of Dimapur, situated in the Himalayan state of Nagaland, India, is currently processing 1:2 tonnes of food waste every month with plans to expand its capacity. Several other reports, on BSF-based food waste management at different scales, are also available (Olivier, 2021; Terrell and Ingwell, 2022), with elaborate do-how details.

REFERENCES

A kind of air drying type black soldier flies larva separator. CN206491189U, 2017.

A kind of automatic feeding of utilization changing food waste and the system of hatching black soldier flies. CN206043152U, 2016.

A kind of black soldier flies box for breeding handling kitchen garbage. CN208159865U, 2018.

A kind of black soldier flies device that utilizes changing food waste cultivation. CN203913000U, 2014.

A kind of black soldier flies larva cultural method. CN109090043A, 2018.

A kind of black soldier flies rotate feed kitchen garbage device. CN206747233U, 2017.

A kind of changing food waste raises black soldier flies Full-automatic uniform cloth system. CN206051041U, 2016.

A kind of drum-type black soldier flies polypide larva screening machine. CN207135997U, 2017.

A kind of method and application using remaining meal rubbish raising black soldier flies larva. CN109757440A, 2019.

A kind of method that black soldier flies larva is cultivated using kitchen leftover base-material. CN107549126A, 2017.

A kind of method that flat angle stratiomyiid of speck converts chicken manure with microbial association. CN106810301A, 2017.

A kind of multilayer cultivation apparatus of feces of livestock and poultry raising black soldier flies. CN208159863U, 2018.

A kind of operation room using black soldier flies processing kitchen garbage. CN208390638U, 2018.

A kind of production system using black soldier flies processing kitchen garbage. CN208390636U, 2018.

A method of it is handled using black soldier flies and utilizes kitchen garbage and stalk. CN110201972A, 2019.

A method of protein feed is prepared using black soldier flies processing pig manure. CN109007284A, 2018.

Aldana, J., Quan, E., Vickerson, A., Marchant, B., Kaulfuss, O. and Radley, R., Enterra Feed Corp, 2016. *Contained systems to provide reproductive habitat for Hermetia illucens.* U.S. Patent 9,510,572.

Alrayya, F.E., 2019. *System and method for breeding flies, particularly black soldier flies (BSFs).* U.S. Patent Application 16/292,392.

Amrul, N.F., Ahmad, I.K., Basri, N.E.A., Suja, F., Jalil, N.A.A. and Azman, N.A. 2022. A Review of Organic Waste Treatment Using Black Soldier Fly (Hermetia illucens). *Sustainability sSwitzerland), 14*(8), p.4565. doi: 10.3390/su14084565.

An automatical conveyor system for mass food waste using bsf. KR101339105B1, 2013.

Antonis, M., 2013. Waste Management 2030-Waste Management World. International Solid Waste Association. https://www.waste-management-world.com/articles/print/volume-11/issue-2/features/waste-management-2030.html. Accessed 17 september 14.

Apparatus for breeding lava of black soldier fly. KR101762186B1, 2016.
Apparatus for eclosion and oviposition of black soldier fly. KR101846282B1, 2017.
Apparatus for processing organic waste using black soldier fly larvae. WO2014137179A1, 2014.
Apparatus for rearing the black soldier fly. KR101123249B1, 2010.
Automatic apparatus for treatment of organic waste using bsf larva. KR20090121779A, 2008.
Awal, M.R., Rahman, M.M., Choudhury, M.A.R., Hasan, M.M., Rahman, M.T. and Mondal, M.F., 2022. Influences of artificial light on mating of black soldier fly (Hermetia illucens)—a review. *International Journal of Tropical Insect Science*, *42*(4), pp.2779–2783.
Bakhsh, N., 2019. *Organic waste treatment system and method*. U.S. Patent Application 16/292,400.
Bandaw, T., and Herago, T. 2017. Review on Abattoir waste management. *Global Veterinaria*, *19*(1), pp.517–524. 10.5829/idosi.gv.2017.517.524.
Bánk, B., 1964. Gleichzeitige Myiasis der Genitalien bei einem Ehepaar. *Gynecologic and Obstetric Investigation*, *157*(2), pp.121–122.
Barroso, F.G., Sánchez-Muros, M.J., Rincón, M.Á., Rodriguez-Rodriguez, M., Fabrikov, D., Morote, E. and Guil-Guerrero, J.L., 2019. Production of n-3-rich insects by bioaccumulation of fishery waste. *Journal of Food Composition and Analysis*, *82*, p.103237.
Beesigamukama, D., Mochoge, B., Korir, N.K., Fiaboe, K.K., Nakimbugwe, D., Khamis, F.M., Subramanian, S., Wangu, M.M., Dubois, T., Ekesi, S. and Tanga, C.M., 2020. Low-cost technology for recycling agro-industrial waste into nutrient-rich organic fertilizer using black soldier fly. *Waste Management*, *119*, pp.183–194.
Bessa, L.W., Pieterse, E., Marais, J. and Hoffman, L.C. 2020. Why for feed and not for human consumption? The black soldier fly larvae. *Comprehensive Reviews in Food Science and Food Safety*, *19*(5), pp.2747–2763. doi: 10.1111/1541-4337.12609.
Black soldier flies cultivation pot and black soldier flies cultivation apparatus. CN207678686U, 2017.
Brian, L., Craig, H., James, L., Lisa, K., Richard, W. and Tim, S., 2013. *Reducing food loss and waste e installment 2 of "Creating a Sustainable Food Future". World Resource Institute*, Washington, DC.
Brits, D., Brooks, M. and Villet, M.H., 2016. Diversity of bacteria isolated from the flies *Musca domestica* (Muscidae) and *Chrysomya megacephala* (Calliphoridae) with emphasis on vectored pathogens. *African Entomology*, *24*(2), pp.365–375.
Cai, M., Zhang, K., Zhong, W., Liu, N., Wu, X., Li, W., Zheng, L., Yu, Z. and Zhang, J., 2019. Bioconversion-composting of golden needle mushroom (Flammulina velutipes) root waste by black soldier fly (Hermetia illucens, Diptera: Stratiomyidae) larvae, to obtain added-value biomass and fertilizer. *Waste and Biomass Valorization*, *10*(2), pp.265–273.
Caprio, F., De Gennaro, D., Gorgone, G. and Roncone, M., Bioconversion Srl, 2016. *Plant and method for organic waste bioconversion and municipal waste biostabilization*. U.S. Patent 9,428,426.
Çetinkaya, M., Ozkan, H., Köksal, N., Coşkun, S.Z., Hacimustafaoğlu, M. and Girişgin, O., 2008. Neonatal myiasis: a case report. *The Turkish Journal of Pediatrics*, *50*(6), pp.581–584.
Cheng, J.Y., Chiu, S.L. and Lo, I.M., 2017. Effects of moisture content of food waste on residue separation, larval growth and larval survival in black soldier fly bioconversion. *Waste Management*, *67*, pp.315–323.
ČiČková, H., Newton, G.L., Lacy, R.C. and Kozánek, M., 2015. The use of fly larvae for organic waste treatment. *Waste Management*, *35*, pp.68–80.
da Silva, G.D.P. and Hesselberg, T., 2020. A review of the use of black soldier fly larvae, Hermetia illucens (Diptera: Stratiomyidae), to compost organic waste in tropical regions. *Neotropical Entomology*, *49*(2), pp.151–162.

Dehghani, R., Mohammadi, M.J., Mostafaii, G., Gilasi, H., Shahrisvand, B., Charkhloo, E. and Hoseindoost, G., 2018. The frequency of stable fly larvae in the process of producing compost from municipal waste. *Fresenius Environ Bull*, *27*, pp.5323–5328.

Demetriou, J., Kalaentzis, K., Kazilas, C., Kunz, G., Muller, B., Mostovski, M.B. and Koutsoukos, E., 2022. An "alien" species on the loose: new records and updated distribution of the black soldier fly Hermetia illucens in the western palearctic.

Device for treating kitchen waste by breeding Hermetia illucens. WO2017173849A1, 2016.

Diener, S., Solano, N.M.S., Gutiérrez, F.R., Zurbrügg, C. and Tockner, K., 2011. Biological treatment of municipal organic waste using black soldier fly larvae. *Waste and Biomass Valorization*, *2*(4), pp.357–363.

Drew, D., Richards, C., Kotze, C., Drew, J., Rudolphe, R. and Kieron, D.U.N.N., Agriprotein Holdings Ltd, 2018. *Method of breeding dipteran insects*. U.S. Patent Application 15/771,927.

Eby, H.J. and Morgan, N.O., US Department of Agriculture, 1977. *Apparatus and method for treating waste products*. U.S. Patent 4,040,810.

El-Massri, M.A., Lakeside Power and Methane LLC, 2019. *Biogas production from excrement*. U.S. Patent Application 16/101,880.

Eriksson, D.I., 2021. Design and evaluation of black soldier fly larvae and frass compost separator - Master thesis in Sustainable Development 2021/53. Department Of Earth Sciences, Uppsala University, Published at Department of Earth Sciences, Uppsala University, Institutionen För Geovetenskaper. 120 pp.

Feces of livestock and poultry cultivates black soldier flies system. CN207444048U, 2017.

Font, A.F., Velasco, S.R., Velasquez, Y., Galinanes, A.L., and Morcillo, A.L., 2015. Equipment for the continuous feeding and self-harvest of larvae of dipterous insects. WO201517345A1.

Fotiadis, F., Entomics Biosystems Ltd, 2020. *Insect Larvae Rearing*. U.S. Patent Application 16/648,200.

Fowles, T.M. and Nansen, C., 2020. Insect-based bioconversion: value from food waste. In *Food waste management* (pp. 321–346). Palgrave Macmillan, Cham.

Furman, D.P., Young, R.D., Catts, E.P. 1959. Hermetia illucens (Linnaeus) as a factor in the natural control of Musca domestica Linnaeus. *Economic Entomology*, 52, pp. 917–921.

Ganda, H., Zannou-Boukari, E.T., Kenis, M., Chrysostome, C.A.A.M. and Mensah, G.A., 2019. Potentials of animal, crop and agri-food wastes for the production of fly larvae. *Journal of Insects as Food and Feed*, *5*(2), pp.59–67.

Gardner, N. 2023. *Black Soldier Fly Poised to Take Flight as Leading Alternative Protein Source. Beta Buzz #3: A Publication for the Insect Farming Industry. (March*), pp. 5–8.

Isibika, A., Vinnerås, B., Kibazohi, O., Zurbrügg, C. and Lalander, C., 2019. Pre-treatment of banana peel to improve composting by black soldier fly (Hermetia illucens (L.), Diptera: Stratiomyidae) larvae. *Waste Management*, *100*, pp.151–160.

It is a kind of to handle kitchen garbage frass separator using black soldier flies larva. CN207978745U, 2018.

Józefiak, D., Mazurkiewicz, J., Rudak, J. and Urbanski, J., Hipromine SA, 2018. *A Modular Lamp System for Insect Breeding, Use Thereof for Stimulation of Insect Reproduction and a Method of Insect Breeding*. U.S. Patent Application 15/748,163.

Jung, S., Jung, J.-M., Tsang, Y.F., Bhatnagar, A., Chen, W.-H., Lin, K.-Y.A., Kwon, E.E. 2022. Biodiesel production from black soldier fly larvae derived from food waste by non-catalytic transesterification. *Energy*, Part A 238, 121700. doi: 10.1016/j.energy.2021.121700.

Kaczor, M., Bulak, P., Proc-Pietrycha, K., Kirichenko-Babko, M., Bieganowski, A. 2023. The variety of applications of hermetia illucens in industrial and agricultural areas—review. *Biology*, *12*(1), p.25. doi: 10.3390/biology12010025.

Kappelt, C.E. and Levenhagen, T.E. (1998). *U.S. Patent No. 5,819,*685. U.S. Patent and Trademark Office, Washington, DC.

Khayrova, A., Lopatin, S., Varlamov, V. 2019. Black soldier fly Hermetia illucens as a novel source of chitin and chitosan. *International Journal of Science and Research, 8*, pp. 81–86.

Kitchen waste treatment system. CN203648978U, 2013.

Kumar, S., Negi, S., Mandpe, A., Singh, R.V. and Hussain, A., 2018. Rapid composting techniques in Indian context and utilization of black soldier fly for enhanced decomposition of biodegradable wastes: A comprehensive review. *Journal of Environmental Management, 227*, pp.189–199.

Kuppusamy, G., Kong, C.K., Segaran, G.C., Tarmalingam, E., Herriman, M., Ismail, M.F., Mehmood Khan, T., Low, L.E. and Goh, B.H., 2020. Hummingbird-leaves-reared black soldier fly prepupae: Assessment of nutritional and heavy metal compositions. *Biology, 9*(9), p.274.

Lalander, C., Diener, S., Zurbrügg, C. and Vinnerås, B., 2019. Effects of feedstock on larval development and process efficiency in waste treatment with black soldier fly (Hermetia illucens). *Journal of Cleaner Production, 208*, pp.211–219.

Lee, S.H., Green Teko Co Ltd, 2011. *Method for rearing soldier flies*. U.S. Patent Application 13/002,024.

Lim, J.W., Mohd-Noor, S.N., Wong, C.Y., Lam, M.K., Goh, P.S., Beniers, J.J.A., Oh, W.D., Jumbri, K. and Ghani, N.A., 2019. Palatability of black soldier fly larvae in valorizing mixed waste coconut endosperm and soybean curd residue into larval lipid and protein sources. *Journal of Environmental Management, 231*, pp.129–136.

Liu, C., Wang, C., Yao, H. and Chapman, S.J., 2020. Pretreatment is an important method for increasing the conversion efficiency of rice straw by black soldier fly larvae based on the function of gut microorganisms. *Science of the Total Environment, 762*, p.144118.

Lolley, M.D., Insecticycle LLC, 2020. *Method, device, and system for use with insect larvae*. U.S. Patent 10,842,138.

Lomonaco, G., Franco, A., De Smet, J., Scieuzo, C., Salvia, R. and Falabella, P., 2024. Larval Frass of Hermetia illucens as organic fertilizer: Composition and beneficial effects on different crops. *Insects, 15*(4), p.293.

Lopes, I.G., Yong, J.W. and Lalander, C., 2022. Frass derived from black soldier fly larvae treatment of biodegradable wastes. A critical review and future perspectives. *Waste Management, 142*, pp.65–76.

Makkar, H.P., Tran, G., Heuzé, V. and Ankers, P., 2014. State-of-the-art on use of insects as animal feed. *Animal Feed Science and Technology, 197*, pp.1–33.

Marchant, B., Vickerson, A., Kaulfuss, O. and Radley, R., Enterra Feed Corp, 2018. *Continuous production system for culturing dipteran insects*. U.S. Patent 10,159,229.

Method for biotransformation of kitchen waste using Hermitia illucens. WO2019056762A1, 2018.

Method for preparing organic fertilizer by utilizing horseshit processed by black soldier flies. CN102329155A, 2011.

Method for processing food waste by using black soldier fly larvae and material formula. CN101889629B, 2010.

Method for processing human excrement by utilizing black solider flies to prepare organic fertilizer. CN102329157A, 2011.

Newton, G.L. and Sheppard, D.C., University of Georgia Research Foundation Inc (UGARF), 2017. *Systems and methods for rearing insect larvae*. U.S. Patent 9,629,339.

Newton, G.L., Sheppard, D.C., Watson, D.W., Burtle, G.J., Dove, C.R., Tomberlin, J.K. and Thelen, E.E., 2005, January. The black soldier fly, Hermetia illucens, as a manure management/resource recovery tool. In *Symposium on the State of the Science of Animal Manure and Waste Management*, San Antonio, TX, pp. 5–7.

Niass, C.T., Thiaw, C., Diome, T. and Sembène, M., 2021. First record of the black soldier fly, Hermetia illucens L, 1758 (Diptera, Stratiomyidae) in central and western Senegal. *Journal of Entomology and Zoology Studies*, *9*(5), pp.32–35

Oliveira, F., Doelle, K., List, R. and O'Reilly, J.R., 2015. Assessment of Diptera: Stratiomyidae, genus Hermetia illucens (L., 1758) using electron microscopy. *Journal of Entomology and Zoology Studies*, *3*(5), pp.147–152.

Olivier, J., 2021. How to build a black soldier fly compost bin. https://www.grit.com/animals/poultry/black-soldier-fly-bin-zm0z21mazbut/. Last accessed August 2024.

Olivier, P.A., 1999. *Device and method for the continuous treatment of waste by means of fly larvae*. U.S. Patent 6,001,146.

Olivier, P.A., 2002. *Method for bio-conversion of putrescent wastes*. U.S. Patent 6,391,620.

Olivier, P.A., 2003. *Apparatus for Bio-conversion of Putrescent Wastes*. U.S. Patent 6,579,713.

Olivier, P.A., 2004. *Disposal apparatus and method for efficiently bio-converting putrescent wastes*. U.S. Patent 6,780,637.

Olivier, P.A., Olivier and Paul A., 1998. *Device and method for the continuous treatment of waste by means of fly larvae*. U.S. Patent 5,759,224.

Organic waste converting system for black soldier flies. CN203327766U, 2013.

Perednia, D.A., 2015. *Rotating feeder bin for growing, feding and harvesting insect larvae*. U.S. Patent Application 14/689,949.

Popa, R. and Green, T.R., Oregon State Board of Higher Education and Portland State University, 2012. *Insect-based removal of organic solutes from liquid*. U.S. Patent Application 13/010,654.

Popa, R. and Nealson, K.H., River Road Research Inc, 2018. *System and method for propagating dipteran larvae*. U.S. Patent Application 16/011,934.

Popa, R. and Nealson, K.H., River Road Research Inc, 2019. *Methods for producing melanin and inorganic fertilizer from fermentation leachates*. U.S. Patent 10,227,626.

Premalatha, M., Abbasi, T., Tabassum-Abbasi, Premalatha, M. and Abbasi, S.A. 2011. Energy-efficient food production to reduce global warming and ecodegradation: The use of edible insects. *Renewable and Sustainable Energy Reviews, 15*, pp.4357–4360.

Rana, K.S., Salam, M.A., Hashem, S. and Islam, M.A., 2015. Development of black soldier fly larvae production technique as an alternate fish feed. *International Journal of Research in Fisheries and Aquaculture*, *5*(1), pp.41–47.

Rehman, K.U., Hollah, C., Wiesotzki, K., Rehman, R.U., Rehman, A.U., Zhang, J., Zheng, L., …, Aganovic, K. 2023. Black soldier fly, Hermetia illucens as a potential innovative and environmentally friendly tool for organic waste management: A mini-review. *Waste Management and Research*, *41*(1), pp.81–97. doi: 10.1177/0734242X221105441.

Roller type black soldier flies biological processor. CN102351394A, 2011.

Ruben, B.N., 2015. *Recycled food processing, products therefrom, and devices useful therein*. U.S. Patent Application 14/574,375.

Salam, M., Shahzadi, A., Zheng, H., Alam, F., Nabi, G., Dezhi, S., Ullah, W., Ammara, S., Ali, N. and Bilal, M., 2022. Effect of different environmental conditions on the growth and development of Black Soldier Fly Larvae and its utilization in solid waste management and pollution mitigation. *Environmental Technology & Innovation*, *28*, p.102649.

Schmitt, E. and de Vries, W., 2020. Potential benefits of using *Hermetia illucens* frass as a soil amendment on food production and for environmental impact reduction. *Current Opinion in Green and Sustainable Chemistry*, *25*, p.100335.

Sheppard, D.C., Newton, G.L., Thompson, S.A. and Savage, S., 1994. A value added manure management system using the black soldier fly. *Bioresource Technology*, *50*(3), pp.275–279.

Siddiqui, S.A., Ristow, B., Rahayu, T., Putra, N.S., Yuwono, N.W., Mategeko, B., Smetana, S., Saki, M., Nawaz, A. and Nagdalian, A., 2022. Black soldier fly larvae (BSFL) and their affinity for organic waste processing. *Waste Management*, *140*, pp.1–13.

Siddiqui, S.A., Süfer, Ö., Çalışkan Koç, G., Lutuf, H., Rahayu, T., Castro-Muñoz, R. and Fernando, I., 2024. Enhancing the bioconversion rate and end products of black soldier fly (BSF) treatment–A comprehensive review. *Environment, Development and Sustainability*, 27, pp.9673–9741, https://doi.org/10.1007/s10668-023-04306-6.

Singh, A., and Kumari, K. 2019. An inclusive approach for organic waste treatment and valorisation using Black Soldier Fly larvae: A review. *Journal of Environmental Management*, *251*, p.109569. doi: 10.1016/j.jenvman.2019.109569.

Smets, R., Verbinnen, B., Van De Voorde, I., Aerts, G., Claes, J. and Van Der Borght, M., 2020. Sequential extraction and characterisation of lipids, proteins, and chitin from black soldier fly (Hermetia illucens) larvae, prepupae, and pupae. *Waste and Biomass Valorization*, *11*, pp.6455–6466.

Surendra, K.C., Tomberlin, J.K., van Huis, A., Cammack, J.A., Heckmann, L.H.L. and Khanal, S.K., 2020. Rethinking organic wastes bioconversion: Evaluating the potential of the black soldier fly (Hermetia illucens (L.))(Diptera: Stratiomyidae)(BSF). *Waste Management*, *117*, pp.58–80.

Suryati, T., Julaeha, E., Farabi, K., Ambarsari, H. and Hidayat, A.T., 2023. Lauric acid from the black soldier fly (Hermetia illucens) and its potential applications. *Sustainability*, *15*(13), p.10383.

Tabassum-Abbasi., Abbasi, T. and Abbasi, S.A., 2016. Reducing the global environmental impact of livestock production: the minilivestock option, *Journal of Cleaner Production* (Elsevier), *112,* pp.1754-766.

Tepper, K., Edwards, O., Sunna, A., Paulsen, I.T., Maselko, M., 2024. Diverting organic waste from landfills via insect biomanufacturing using engineered black soldier flies (Hermetia illucens). *Communications Biology*, *7*(1), p.862.

Terrell, C., and Ingwell, L., 2022. Vegetable Insects - Black soldier fly composting guide. Department of Entomology, Purdue University. https://extension.entm.purdue.edu/publications/E-276/E-276.html. Last accessed August 2024.

Tettamanti, G. and Bruno, D. 2024. Black soldier fly larvae should be considered beyond their use as feedstuff. *Journal of Insects as Food and Feed*, *10*(1), pp.1–7. https://doi.org/10.1163/23524588-241001ED.

The bio combined processing unit of kitchen waste. CN208829512U, 2018.

The cycle formula device of food waste is handled using black soldier flies, CN207911828U, 2018.

The devices and methods therefor of organic waste is handled using black soldier flies. CN108739684A, 2018.

The ecological cultivation method of black soldier flies processing pig manure. CN109006699A, 2018.

Thi, N.B.D., Kumar, G. and Lin, C.Y., 2015. An overview of food waste management in developing countries: Current status and future perspective. *Journal of Environmental Management*, *157*, pp.220–229.

Tinder, A.C., Puckett, R.T., Turner, N.D., Cammack, J.A. and Tomberlin, J.K., 2017. Bioconversion of sorghum and cowpea by black soldier fly (Hermetia illucens (L.)) larvae for alternative protein production. *Journal of Insects as Food and Feed*, *3*(2), pp.121–130.

Tomberlin, J.K., Sheppard, D.C. and Joyce, J.A., 2002. Selected life-history traits of black soldier flies (Diptera: Stratiomyidae) reared on three artificial diets. *Annals of the Entomological Society of America*, *95*(3), pp.379–386.

Tran, C., Le, T.M., Pham, C.D., Duong, Y., Le, P.T.K. and Tran, T.V., 2022. Valorization of black soldier flies at different life cycle stages. *Chemical Engineering Transactions*, *97*, pp.139–144.

United Nations Development Programme (2023). Hail the Black Soldier Fly: turning waste to animal feed. https://www.undp.org/india/stories/hail-black-soldier-fly-turning-waste-animal-feed. Last accessed August 2024.

Urbanski, J., Józefiak, D. and Mazurkiewicz, J., Hipromine SA, 2019. *Apparatus and method for rearing or breeding of insects*. U.S. Patent Application 16/324,325.

Utilize the apparatus and method of black soldier flies processing kitchen garbage. CN107497833A, 2017.

Utilize the equipment of black soldier flies processing kitchen garbage. CN207271797U, 2017.

Utilize the method for black soldier flies processing humified material. CN108094692A, 2017.

Varelas, V., 2019. Food wastes as a potential new source for edible insect mass production for food and feed: A review. *Fermentation*, *5*(3), p.81.

Vickerson, A., Radley, R., Marchant, B., Kaulfuss, O. and Kabaluk, T., ENTERRA FEED Corp, 2017. *Hermetia illucens frass production and use in plant nutrition and pest management*. U.S. Patent 9,844,223.

Villamar, C.R., 2017. *System and method for solar greenhouse aquaponics and black soldier fly composter and auto fish feeder*. U.S. Patent 9,585,315.

Wang, Q., Ren, X., Sun, Y., Zhao, J., Awasthi, M.K., Liu, T., Li, R. and Zhang, Z., 2021. Improvement of the composition and humification of different animal manures by black soldier fly bioconversion. *Journal of Cleaner Production*, *278*, p.123397.

Wang, Y.S. and Shelomi, M., 2017. Review of black soldier fly (*Hermetia illucens*) as animal feed and human food. *Foods*, *6*(10), p.91.

Whitaker, K.C.O. and Hillmann, P.S., Entocycle Ltd, 2020. *Apparatus and methods for production of dipteran insects*. U.S. Patent Application 16/647,356.

With the device of feces of livestock and poultry cultivation black soldier flies. CN207322428U, 2017.

With the system of feces of livestock and poultry cultivation black soldier flies. CN207544109U, 2017.

Wooley, K. and Tibbetts, C. 2023. Transforming flies into degradable plastics. In: *American Chemical Society Meeting Fall 2023*, American Chemical Society, ACS Publication, San Francisco. https://www.acs.org/pressroom/newsreleases/2023/august/transforming-flies-into-degradable-plastics.html (Last accessed July 2025).

Xia, J., Ge, C., Yao, H. 2021. Antimicrobial peptides from black soldier fly (Hermetia illucens) as potential antimicrobial factors representing an alternative to antibiotics in livestock farming. *Animals*, *11*(7), 1937. doi: 10.3390/ani11071937.

Xiong, A., Ruan, L., Ye, K., Huang, Z. and Yu, C., 2023. Extraction of chitin from black soldier fly (*Hermetia illucens*) and its puparium by using biological treatment. *Life*, *13*(7), p.1424.

Yang, S., 2019. Explanation of an industrial application of black soldier fly converting food waste into valuable products. https://www.evoconsys.com/blog/explanation-of-an-industrial-application-of-black-soldier-fly-converting-food-waste-into-valuable-products. Last accessed August 2024.

16 A Novel, Versatile, Effective, Yet Inexpensive System for Carrying out BSF-Mediated Zoocomposting of Difficult-to-Treat Food and Abattoir Waste

ZOCOFA

16.1 INTRODUCTION

In the previous chapter, we had described the unique role that is played by black soldier flies (BSF) in zoocomposting food and abattoir waste, which can be effectively handled neither by earthworms nor by termites. Nor can it be easily composted, anaerobically digested, or converted to useful products. The chapter had also dwelt at length on the criteria for designing BSF-based zoocomposting units and the approaches used so far in developing various types of systems that have been reported in the prior art. The assessment indicates that despite a plethora of proposed designs, which can effectively handle manures and several other types of organic waste, there are no systems reported in the prior art that can effectively and efficiently handle cooked/uncooked food waste, abattoir waste, and fruit and vegetable waste. Also, most reported systems are highly instrument-intensive and energy-intensive, requiring skilled personnel to control and operate. This makes the systems expensive and unaffordable, especially to the developing world, which harbours most of the world's population and suffers the most from the pollution caused by these wastes. The BSF-mediated zoocomposting system developed by the authors, which is at once versatile and effective yet inexpensive, is acronymed ZOCOFA (Zoo COmposting of Food and Abattoir waste).

DOI: 10.1201/9780429331886-19

16.2 DRAWBACKS IN THE PRIOR ART AND HOW THE DRAWBACKS HAVE BEEN OVERCOME BY THIS

1. The major organic wastes which have been earlier explored for treatment with BSF are mostly cattle and poultry manure, semi-digested agricultural wastes like coconut waste, soybean curd, maize barn, grasses, etc., and brewery and distillery spent grain. Much less work exists on fruit and vegetable waste, human faecal waste and sludge, and fish waste (Cickova et al., 2015; Abbasi, 1999, 2018; Kumar et al., 2018; Fowles & Nansen, 2020). Even those processing this semi-wet waste are compelled to mix it with dry ingredients (hay, coco pith) and microbial inoculum (yeast, bacteria) without which the rate of zoocomposting remains very slow. Apart from it, the food waste before bioprocessing is usually pulverized, disinfected, dewatered, and treated. These material-intensive and energy-intensive steps make the overall system cumbersome and costly. There is no prior art in which blends of cooked / uncooked food waste, abattoir waste, and fruit/vegetable waste – which, together, constitute three of the most problematic biowastes – have been treated simultaneously to achieve process synergy and optimality.
2. The systems that have been developed so far are highly mechanized, usually consisting of separate units for various stages of the bioprocess, viz mating, oviposition, rearing, migration, collection, and harvesting. Each of these steps requires expensive gadgets and controls, and needs substantial quantities of energy to run. Moreover, they require skilled and trained people to operate, adding to the costs. These factors put the existing BSF-based zoocomposting units beyond the reach of not only laypersons but also municipal authorities of developing countries. In contrast, the proposed system is modular, simple, and inexpensive yet very effective.

16.3 SUMMARY OF THE INVENTION

The uniqueness of the present invention is that, for the very first time, the processing of cooked/uncooked food wastes, abattoir wastes, and fruit/vegetable wastes together in a rapid and efficient yet inexpensive manner has been achieved. For this, novel bioreactors which deploy naturally and freely available BSF have been designed, tested over several years, and refined. Processing all the abovementioned three waste types singly is fraught with a series of problems because each individual waste stream is either so excessively rich in some ingredients that it upsets the feeding larvae, while being short of some other ingredients. We have got around this problem by developing a blend of these waste types in a manner that the shortcoming of one becomes the strength of the other, enabling rapid and efficient zoocomposting of all. The waste processing in the bioreactors is synchronized with the life cycle of the BSF so that the former follows the latter to achieve the best results in the shortest possible time. This makes the process cost effective and easy to operate. The system is able to reduce the bulk of the food waste by 75%–80% within a period of 20–25 days. About 25% of the original waste mass gets converted to frass, and about 20% into pre-pupae. The pre-pupae make excellent poultry feed, while frass – which is the material left after processing of the wastes – contains larval excrements, moultings, and undigested materials. It is an excellent organic fertilizer. The fertilizer potential of frass has been assessed on a field scale, by testing it on three seasonal vegetables, lady's finger, cluster bean, and spinach, and comparison was made with other organic fertilizers and synthetic fertilizers. It has been found that plants amended with frass showed better germination, growth, and yield than plants grown on other fertilizers.

16.4 NOVELTY AND DISTINGUISHING FEATURES OF THE INVENTION

The most significant distinguishing features of the invention are:

1. At the heart of the system is the ability to, first time ever, totally utilize the three most difficult waste streams – cooked/uncooked food waste, abattoir waste, and fresh fruit/vegetable waste in generating fertilizer and feed. Food waste is high in oils and spices. When treated in any system, it leads to foul smell and anaerobic conditions, which are detrimental for the larvae, thereby fouling the bioprocess. Likewise, vegetable/fruit waste and abattoir waste, when treated by BSF in pre-existing systems, take a longer time and produce smaller-sized larvae. In the presently developed system, all these wastes, when fed in a 1:1:1 ratio, are treated rapidly and effectively. The shortcoming of one becomes the strength of the other, and thus, processing is done in the reactor at a faster rate without the use of pre-processing and pre-treatment of the waste. This has great potential in populous developing countries like India, which generate massive quantities of these wastes but have no affordable method to treat them.
2. The bioreactor is made using low-cost materials and requires low maintenance. The unique configuration of the reactor is self-sustaining and is able to accomplish oviposition, larval hatching, rearing, migration, and waste treatment in a single unit.
3. Unlike other systems, it does not require any form of mechanization, aerators, lamps, environmental control systems, etc. There is optimal space utilization and ease of expansion.
4. The wastes put into the reactor do not require any sort of pre-treatment such as grinding, pasteurization, and chemical and microbial treatment, as is mandated in other BSF systems.
5. There is no need to mix drying agents with the waste, as necessary in prior art.
6. The bioprocessing is done using native BSF and their larvae. As BSF are largely available in a tropical climate, they are easily attracted to the reactors. This reduces the cost of buying and culturing the flies separately.
7. When reactors are kept in a well-ventilated room, the whole life cycle of the flies continues to take place along with waste treatment.
8. Harvesting, along with separation of the frass and the pre-pupae, is done by utilizing the pre-pupae's sensitivity to light. The pre-pupae self-migrate out

of the reactor and the frass. There is no compulsion to use artificial light and mechanized harvesters.

9. The ability to treat wastes along with running the whole life cycle of the native BSF in a single cost-effective reactor makes it affordable for marginalized populations and small-scale farmers to adopt this technology for their economic benefit. This is very crucial for developing and underdeveloped countries.
10. The system is a zero-waste generating process. In addition, useful byproducts such as frass for use as fertilizer and pre-pupae for poultry and fish feed are generated.

16.5 DETAILED DESCRIPTION

The features of design and operation that bestow unprecedented novelty and usability to this process are detailed below.

There are two embodiments of the specially designed bioreactor. Both the embodiments can be put singly or stacked one above the other to form a unit of three or four. Both the embodiments contain leachate collection systems, larval migration and collection systems, and upper lids with attached egg laying provision for BSF to lay eggs.

The first embodiment, S1, is a horizontal barrel-shaped reactor, made up of high-density polyethylene (2–3 mm thick) sheets. A typical reactor has a length of 150 cm, diameter of 30 cm, and height of 30 cm. The wastes are filled in such reactors up to a height of 10–15 cm from the bottom of the barrel, so that they can be acted upon by the BSF without the formation of anaerobic conditions (Figure 16.1). The bottom of the reactor has a cut-out of 120 cm (length) by 10 cm (breadth), overlain

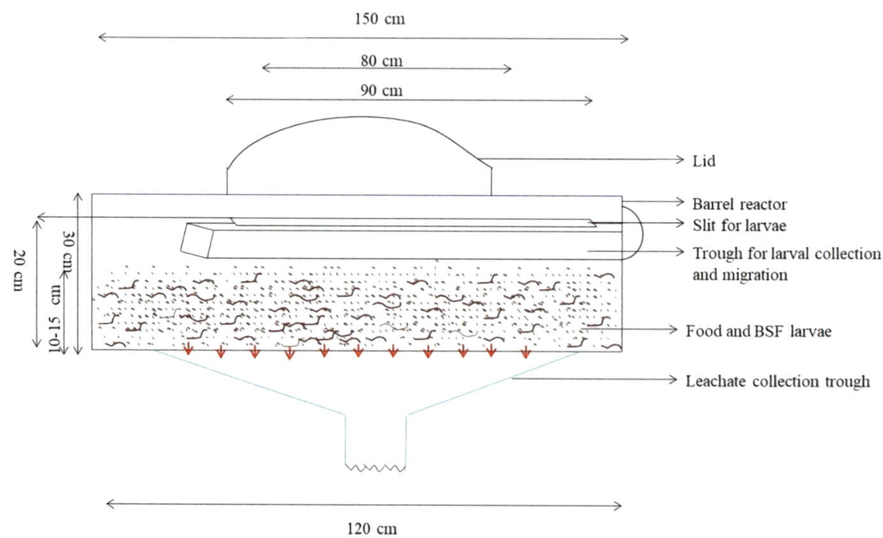

FIGURE 16.1 Embodiment S1: Side view of the reactor showing larval and leachate collection systems.

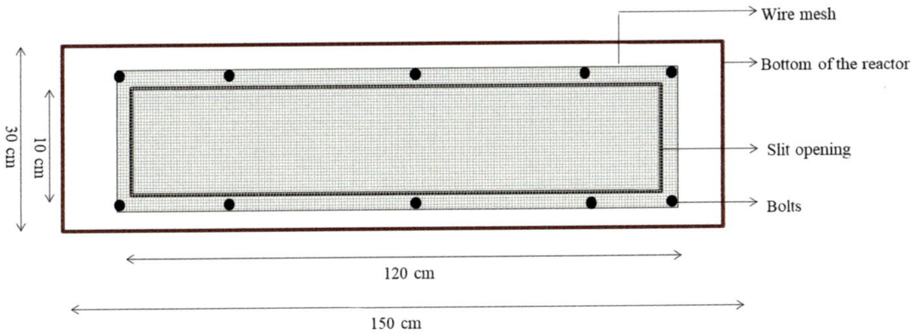

FIGURE 16.2 Embodiment S1: The reactor bottom, showing cut-out overlain with the bolted-on wire mesh.

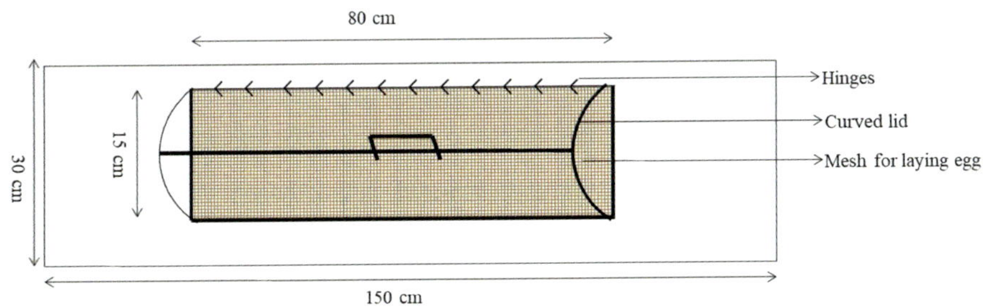

FIGURE 16.3 Embodiment S1: Top view of the reactor showing the lid and the mesh trough for egg laying.

by a plastic-coated steel mesh of 0.4–0.5 mm diameter, for leachate removal and air circulation (Figure 16.2). This mesh is bolted down securely to prevent gaps between the mesh edges and the reactor from which the larvae could escape. The leachate is collected into a slightly inclined trough, which pours down into a common gutter of 10 cm width and then collected in a leachate collection vessel (Figure 16.1). Two slits of 90 cm length and 2 cm breadth, with an attached flap, are provided on either side of the barrel, at a height of 20 cm from the bottom. The flaps open into troughs of 10 cm width (Figure 16.1). The pre-pupal stage larvae migrate from the reactor through the slit openings, over the flap and into the trough. As the larvae are photophobic, the trough is covered and constructed in a way that does not require artificial light to promote migration. The top of the reactor has an opening of 80 cm (length) by 15 cm (breadth) covered by a lid (Figure 16.3). This opening serves several purposes. Though it, the feed is charged into the reactor, and the frass is removed. It also provides an opening for entry point for the native female BSF to lay their eggs on the feed. The semi-circular lid that covers the opening is attached to the reactor with hinges and is positioned such that when closed, it leaves a gap of 10 cm on either end (Figure 16.4). This gap is provided in order for the female BSF to enter the reactor and lay eggs on the shallow mesh trough. The trough is attached to the lid in a way that when the lid is closed, it is positioned just under the opening of the reactor.

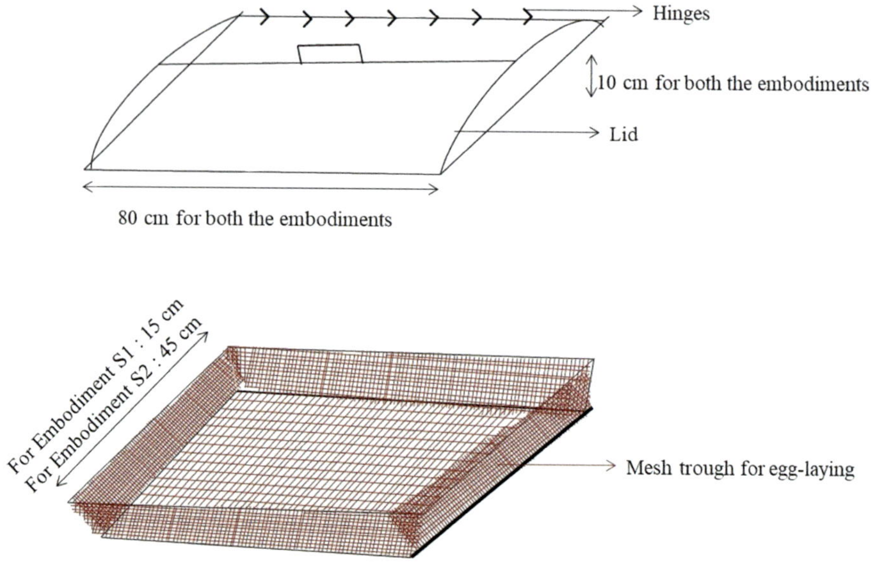

FIGURE 16.4 Lid covering the opening at the top of the reactor and attached egg laying mesh trough in both the embodiments of the reactor.

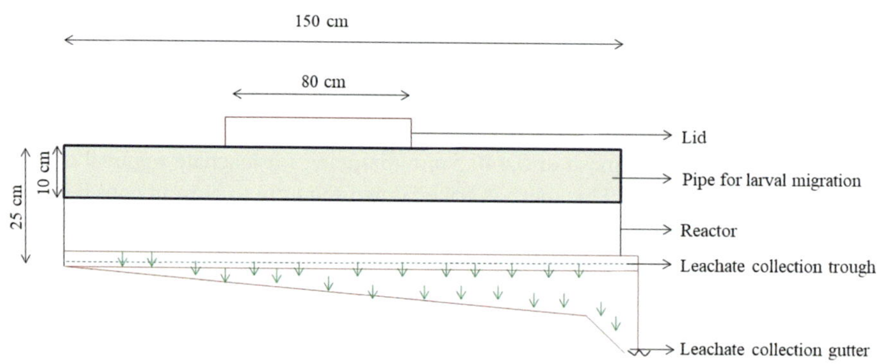

FIGURE 16.5 Embodiment S2: Long side view of the reactor.

The mesh is made of polyester having perforations of 1mm in diameter. The BSF females enter the reactor and lay the eggs on the net. The advantage of this arrangement is that once the eggs hatch, the larvae fall directly into the underlying waste. This arrangement eliminates the need for separate mating and culturing units that have been essential in the prior art.

In the second embodiment, S2, the reactor is shaped like a rectangle with slanting side walls (Figures 16.5–16.7). The reactor, typically, made up of high-density polythene sheets of 2–3 mm thickness, has a length of 150 cm, breadth of 90 cm, and height of 25 cm. The longer side walls are inclined at an angle of 45° to facilitate the ease of movement of the larvae at the migration stage. The depth of the feed in the

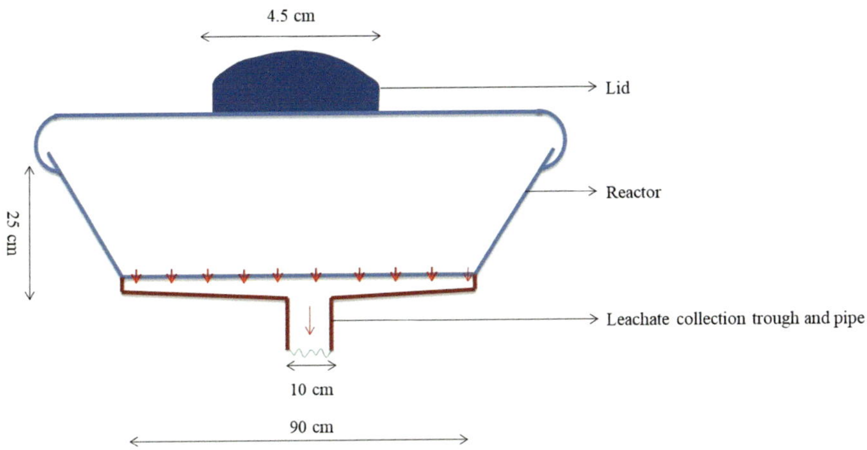

FIGURE 16.6 Embodiment S2: Side view of the reactor, showing the leachate collection system.

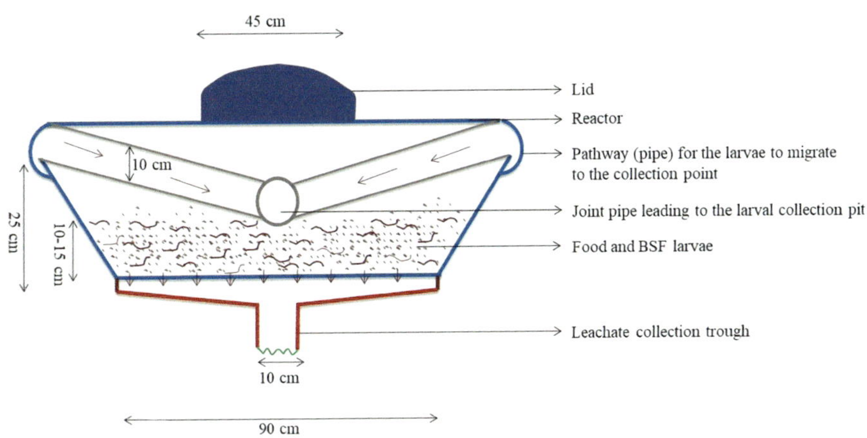

FIGURE 16.7 Embodiment S2: Side view of the reactor, showing the larval collection system.

reactor is kept between 10 and 15 cm for better aeration and feeding by the larvae (Figure 16.7). The bottom of the reactor is perforated, with holes of 3 cm in diameter, for draining leachate. The distance between the holes is kept at 10 cm. The bottom of the reactor is covered with a plastic-coated steel mesh of 0.4–0.5 mm diameter, to prevent the escape of larvae from the bottom while simultaneously allowing free passage for the drainage of the leachate (Figure 16.8). The leachate drips into a slightly inclined trough and then pours down into a 10 cm wide common gutter (Figures 16.6 and 16.7). It is eventually collected in a leachate collection vessel. Both the longer upper edges of the slanted walls are encased in a hollow pipe of 10 cm diameter, to facilitate the migration of the larvae. Both these pipes converge into a common joint

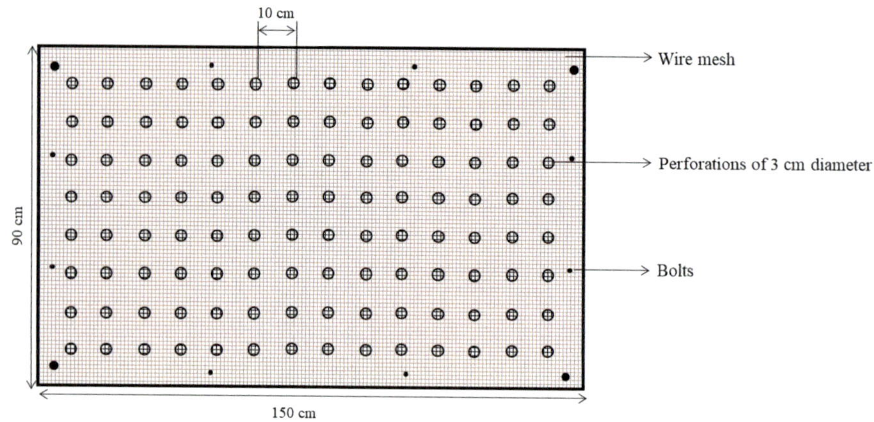

FIGURE 16.8 Embodiment S2: Top view of the reactor bottom, showing the perforated bottom overlain with a mesh.

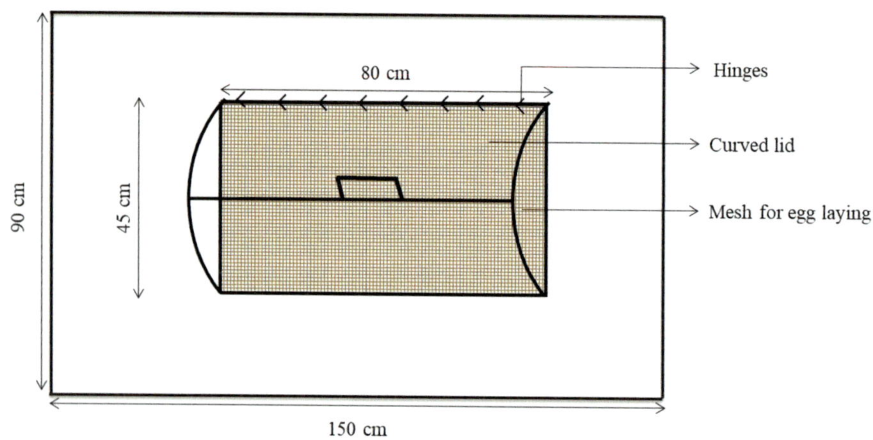

FIGURE 16.9 Embodiment S2: Top view of the reactor, showing the lid and egg laying mesh.

pipe leading to the larva collection pit. This system does not require artificial lighting for the larval migration (Figures 16.5 and 16.7). The reactor top has an opening of 80 cm (length) and 45 cm (breadth). This serves the triple purpose of feed inlet, frass removal, and a gateway to enable the native female BSF to go inside the reactor to lay eggs (Figure 16.9). Similar to Embodiment 1, the opening is covered by a semi-circular lid attached by hinges. The lid, when closed, leaves a 10 cm gap on either side. Through this gap, the female BSF enters the reactor and lays eggs on the mesh trough attached to the lid. The trough lowers into the reactor when the lid is closed. It is made of polyester net having openings of 1 mm in diameter (Figure 16.4). When the eggs are laid on the mesh hatch, the larvae fall directly into the feed below. This set-up also eliminates the need for separate provisions for BSF mating and culture, which has been essential in the prior art.

The reactors need to be placed inside a well-ventilated room with adequately large windows, allowing for abundant sunlight to come in. The windows need to be covered with PVC-coated fibreglass nets having holes of 1 mm in diameter. For allowing the native BSF to enter the room at the beginning of the lifecycle, the nets are kept slightly open. The presence of large windows also provides sunlight for the BSF's required photoperiod. This reduces the use of artificial lighting for mating. Some potted plants with large leaf areas are kept inside the room to provide surfaces for the BSF mating and resting. To facilitate pupation, containers filled with dry sand are kept inside the room. Some of the larvae collected in the larval collection vessel are transferred to the pupal container for pupation.

With these arrangements, the entire lifecycle of the BSF is controlled inside a single room, which also houses the bioreactors. This has the potential to reduce the cost of the whole process very substantially. For the separation of larvae from the frass, the harvest from the reactor is placed in a harvesting container, which is a large tray lined with a black polythene sheet. The larvae self-migrate under the sheet, as they are photophobic and thus separate out from the frass. As a result, frass and larvae are collected separately without any mechanized intervention. The young larvae are put back into the reactor for feeding, and the ones in the pre-pupal stage are ready for use as animal feed, for bio-oil extraction, etc. The collected frass is kept for air-drying for a period of 1–2 days. This leads to the maturation of the frass, which can then be utilized as an organic fertilizer.

16.5.1 Priming of the Reactor

As the reactors depend upon native BSF, the reactors are first started with only fresh fruit wastes. The smell attracts the native female BSF to come in and lay eggs on the egg laying mesh attached to the lid of the reactor. Once the eggs are laid, the reactor can be fed other wastes.

16.5.2 Reactor Operation and Production of Larvae and Frass

The reactors can be operated in two ways. The first is to run the reactor in a pulse-fed continuous operation. The reactors are fed with cooked food waste, abattoir waste, and fruit/vegetable waste in a 1:1:1 ratio, on a dry weight basis. The feed is put at a rate of 10 kg/day (fresh weight), every alternate day. This is done until 20–25 days of reactor operation, as by that time most of the larvae would have gone to the last instar stage. Then the reactors are stirred and kept open for drying the frass. After 2–3 days, the frass and leftover larvae in the reactor are harvested and kept in the harvesting container, where the larvae separate out from the frass. The frass thus collected is air dried and stored in airtight containers. The matured larvae are taken to be used as feed, for biofuel production, and some for pupation. The young larvae, along with a small amount of frass, are put back into the reactor to start the process again.

The second operational strategy is to run the reactor in a continuous manner for a period of time or until the reactor is not filled with the frass. This period of time may take from 6 months to 1 year. The reactors are fed with cooked food waste, abattoir waste, and fruit/vegetable waste in a 1:1:1 ratio, on a dry weight basis. The feed is put at a rate of 10 kg/day (fresh weight), every alternate day. The pre-pupae self-migrate out from the reactor into the larval collection container. Out of the total larvae collected, some are put into the pupation container, and others are taken to be used as feed. When the reactor is filled with the frass, it is harvested in the same manner as in the first option.

REFERENCES

Abbasi, S.A., 1999. Water quality indices – state of the art. *Journal of the Institution of Public Health Engineers*, (1), pp.13–24. https://www.researchgate.net/publication/317824095_Water_quality_indices_-_state_of_the_art

Abbasi, S.A., 2018. The myth and the reality of energy recovery from municipal solid waste. *Energy, Sustainability and Society*, *8*(1), pp.1–15, doi: 10.1186/s13705-018-0175-y.

ČiČková, H., Newton, G.L., Lacy, R.C. and Kozánek, M., 2015. The use of fly larvae for organic waste treatment. *Waste Management*, *35*, pp.68–80.

Fowles, T.M. and Nansen, C., 2020. Insect-based bioconversion: value from food waste. In: Närvänen, E., Mesiranta, N., Mattila, M., Heikkinen, A. (Eds.), *Food Waste Management*. Palgrave Macmillan, Cham, pp. 321–346.

Kumar, S., Negi, S., Mandpe, A., Singh, R.V. and Hussain, A., 2018. Rapid composting techniques in Indian context and utilization of black soldier fly for enhanced decomposition of biodegradable wastes: A comprehensive review. *Journal of Environmental Management*, *227*, pp.189–199.

Part IV

Zoocomposting with Termites

Termigradation

17 Termites and Their Unique Ability to Biodegrade Ligninous Waste and Fix Nitrogen

17.1 INTRODUCTION

Few animals are more dreaded and loathed by laypersons than termites. They are dreaded because termites can eat through not only precious furniture, wooden artefacts, books, clothing, etc, but even the woodwork of entire buildings – making them hallow and worthless within days. And they are loathed because the termites do it so silently and surreptitiously that often one isn't able to perceive the damage before it is too late.

But, as is true with each and every other species in the world, the adverse impacts of termites are felt by human beings not because termites interfere with human existence. It is the other way – due to the massive encroachments of human beings on forests and other land areas, termites have been deprived of their main source of food: ligninous debris of trees and shrubs. They are also deprived of open spaces needed to establish their nests. This forces them to eat whatever they can of the ligninous and cellulosic material they can reach in human dwellings.

But even when we go by the general human perspective of a "pest", only about 10% of all termite species have been found to be a nuisance. The rest 90% are not capable of inflicting any significant damage to human dwellings (Premalatha et al., 2013).

Otherwise, left to themselves, all species of termites together perform two of the most essential and daunting of ecosystem services: (a) biodegrading ligninous material like stems and twigs of trees and shrubs which most other animals can not biodegrade and (b) soil-working – comminuting the soil around them, enriching it with nitrogen and other nutrients, bringing its pH to as near neutral as they can, and moving it about to enhance its water holding capacity, oxygenation, and overall rejuvenation. As we had elaborated in Chapter 1, most animals contribute to biodegrading waste and processing soil in one way or the other, but termites do both at scales that are matched only by two other groups of animals – earthworms and ants. It is for this reason that the trio has been given the sobriquet of "soil engineers". But even within this exclusive list, termites are unique as they possess virtues the other two groups do not possess – the ability to biodegrade lignin and assist nitrogen fixation. They do have one thing in common with ants: They are as highly eusocial as ants, with a very rigid and well-defined hierarchical system of position, duties, and responsibilities.

DOI: 10.1201/9780429331886-21

This chapter presents an overview of termites and their special attributes, which are relevant to their possible use in zoocomposting. It is a prelude to Chapter 18, which provides the gist of these authors' efforts in harnessing termites and using them in the zoocomposting of ligninous and other hard biowaste.

17.2 THE CHALLENGES OF BIOCOMPOSTING LIGNINOUS AND "HARD" BIOWASTE

Biomass, which is high on volatile solids (VS) but low on lignin, such as animal manure and leaves of several plants, is amenable to fairly rapid biodegradation under ambient conditions. Due to this feature, such biomass can be biocomposted as well as biodegraded aerobically or anaerobically to generate solid/liquid/gaseous biofuels (Osman et al., 2021; Singh et al., 2023).

But biomass high in lignin – such as branches and stems of several plants, maize stalks, bagasse, coconut shells, leaves of xerophytes like prosopis (*Prosopis juliflora*) (Figure 17.1), and other "hard" carbonaceous waste such as thermocol and paper cups – resists biodegradation. It can not be biocomposted efficiently either through direct composting or earthworm (EW)-mediated vermicomposting. Nor can it be processed effectively by BSF or other animal species explored so far in biocomposting (Srivastava et al., 2012).

As much as 150 billion tonnes of lignin is generated per annum (Khan et al., 2022), and this figure is still to peak (Cajuko et al., 2021, He et al., 2017). In the absence of commercially viable methods to valorize ligninous waste, it is either disposed of untreated or burnt as a solid fuel (Sajjadi et al., 2021; Al-Tohamy et al., 2022). If bioprocesses can be developed for product-specific conversion of lignin into

FIGURE 17.1 The leaves of prosopis are deeply resistant to biodegradation.

useful chemicals, much benefit can be derived from the presently valueless ligninous waste (Nguven et al., 2021).

The termite gut is known to possess an extraordinarily rich and diverse microflora – encompassing bacteria, yeast, and other eukaryotic symbionts (Ali et al., 2022). Some of these are believed to be responsible for the special ability of termites to biodegrade lignin. It is being increasingly acknowledged that there is a great abundance of species, enzymes, and genes in the termite "gut digestome" (Sun et al., 2013; Rajeswari et al., 2021). This makes it possible that a complex web of interactions between these constituents exists, which bestows upon termites their ability to biodegrade lignin. This realization has led in recent years to intense interest and effort in extracting, isolating, and exploring the constituents of the termite gut. There is a growing body of evidence that underscores the importance of symbiotic yeasts, their extensive distribution across the termite gut, and the role they play in termite–symbiont interactions (Ali et al., 2022).

17.3 TERMITES: SOME RELEVANT FACTS

Termites are believed to be the oldest of all social insects. While cockroaches, with whom termites are more closely related taxonomically than they are to any other group of insects, have been present for 250 million years, termites are estimated to have been around for 220 million years (Pearce, 1997). Presently, as many as 2,500 species of termites are known. They are present in all the continents, except for very cool regions of the world (Figure 17.2).

Some species built subterranean nests which are entirely underground, many make subterranean nests with aboveground mounds, and some make arboreal nests well above the ground (Figure 17.3).

Even though termites are also frequently referred to as "white ants", they are not only different from ants but are wary of ants because the latter lose no opportunity to raid termite mounds and make meals of the termite workers. Ants have a thin waist

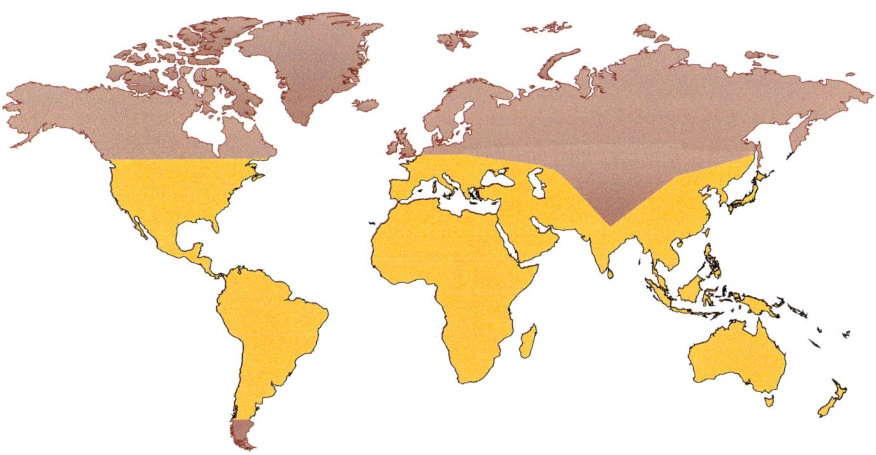

FIGURE 17.2 Termites ☐ are present in all but very cold regions of the Earth.

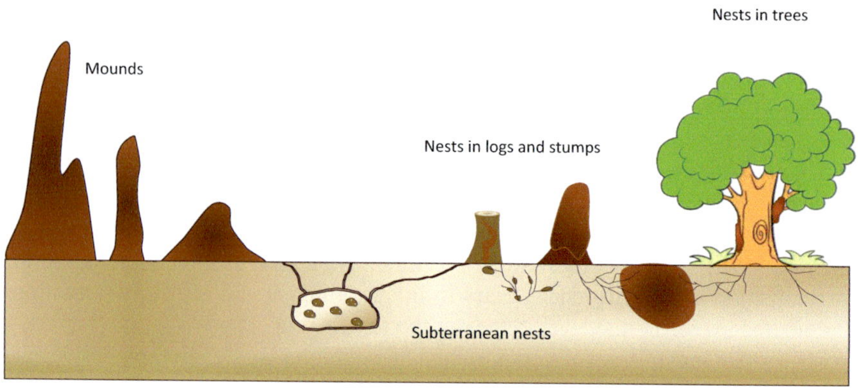

FIGURE 17.3 Different types of nests built by different termite species.

FIGURE 17.4 White-bodied termite workers on a termite mound.

and elbowed antennae, which are absent in termites. And while the males of ants and other social insects die after mating, the termite kings remain with the queen all their lives.

Termite workers are wingless and sexually incapable. Except for the members of Hodotermitidae, all termite workers are blind. Their abdominal sclerotization is poor, which causes them to become white (Figure 17.4). Termite workers work very hard and long, almost incessantly, tunnelling and collecting food, building and maintaining their nest, feeding and grooming the young, and serving the higher castes of soldiers, king, and queen (Figure 17.5).

FIGURE 17.5 A termite queen with king, soldier, and worker. (https://commons.wikimedia.org/wiki/File:Termite_queen_1.jpg.)

Termite soldiers, as the title suggests, are meant to smell trouble and mount a defence. They have a longer and wider head than that of the workers, coloured yellow to brown, with more muscle than possessed by the workers. Soldiers of some species have a long, nose-like modification, from which they are able to fire toxic glandular secretions with the intent of overwhelming an attacker. Depending on the genera, soldiers are either all male, all female, or a 50:50 mixture of both.

The reproduction in termites occurs with winged reproductives, male and female (Figure 17.6), who emerge from their nests to form swarms, preferably around light sources (Figure 17.7). Humid, rainy conditions are among the signals alates are believed to be following in the timing of their emergence. Upon landing, the alates of some species shed their wings (Figure 17.8) directly, while the alates of other species do it after contact with the opposite sex. As soon as the mates choose each other, they dig out tunnels in soil, or in crevices and cracks available in wood, to protect themselves from predators as well as to construct the royal chamber or the copularium. This effort takes several days before the "marriage" is consummated. Since it is much easier and quicker to dig into soil or wood that has been moistened, the rainy season is the one preferred by the alates to emerge.

In due course, termite queens become egg-laying machines and become several times larger in size than the king, the soldiers, and the workers (Figure 17.5).

FIGURE 17.6 A typical winged reproductive (alate). (https://commons.wikimedia.org/wiki/File:Flying_Termite.jpg.)

FIGURE 17.7 Winged termite reproductives (alates) swarming in quest of mates. (https://commons.wikimedia.org/wiki/File:Winged_termites_swarming_IMG_20201021_191840.jpg.)

FIGURE 17.8 A dealate among wings shaded by him/her and others. (https://commons.wikimedia.org/wiki/File:Wings_of_Winged_Termite_swam_from_ValparaiJEG0102.jpg.)

Lower termites – who can not themselves digest lignin but rely on particular species of protozoa living in their guts to process lignin ingested by the termites – are slower in egg production than the higher termites. In the latter, egg production by the queens can occur explosively. For instance, queens of the genus *Macrotermes* can produce 30,000–40,000 eggs per day, adding to tens of millions of eggs per year. This goes on throughout the queen's lifetime, which spans 15–20 years but can be as long as 50 years.

Figure 17.9 illustrates the termite life cycle.

17.4 THE ROLE AND THE POTENTIAL OF TERMITES

Termites are not only among those very few animals who can feed upon lignin and biodegrade it (Geib et al., 2008; Rajan et al., 2022), they are also among very few animals who are capable of fixing nitrogen with the help of nitrogen-fixing bacteria they carry in their guts (Kong and Varma, 2006; Bignell et al., 2011). Wood wasps, beetles, and wood-feeding roaches are among the other animals that host bacterial communities capable of degrading lignin (Luo et al., 2019; Tokuda et al., 2019) and harbour nitrogen-fixing microflora. But the scale at which termites operate makes the role of other insect-eating nitrogen-fixing animals appear insignificant in comparison. Termites form as much as 40%–60% of belowground biomass in some biotopes and about 10% of all the world's zoomass (Loveridge & Moe, 2004; Plunketts, 2021). Their total combined weight has been estimated as 445 million tonnes, which is about 100 million tonnes more than the weight of the 8 billion humans (UN, 2022) presently in the world.

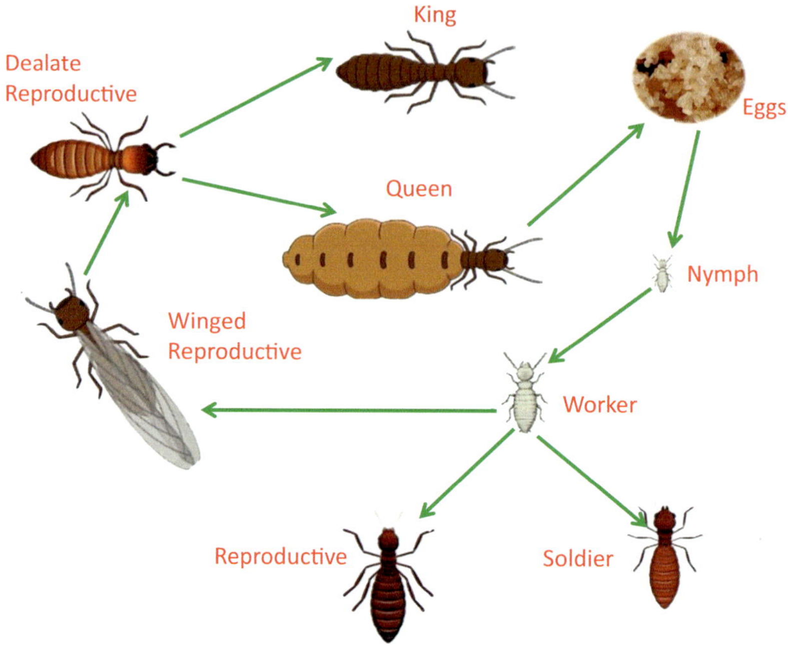

FIGURE 17.9 The termite life cycle.

To keep stable and rising the multi-trillion termite population that makes up a large fraction of all animal biomass, termite queens (Figure 17.10) not only produce, as mentioned earlier, up to 10 million eggs per year, they also have long life-spans during which they initiate and populate several termite colonies (Holt & Lapage, 2000). The overwhelming majority of offspring that come out of the termite eggs are workers. As stated earlier, workers of most species happen to be blind and, barring a few exceptions, terminally sterile. Additionally, they keep working relentlessly, around the clock, ingesting and comminuting plant debris with soil, making "sheathings" or galleries with soil of similar particle sizes to facilitate movement, and constructing nests. In the course of this, they keep moving and enriching the soil while also generating pores in it to enhance its aeration and its water retention ability (Joquet et al., 2016). They enrich soil nitrogen and make several other plant nutrients bioavailable, besides performing numerous other ecosystem services (Joquet et al., 2011, 2016). The impact of termite activity on the spatial heterogeneity of biotopes can be so powerful that it can last for years to centuries (Erens et al, 2015; Joquet et al, 2016). It is this effort, massive in scale as well as effect, which has led to termites being recognized as "soil engineers", alongside ants and earthworms.

The additional ability of termites to digest lignin is of overriding significance because if the microorganisms and biomolecules responsible for it can be made to biocompost ligninous waste at economically viable rates, a very large fraction of biomass that is presently unutilizable in biofuel production can become utilizable. Very extensive research in recent years has been focused on this aspect of termites

FIGURE 17.10 A termite queen with workers in attendance.

(Khan & Ahmad, 2018; Rajan et al., 2022), but a commercially viable process is yet to be developed.

So critical is the role of termites that, but for them, all the soil in the world will not only become barren and infertile but will also get sealed off by undecaying wood, making all agriculture impossible (Janusz et al., 2017). Indeed, but for the leading role of termites in biocomposting ligninous material, the other two "soil engineers" – ants and earthworms – will also become ineffectual.

17.5 TERMITES: ARCHITECTS EXTRAORDINARY

Having already received universal acclaim as one of nature's three great "soil engineers", termites in more recent years are attracting attention also as nature's unique construction engineers and architects. Different species of termites construct nests of different sizes, shapes, and locations (Figures 17.11 and 17.12). These nests may, from the outside, look like mounds of compacted soil, but their insides are architectural marvels for their functionality, eco-sensitivity, and optimality. Considering that these nests have been built by workers who are just a few millimetres long, high, and wide, weighing a few milligrams, and are blind and mute, the precision, quality, and size of the constructions that termites achieve are mind-boggling. The cathedral termites (*Nasutitermes triodiae*), which are found in Australia's states of Queensland and Western Australia, build some of the tallest structures that non-humans are known

FIGURE 17.11 Termite mounds can vary in size from minuscule to tiny (a), large (b), to very large (below left and right (c, d)).

to build. Their nests can be as tall as 30 ft. If we take the ratio of an average termite worker's height and that of the 30 ft tall nest the workers build, and calculate the height of a building of the same ratio but human beings as workers, the resulting building will be 5 km tall, much taller than the tallest building in the world!

And termite-constructed "buildings" are also incredibly strong. If human-constructed buildings have a safety ratio of 1–3, the ratio for termite buildings is 50 and above!

Even more remarkably, this engineering feat is accomplished with materials no more sophisticated and/or costly than the surrounding soil. Termite workers pick soil particles of desired water content and cement them together with their saliva (Krishnan, 2020). The resultant building blocks, or "termite bricks", are called "bolus". By very fine selection of appropriate soil particles and by precise bolus formation, the thickness and the porosity of the termitaria walls are controlled, which provides proper air-circulation, oxygen flow, pathway networks, and other sub-structures for ensuring the comfort as well as the safety of the termitaria inhabitants.

FIGURE 17.12 Termite nests can take interesting shapes: mushroom (a), graveyard epitaphs (b), and sculptures (c and d).

- As we realize the uniqueness and the utility value of the termite architecture, we are making efforts to mimick its special features. A famous example is the Eastgate Centre in Harare, Zimbabwe (Figure 17.13), designed by Mick Pearce and Arup Associates. It employs a passive cooling system, which is inspired by the termites. With it, the building achieves 90% less energy for ventilation than conventional buildings of the same size.

How strong and durable the termite nests and the overlaying mounds are can be realized from a 2018 report (Wilkinson et al., 2018) on a gigantic network of 200 million termite mounds that was discovered in Brazil. Covering an area of 230,000 km^2, which is as large as that of Great Britain, the mounds were built 700–4,000 years ago by a single termite species, *Syntermes dirus*. In the process, about 10.1 km^3 of soil, equivalent to 4,000 Great Pyramids of Giza, was excavated (Hsu, 2018). The vast network of mounds can be spotted on Google Earth, but at the ground level, it is largely hidden from view by semi-arid, thorny-scrub forests. The mounds form a regular pattern, with about 20 m separating one from the other (Figure 17.14).

FIGURE 17.13 The Eastgate Centre at Harare, Zimbabwe, the architecture of which has been inspired by the unique features of the termite nests, and which achieves huge savings on air-conditioning. Credit: damien_farrell, P1000957@flickr, Public domain (https://www.flickr.com/photos/92094658@N00/7661976432/in/photolist-2izidMH-cF4CDW-5NGg5d-4CVyxZ-2kKXqsj-4CZPah-4CZP8f-ayVvjg-gQXed-52bqi8-2psPQdn-2psUs7n-2psPJCj-2psHCjy-2psWCpb-2psUrCm.)

FIGURE 17.14 Systematically constructed termite mounds, discovered in Brazil.

When the best of human-constructed buildings can not survive more than a few hundred years without periodic reinforcement and renovation, for these mounds to have survived for up to 4,000 years – braving the innumerable calamities of storms, floods, droughts, hurricanes, etc that must have assaulted them – is an unbelievable feat of construction engineering. Interestingly, *S. dirus* hasn't yet called it a day, and is still constructing new mounds in that area.

17.6 TERMITES AS DELICACIES

Humankind has evolved as an entomophagous (insect-eating) species, and termites appear to be a preferred source of food for human beings for over a million years (Blackwell and d'Errico, 2001; van Huis, 2003; Tabassum-Abbasi et al., 2016). There is evidence that the predecessor of us, *Homo sapiens – Australopithecus robustus* – used bones and twigs for harvesting termites from their nests. Even now, termites are among the world's most widely eaten insects, next only to locusts, especially in Africa (Figures 17.15 and 17.16), Asia, and South America (De Foliart, 2002; Jongema, 2017; Siddiqui et al., 2023). Termites are rich in protein (34%–40%

FIGURE 17.15 Cooked termites as a delicacy. (https://commons.wikimedia.org/wiki/File:A_young_boy_eating_cooked_termites.jpghttps://commons.wikimedia.org/wiki/File:A_young_boy_eating_cooked_termites.jpg.)

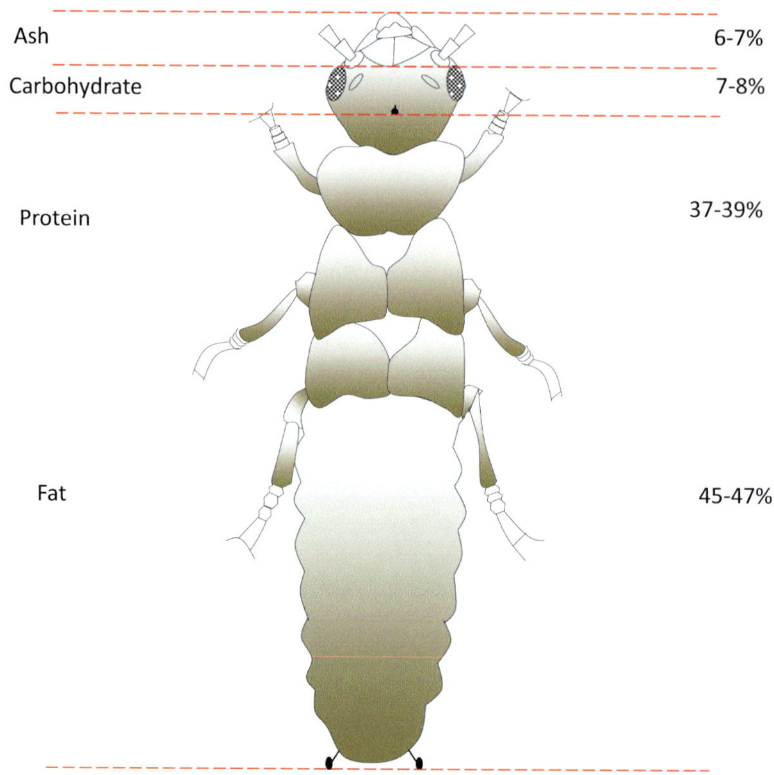

FIGURE 17.16 Indicative nutrient composition of termites and the body parts in which each is predominantly concentrated.

by weight), vitamins (especially A, B, and C), and minerals (especially iron, zinc, and calcium). Winged termites and termite queens are among the elite constituents of termite-based human food. Termite powder is among the prominent constituents of folk medicine (Mahapatro, 2015).

REFERENCES

Ali, S.S., Al-Tohamy, R., Mohamed, T.M., Mahmoud, Y.A.G., Ruiz, H.A., Sun, L. and Sun, J., 2022. Could termites be hiding a goldmine of obscure yet promising yeasts for energy crisis solutions based on aromatic wastes? A critical state-of-the-art review. *Biotechnology for Biofuels and Bioproducts*, *15*(1), p.35.

Al-Tohamy, R., Ali, S.S., Li, F., Okasha, K.M., Mahmoud, Y.A.G., Elsamahy, T., Jiao, H., Fu, Y. and Sun, J., 2022. A critical review on the treatment of dye-containing wastewater: Ecotoxicological and health concerns of textile dyes and possible remediation approaches for environmental safety. *Ecotoxicology and Environmental Safety*, *231*, p.113160.

Backwell, L. and d'Errico, F. 2001. *Early hominids ate termites. News24, Trusted news. First.* https://www.news24.com/news24/xarchive/archive/early-hominids-ate-termites-20010116. Last accessed November 2024.

Bignell, D.E., Roisin, Y. and Lo, N., 2011. *Biology of Termites: A Modern Synthesis*, Springer Dordrecht, XIV, p.576. https://doi.org/10.1007/978-90-481-3977-4.

Cajnko, M.M., Oblak, J., Grilc, M. and Likozar, B., 2021. Enzymatic bioconversion process of lignin: Mechanisms, reactions and kinetics. *Bioresource Technology, 340*, p.125655.

De Foliart, G.R., 2002. *The Human Use of Insects as a Food Resource: A Bibliographic Account in Progress*. University of Wisconsin, Madison.

Erens, H., Mujinya, B.B., Mees, F., Baert, G., Boeckx, P., Malaisse, F. and Van Ranst, E., 2015. The origin and implications of variations in soil-related properties within Macrotermes falciger mounds. *Geoderma, 249*, pp.40–50.

Geib, S.M., Filley, T.R., Hatcher, P.G., Hoover, K., Carlson, J.E., Jimenez-Gasco, M.M., Nakagawa-Izumi, A., Sleighter, R.L. and Tien, M. 2008. *Lignin degradation in wood-feeding insects. Proceedings of the National Academy of Sciences of the United States of America, 105*(35), pp.12932–12937. https://doi.org/10.1073/pnas.0805257105.

He, W., Gao, W. and Fatehi, P., 2017. Oxidation of kraft lignin with hydrogen peroxide and its application as a dispersant for kaolin suspensions. *ACS Sustainable Chemistry & Engineering, 5*(11), pp.10597–10605.

Holt, J.A. and Lepage, M., 2000. Termites and soil properties. In: Abe, T., Bignell, D.E., Higashi, M. (Eds), *Termites: Evolution, Sociality, Symbioses, Ecology*, Springer, US pp.389–407.

Hsu, C. 2018. *An impromptu meeting in a remote Brazilian town led to a study that made headlines around the world*. https://www.buffalo.edu/news/releases/2018/12/027.html. Last accessed November 2024.

Janusz, G., Pawlik, A., Sulej, J., Świderska-Burek, U., Jarosz-Wilkołazka, A. and Paszczyński, A., 2017. Lignin degradation: Microorganisms, enzymes involved, genomes analysis and evolution. *FEMS Microbiology Reviews, 41*(6), pp.941–962.

Jongema, Y. 2017. *List of Edible Insects of the World*. https://www.wur.nl/en/research-results/chair-groups/plant-sciences/laboratory-of-entomology/edible-insects/worldwide-species-list.htm. Accessed 09 October 2022.

Jouquet, P., Bottinelli, N., Shanbhag, R.R., Bourguignon, T., Traoré, S. and Abbasi, S.A., 2016. Termites: The neglected soil engineers of tropical soils. *Soil Science, 181*(3/4), pp.157–165.

Jouquet, P., Traoré, S., Choosai, C., Hartmann, C. and Bignell, D., 2011. Influence of termites on ecosystem functioning. Ecosystem services provided by termites. *European Journal of Soil Biology, 47*(4), pp.215–222.

Khan, M.A. and Ahmad, W. eds., 2018. *Termites and Sustainable Management, Volume-1 – Biology, Social Behaviour and Economic Importance*. Springer, Cham, Switzerland. https://doi.org/10.1007/978-3-319-72110-1.

Khan, R.J., Lau, C.Y., Guan, J., Lam, C.H., Zhao, J., Ji, Y., Wang, H., Xu, J., Lee, D.J. and Leu, S.Y., 2022. Recent advances of lignin valorization techniques toward sustainable aromatics and potential benchmarks to fossil refinery products. *Bioresource Technology, 346*, p.126419.

Kong, H. and Varma, A., 2006. *Intestinal Microorganisms of Termites and Other Invertebrates, Soil Biology*, 6, Springer, Berlin, Heidelberg, XXIV, p.484. https://doi.org/10.1007/3-540-28185-1.

Krishnan, A. 2020. *Architectural secrets of termite mounds, Mongabay*. https://india.mongabay.com/2020/10/architectural-secrets-of-termite-mounds/. Last accessed November, 2024.

Loveridge, J.P. and Moe, S.R., 2004. Termitaria as browsing hotspots for African megaherbivores in miombo woodland. *Journal of Tropical Ecology, 20*(3), pp.337–343.

Luo, C., Li, Y., Chen, Y., Fu, C., Long, W., Xiao, X., Liao, H. and Yang, Y., 2019. Bamboo lignocellulose degradation by gut symbiotic microbiota of the bamboo snout beetle *Cyrtotrachelus buqueti*. *Biotechnology for Biofuels, 12*(1), pp.1–16.

Mahapatro, 2015. Ethno-entomology: Termites as food and medicine, *Current Biotica*, *9*(1), pp.5–8.

Nguyen, L.T., Phan, D.P., Sarwar, A., Tran, M.H., Lee, O.K. and Lee, E.Y., 2021. Valorization of industrial lignin to value-added chemicals by chemical depolymerization and biological conversion. *Industrial Crops and Products*, *161*, p.113219.

Osman, A.I., Mehta, N., Elgarahy, A.M., Al-Hinai, A., Al-Muhtaseb, A.A.H. and Rooney, D.W., 2021. Conversion of biomass to biofuels and life cycle assessment: A review. *Environmental Chemistry Letters*, *19*, pp.4075–4118.

Pearce, M.J., 1997. *Termites: Biology and Pest Management*. Cab International, Wallingford.

Plunkett, 2021. *Twenty unbelievable facts about termites, Blog 20*. https://plunketts.net/blog/20-unbelievable-facts-termites. Last accessed September 2022.

Premalatha, M., Tauseef, S.M., Abbasi, T. and Abbasi, S.A., 2013. A website to showcase the human-friendly aspects of termites. *Current Science*, *105*(11), pp.1459–1460.

Rajan, R., Sudhakaran, A., Ravindranath, A., Sivalingam, R., and Kumar, R., 2022. Isolation, screening and identification of lignin degraders from the gut of termites *odontotermes obesus*. *Journal of Pure and Applied Microbiology*, *16*(3), pp.1696–1704.

Rajeswari, G., Jacob, S., Chandel, A.K. and Kumar, V., 2021. Unlocking the potential of insect and ruminant host symbionts for recycling of lignocellulosic carbon with a biorefinery approach: A review. *Microbial Cell Factories*, *20*(1), p.107.

Sajjadi, M., Ahmadpoor, F., Nasrollahzadeh, M. and Ghafuri, H., 2021. Lignin-derived (nano) materials for environmental pollution remediation: Current challenges and future perspectives. *International Journal of Biological Macromolecules*, *178*, pp.394–423.

Siddiqui, S.A., Fernando, I., Saraswati, Y.R., Rahayu, T., Harahap, I.A., Yao, Q., Nagdalian, A., Blinov, A. and Shah, M.A., 2023. Termites as human foods—A comprehensive review. *Comprehensive Reviews in Food Science and Food Safety*, *22*(5), pp.3647–3684.

Singh, A., Prajapati, P., Vyas, S., Gaur, V.K., Sindhu, R., Binod, P., Kumar, V., Singhania, R.R., Awasthi, M.K., Zhang, Z. and Varjani, S., 2023. A comprehensive review of feedstocks as sustainable substrates for next-generation biofuels. *BioEnergy Research*, *16*, pp.105–122.

Srivastava, S., Mudgal, V. and Jain, R.K., 2012. Lignin-its role and importance in animal nutrition. *International Journal of Livestock Research*, *2*(1), pp.7–23.

Sun, J., Ding, S.Y. and Peterson, J.D. eds., 2013. *Biological Conversion of Biomass for Fuels and Chemicals: Explorations from Natural Utilization Systems*. Royal Society of Chemistry, UK.

Tabassum-Abbasi., Abbasi, T. and Abbasi, S.A., 2016. Reducing the global environmental impact of livestock production: The minilivestock option, *Journal of Cleaner Production*, *112*, pp.1754–766.

Tokuda, G., 2019. Plant cell wall degradation in insects: Recent progress on endogenous enzymes revealed by multi-omics technologies. *Advances in Insect Physiology*, *57*, pp.97–136.

United Nations (UN), 2022. World population to reach 8 million in November 2022. https://www.un.org/en/desa/world-population-reach-8-billion-15-november-2022#:~:text=The%20global%20population%20is%20projected,today%20on%20World%20Population%20Day. Last accessed September 2022.

Van Huis, A., 2003. Insects as food in sub-Saharan Africa. *International Journal of Tropical Insect Science*, *23*(3), pp.163–185.

Wilkinson, B., 2018. Ancient termite megapolis as large as Britain found in Brazil, CNN travel. https://edition.cnn.com/travel/article/termites-brazil-old-intl/index.html. Last accessed November 2024.

18 Opening a New Frontier in Bioprocess Technology
Direct Use of Termites in Biodegrading Highly Ligninous Waste

18.1 INTRODUCTION

As brought out in the previous chapter, termites are not only among the very few animals capable of degrading lignin (Table 18.1), but they are also by far the most numerous and effective.

However, it is not possible to use termites in the same way as earthworms can be used in controlled biocomposting because termites are rigidly eusocial and their workers do not normally breed in captivity. These constraints have, till now, prevented the controlled use of termites in waste treatment. This chapter is a recapitulation of an 18-year-long study in which attempts were made to harness termites for controlled degradation of ligninous and other hard biowaste. It describes successful efforts in utilizing termites in *in situ* termireactors (ISTs) as well as developing termite-based "bioengines" which are capable of processing specific wastes with due process monitoring and control. The work has been granted a patent and, perhaps, opens a new frontier in bioprocess technology, which can be called "termigradation".

TABLE 18.1
Insects and Other Invertebrates Explored so far Towards Degradation of Solid Waste: All Available Reports Are on Studies Done at Laboratory Scale

Species	Common Name	Targeted Solid Waste	References
Calliphora vicina	Blue blow fly	Slaughter waste	Barroso et al. (2013) and Pastor et al. (2015)
Chrysomya megacephala	Oriental blue fly	Restaurant waste, manure slaughter wastes	Li et al. (2012), Barroso et al. (2013), Yang and Liu (2014), Cickova et al. (2015), and Pastor et al. (2015)
Conicera tibialis	Phorid fly or scuttle fly	Animal carcasses	Sharanowski et al. (2008)
Coproica ferruginata	Lesser dung fly	Solid urban waste	Morales and Wolf (2010)

(*Continued*)

DOI: 10.1201/9780429331886-22

TABLE 18.1 (*Continued*)
Insects and Other Invertebrates Explored so far Towards Degradation of Solid Waste: All Available Reports Are on Studies Done at Laboratory Scale

Species	Common Name	Targeted Solid Waste	References
Curtonotum helvum	Curtonotid fly	Solid urban waste	Morales and Wolf (2010)
Cydia pomonella	Codling moth	Starch and cheese waste, waste water sludge	Meats et al. (2004), Brar et al. (2008), and Fowles and Nansen (2019)
Desmometopa varipalpis	Free loader fly	Solid urban waste	Morales and Wolf (2010)
Drosophila melanogaster	Fruit fly	Urban waste, wastewater	Wheeler (1987) and Clark et al. (2022)
Galleria mellonella	Greater wax moth larvae	Polyethylene polystyrene	Billen et al. (2020)
Lucilia sericata	Common green bottle fly	Carrion, meat, manure, slaughter waste	Nuove et al. (1995), Marchaim et al. (2003), Yehuda et al. (2011), Barroso et al. (2013), Cickova et al. (2015), and Pastor et al. (2015)
Musca autumnalis	Face fly	Cattle manure	Koo et al. (1980), Arends and Wright (1981), and Cickova et al. (2015)
Musca domestica	House fly	Manure, garbage, food wastes, restaurant waste, corn sillage, sawdust, gibberlin fermentation residue, agro-processing waste products	Dordevic et al. (2008), Pastor et al. (2011), Wang et al. (2013), Cickova et al. (2015), Pastor et al. (2015), Roffeis et al. (2015), Vanzantan et al. (2015), Yang et al. (2015), Jiang et al. (2017), Niu et al.,(2017), Bloukounon-Goubalan et al. (2020), Ganda et al. (2019), Fowles and Nansen (2019), and Miranda et al. (2021)
Ornidia obesa	Syrphid fly	Solid urban waste	Morales and Wolf (2010)
Orthaga aenescens	Shout moth	Solid urban waste	Morales and Wolf (2010)
Protaetia brevitarsis	Oriental edible beetle larvae	plant matter	Li et al. (2019)
Protophormia terraenovae	Blue-bottlefly	Slaughter house waste	Barroso et al. (2013) and Pastor et al. (2015)
Sarcophaga carnaria	Common flesh flies	Decaying meat	Pape (1987), Yehuda et al. (2011), Cickova et al. (2015), and Pastor et al. (2015)

(*Continued*)

TABLE 18.1 (*Continued*)
Insects and Other Invertebrates Explored so far Towards Degradation of Solid Waste: All Available Reports Are on Studies Done at Laboratory Scale

Species	Common Name	Targeted Solid Waste	References
Sarcophaga peregrina	Flesh fly	Organic waste, defatted restaurant waste, faeces, decaying meat	Yang et al. (2012), Cickova et al. (2015), and Pastor et al. (2015)
Stomoxys calcitrans	Stable fly	Manure	Mramba (2007), Dehghani et al. (2018)
Teleogryllus testaceus	Cambodian field cricket	Agriculture waste, plants, spent grain, weeds	Melillo (2013), Miech et al. (2016), and Fowles and Nansen (2019)
Tenebrio molitor	yellow mealworm	Wheat straw, vegetable waste, corn stove, crop waste, plastic waste (polyurethane, polyethylene, polylactic acid, polystyrene)	Lush (1943), Meyer et al. (2012), Li et al.,(2013), Wang et al. (2017), Yang et al. (2018, 2019), Fowles and Nansen (2019),Billen et al. (2020), He et al. (2021), and Bulak et al. (2021)
Themira annulipes	Black scavenger fly	Solid urban waste	Morales and Wolf (2010)
Zophobas atratus	Super worm	Plastics such as polystyrene, polyethylene	Yang et al. (2020)
Zophobas morio	Super worm	Spent grain, distillers' grain, baking remains, plastics such as polystyrene, polyethylene	Van Broekhoven et al. (2015), Harsanyi et al. (2020), Peng et al. (2020), Yang et al. (2020), and Rumbos and Athanassiou (2021)

18.2 PREVIOUS ATTEMPTS AT EMPLOYING TERMITES FOR SOLID WASTE TREATMENT

Even though, from time to time, termites have been proposed for direct use in biocomposting lignocellulosic waste (Haritos et al., 1994; Myles 1995; Toninger, 1997), the prospect has not been intensively explored for its feasibility. The first recorded attempt to employ termites in handling solid waste was made by Severtson (2006). It consisted of two parts. One part was carried out within the confines of a laboratory, which had a naturally developed colony of *Coptotermes acinaciformis* existing there. Plastic containers were filled with different types of paper and were connected to the termite colony via pipes. The rate of consumption of the paper by the termites was assessed. In the other part, carried out in the open, efforts were made to create "hubs" of termite activity by putting 44-gallon (about 100 L) drums filled with wood, soil, and grass to attract termites. Waste paper was put in plastic bags, which were connected to the hubs so that termites could move from the drum to the plastic bag to feed upon the paper.

The authors found that in both embodiments of their experiments, even though termites did consume some paper waste but their activity was extremely slow, and in the majority of trials, complete decomposition of paper waste was not witnessed even after 20 weeks. Second, the design of their experiments was such that no process control was possible. Third, waste paper is not a substrate problematic to dispose of. Most of it is reutilized by paper mills after pulping it. It can also be a good feed for energy-producing incinerators.

In parallel to Severtson (2006), the present authors and coworkers (Ganesh et al., 2007) began their studies, an overview of which is contained in this paper. A few years later, Lenz et al. (2011) reported a study, carried out in a manner similar to vermicomposting, in which groups of *C. formosanus* and *Reticulitermes speratus* were kept for 4 and 8 weeks with 10 different types of paper as their food source. As can be expected from the way termites function in nature, their survival when confined in a vessel, cut-off from their colonies, was low, and most groups on recycled office paper were dead after 8 weeks. The study reconfirmed that termites cannot be used for processing waste in ways in which other animals can be, especially earthworms.

18.3 THE WORK OF THESE AUTHORS

The work of these authors, an overview of which is presented in this paper, was prompted by two considerations:

a. Despite very intense and persistent efforts by scientists across the world, use of microflora or biomolecules isolated from the termite gut for biocomposting lignin is still far away from becoming utilizable at a large scale.
b. If ways can be found to harness termites in a way that they can be used to biocompost ligninous waste in a controlled fashion, most of the costs and efforts associated with the obtaining/culturing of lignin-degrading substances can be saved. In other words, termites can be directly used as bioreactors.

An overview of the challenges faced by the present authors in the course of making controlled use of termites in biocomposting, and the efforts made to overcome them, is presented in this report.

18.4 THE SPECIAL CHALLENGES ASSOCIATED WITH THE USE OF TERMITES IN BIOCOMPOSTING

Controlled use of termites poses special challenges, as explained below.

The first challenge is that, unlike earthworms and a lot many other bioagents, in termites, only one royal pair has the ability to reproduce, and all workers are sterile.

Hence, termites cannot be cultured in the way other organisms can be, viz, by giving a few individuals of mixed sexes the feed and the living conditions suitable to them and letting them reproduce.

Second, it is not even possible to run a reactor for any reasonable length of time by capturing a few thousand termite workers and letting them feed upon the desired waste because, after being cut-off from their colony, the workers die quickly. This happens possibly because it disrupts the very subtle and intricate communication that exists across the termite hierarchy (workers, soldiers, alates, royal pair) and which is based on environmental cues and biomolecules.

These two constraints make it necessary that (a) either the termite colonies already functioning in nature are utilized for somehow drawing workers from there to the solid waste that is aimed to be biocomposted, and/or, (b) captive termite colonies are developed within the confinement of work sheds which can then be made to feed predominantly on the waste to be biocomposted. Both these challenges were addressed by the authors as described below.

18.5 THE PRESENT WORK

18.5.1 Identifying Termite Species to Be Worked With

As the authors began efforts to search for solutions to both problems, a new challenge surfaced. It was to ensure that only those termite species should be explored that have been neutralized in the area where the authors work – the city of Puducherry and its outskirts. No new species was to be brought in to prevent the risk of it becoming invasive and colonizing. To ensure this, elaborate surveys were done of the termifauna of the region, as detailed elsewhere (Kaur et al., 2010; 2013; 2016), to shortlist species of relevance to the present study. Separate and mutually independent surveys were carried out by the method of randomly arranged transects (Jones & Eggleton, 2000) and baits of different types (Abensperg-Traun, 1993; Jones et al., 2005).

Together, the surveys revealed the presence of 13 species (Figure 18.1). Of these, only one – *Eremotermes paradoxalis* Holmgren – is predominantly a soil feeder, while the rest 12 are essentially wood/litter feeders. The termifauna was dominated by *Hypotermes obscuriceps* Wasmann, which constituted 52% of all termite population. The soil feeder *Eremotermes paradoxalis* formed only 0.7% of the population, just a fraction more than the least populous *Microcerotermes pakistanicus* Akhtar.

After *H. obscuriceps, Macrotermes convulsionarius* Konig and *Odontotermes anamallensis* Holmgren were the next most abundant, forming 25% and 12% of the termite population. Together, these three species constituted 87% of all the termifauna of the region. Comparison with the past studies reported by others (Zeidler et al., 2002; Carrijo et al., 2009; Dosso et al., 2010; Hemachandra et al., 2010) and the resulting values of the indices of Simpson, Shannon, and Pielon worked out by us indicated that the termifauna in the present study area had comparable or better diversity /evenness than the areas studied by other authors.

After having identified the species that have been neutralized in the study area, and cataloguing their feeding preference (Kaur et al., 2016), it became clear that the role of the three leaf litter-cum-wood feeder species *H. obscuriceps, M. convulsionarius,* and *O. anamallensis*, which, together, constituted 87% of the termite population, was

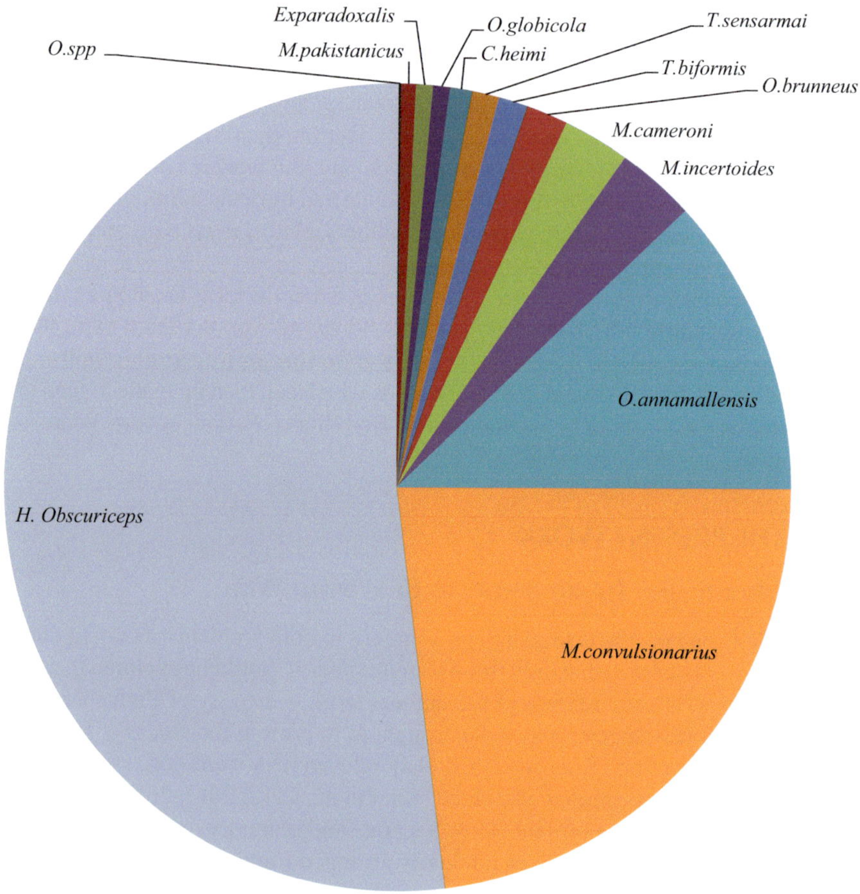

FIGURE 18.1 Relative abundance of termite species in the study area.

likely to be pivotal. It appeared more so because these species belonged to the feeding group II as per the classification of Donovan et al. (2001), representing wide-ranging feed acceptance, including dead wood, grass, leaf litter, and micro-epiphytes. For this reason, they appeared more likely to accept ligninous waste of different types than the more specialized or less predominant termite species.

18.6 APPROACHES TO THE DESIGN OF THE TERMIREACTORS

Two ways appeared possible with which termites could be used in a controlled fashion to process ligninous and other forms of hard biowaste (LHBW). One was to a) design reactors which can be filled with LHBW of different types and placed near naturally occurring termite mounds of desired species, and b) find ways by which termites can be attracted to those termireactors so that the termicomposting of LHBW could be achieved speedily. These types of systems were identified as "ISTs".

The second option was to develop captive termite colonies of appropriate species within close confines and make the termites feed exclusively upon LHBW. The authors felt that such captive termite colonies would be like bioengines, supplying termite workers in a sustainable fashion; hence, these systems were named "bioengine-based termireactors" (BBTs).

18.7 THE DESIGN AND THE PERFORMANCE EVALUATION OF THE ISTs

18.7.1 The Design Approach

Three embodiments of the ISTs were designed, based on the nature of the substrate to be processed and the termite species to be harnessed for carrying out the "termigradation". The designs were based on the following criteria:

1. The reactor geometry should be so matched with the termite behaviour that it enables as complete utilization of the substrate as possible in the shortest possible time.
2. The reactors should be easy to move around, charge, and install.
3. The termigraded product should be easy to remove from the reactors.
4. Provision should be made for non-destructive separation of the termites still present in the substrate.

In addition, a system of trails was developed that could be attached to all three embodiments to enhance the prospects of scout workers reaching the IST.

18.7.2 The Three Embodiments Designed Based on the Above Approach

Three versions or embodiments were designed on the basis of the objectives itemized above.

18.7.2.1 Embodiment 1: Multi-Module IST, Suitable for Harnessing Subterranean Termites

This embodiment of IST consists of several modules (Figure 18.2a); the number of modules to be used can vary with the quantity of substrate to be processed. Each module (Figure 18.2b) is made up of galvanized iron wire mesh. A protective covering, equipped with a handle, serves as a lid to the module. The lid is lifted to charge (feed) the module and then replaced. All the modules of the reactor are thus charged, and then ported to the site and installed in below-ground pits.

After the designated termigradation time, the modules are lifted out and placed in a carry case (Figure 18.2c). The carry case consists of sliding, removable panels on the sides with a larger surface area of the case and two fixed panels on the sides with a smaller surface area. The purpose of the carry case is to transport the module – which is made of wire mesh on its five sides – in a way to prevent termites from falling out of the container.

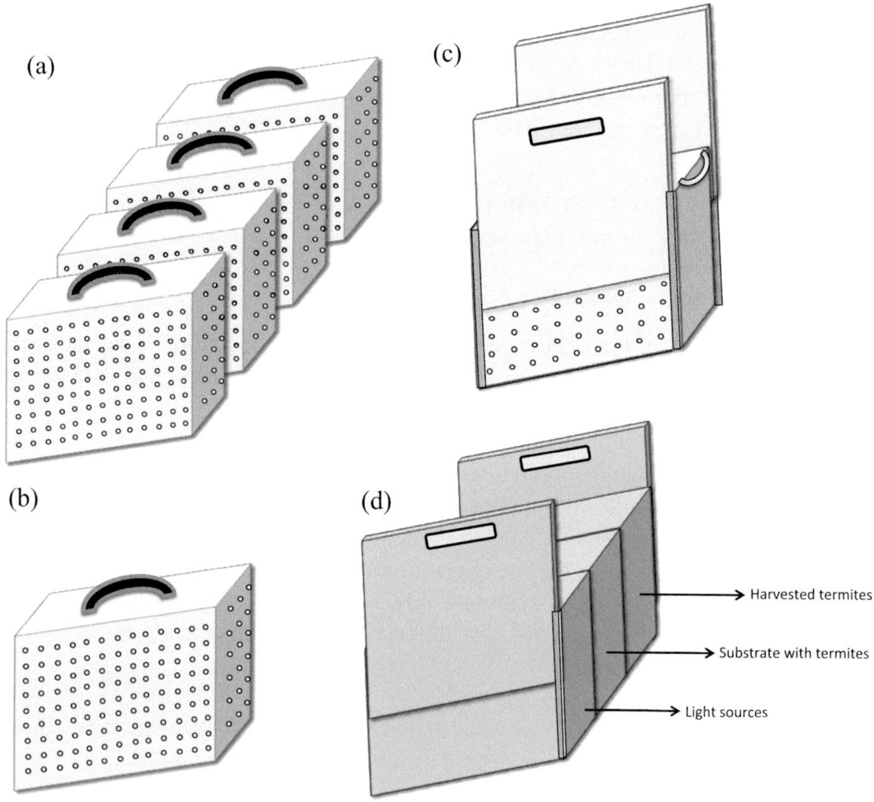

FIGURE 18.2 A typical train of IST modules (a); an individual module (b); a carry case with a module inside (c); a typical harvester assembly (d).

Figure 18.2d shows the harvester assembly, consisting of two components. The first one fits into one of the larger surface area sides of the reactor module and carries a light source. The other one fits on the other larger surface area side of the reactor, and contains cardboard. Once fitted into the harvester assembly, the two sliding panels of the carry case are removed, and the light source is turned on. This causes the termites present in the reactor module to move away from the light source towards the darker zone, which is the cardboard substrate present on the other side of the reactor module. The light source is left switched on till the animal migration is complete. The removable panels are then repositioned in the carry case, and the harvester assembly is disengaged from the module. The termigraded product is harvested from the reactor module by lifting the lid and emptying the reactor contents.

18.7.2.2 Embodiment 2: IST for Termites Which Forage on, or just Below, the Soil Surface

The termireactor has been sized to maximize surface area/volume ratio within the limiting condition that adequate substrate depth should be maintained to provide

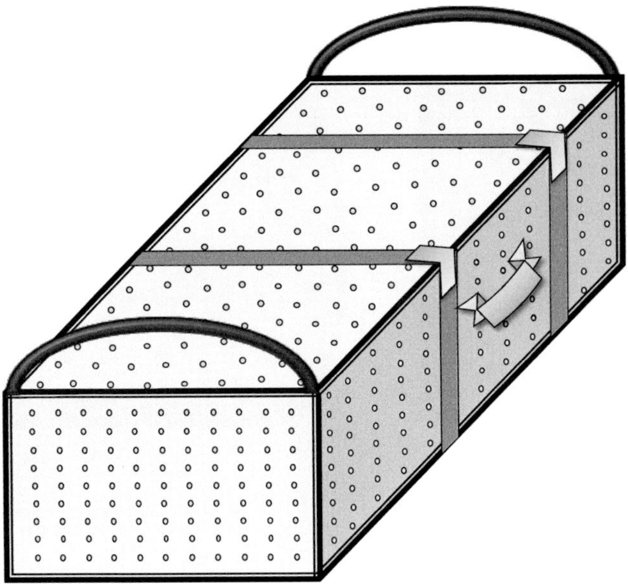

FIGURE 18.3 ISTs appropriate for using termites that forage on, or just below, the soil surface.

the termites with adequate humidity as well as shading from sunlight (Figure 18.3). The reactor is made of wire mesh, with metal angles placed at strategic intervals to provide the scaffolding needed to prevent the wire mesh from getting bent. Both the upper and the lower faces of the termireactor serve as lids and can be unfastened and swung open at the hinges to charge/empty the reactor. Two handles are provided for facilitating the transportation of the reactor.

18.7.2.3 Embodiment 3: General Purpose IST, Which Can Be Used for Both Above-Ground and Subterranean Foragers

The third embodiment is for situations in which several species of termites are active in the area around the IST. It can handle species that forage above-ground as well as the ones that are subterranean foragers (Figure 18.4). It is a cylindrical vessel made of wire mesh with lids on both sides. A pair of such vessels are to be used at each site, both filled with LHBW. Of these, one is oriented vertically (Figure 18.4a) and buried in the soil for the use of subterranean species. The other is kept sideways (Figure 18.4b) to facilitate access to above-ground foragers. Figure 18.5 illustrates how the reactor when used with subterranean termites (vertical orientation) is charged, installed, and emptied.

18.7.3 Attachment of Trails

In nature, termite scouts keep exploring food sources in the area around their colony. Upon finding a viable source, they generate pheromone trails between the termite colony and the food source so that other workers can get attracted to the source and

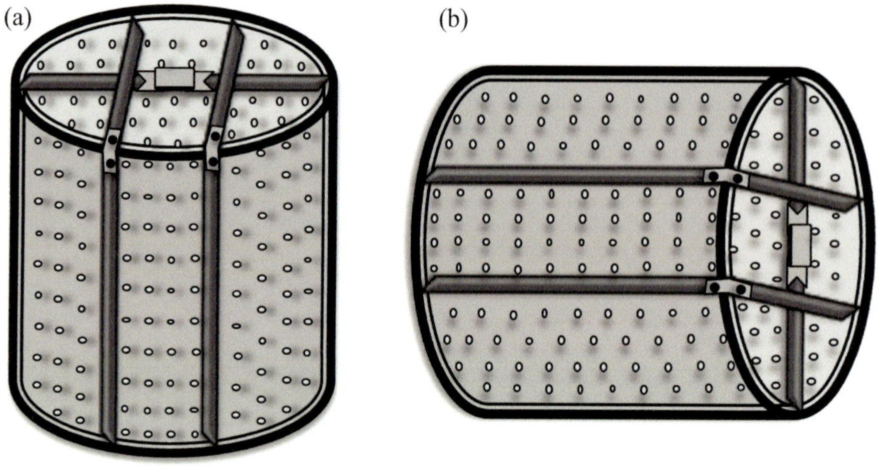

FIGURE 18.4 Vertical and sideways orientation of ISTs depending on the nature of the termites to be attracted.

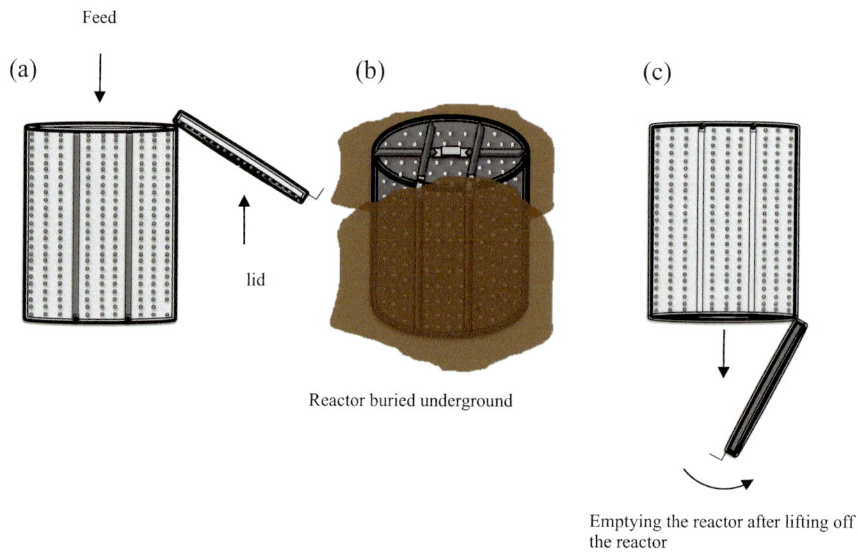

FIGURE 18.5 Arrangement for utilizing subterranean termites in ISTs.

reach it by following the trails. The present authors thought that if termites can be induced to make pheromone trails leading to ISTs from all directions, there will be a greater probability of the foragers reaching the ISTs from all sides. To enable this, pipes filled with waste paper, which led to the ISTs, were laid out projecting radically outwards. Experiments with ISTs placed with or without trails showed significantly better performance of the former compared to the latter. Hence, on the basis of this, provision has been made in all the IST embodiments described above for "trails"

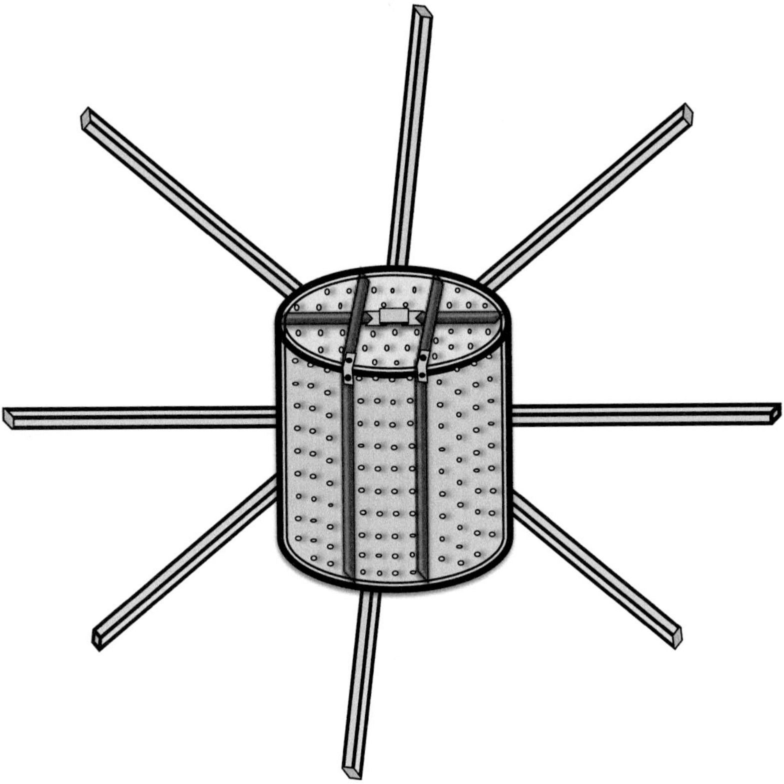

FIGURE 18.6 A typical system of trials attached to an IST.

radically projecting outwards from all sides (Figure 18.6). These trails are made up of narrow pipes in which waste paper, or similar material (like sawdust) known to be consumed by most termite species, is partially filled to attract scout termites. The scouts forage through the bait to create strong pheromone trails, which then attract termites from all sides. Figure 18.6 illustrates this provision with respect to the third embodiment of the ISTs.

Subsequent work by the present author's group (Kaur et al., 2018) has shown that trails of 2-phenoxyethanol, comprising its 0.0005% solution in ethanol, attract termite foragers towards food sources to which the trails lead. Hence, an ethanolic solution of 2-phenoxyethanol can be used to lay trails for guiding and controlling the movement of the termites to-and-fro termireactors.

ISTs were used to termicompost ligninous parts of weeds parthenium (*Parthenium hysterophorus*), nightshade (*Solanum elaeagnifolium*), prosopis (*Prosopis juliflora*), ipomoea (*Ipomoea carnea*), and lantana (*Lantana camara*) (Anantharaju et al., 2017a,b; Patnaik et al., 2022; Kaur et al., 2022). They employed ISTs of different designs and of capacities ranging from 5 to 50 kg. In all cases, 40%–50% of the substrate was termicomposted within 3–4 weeks. ISTs with trails performed significantly better than ISTs without trails. Two interesting observations were made in all cases:

a. The rate of termicomposting went down as the quantity of the substrate remaining in the ISTs went down. It appeared that as the food source (the substrate being termicomposted) got depleted, the termite scouts gave appropriate signals, which led to a reduction in the number of incoming foragers being summoned. It also indicated that if the ISTs are replenished with fresh substrate once every 30 days or so, the rate of termicomposting can be maintained at a high.
b. Unlike vermicomposting in which 50%–60% of the substrate is converted to vermicast – the rest getting metabolized into warm zoomass and CO_2 (Banupriya et al., 2022) – there is little residue left after termites have fed upon the substrate. It is partly because termites utilize the nutrients present in the substrate more efficiently than most other animals do, generating very little excreta. In other words, most of the ingested substrate gets converted into termite zoomass. Termites are also known to carry those bits of the substrate that they do not ingest to the termitaria where fungus gardens are maintained for growing fungi as a termite feed. The bits taken there get upgraded as food by the fungal action. Due to these reasons, termicomposting does not generate termicast in the way vermicomposting produces vermicast. The processing of waste by termites is more in the nature of "assimilative disposal", in which the waste is assimilated with nature by the termite action. There is no termicast left by the termites to harvest.

18.8 THE DESIGN AND PERFORMANCE EVALUATION OF THE BBTs

Given that the BBTs require development of captive termite colonies, attention was focused on the challenges and opportunities associated with the initiation and the development of termite colonies within confines like the typical worksheds that are used in the processing of designated solid waste.

18.8.1 Initiation of Termite Colonies

18.8.1.1 Role of Alates

All termite colonies are started in nature by the mating between winged reproductives (alates), which emerge from pre-existing colonies at certain times of the year (Arora & Pajni, 2008; Nasir & Akhtar, 2011; Premalatha et al., 2011). The alates swarm towards nearby sources of light, enabling a large number of individuals to come together and facilitate mating. Once a pair decides to mate, it moves to preferred niches (which differ from species to species), generally formed by moist soils. The oviposition is then initiated, leading to colony development.

The present authors decided to initiate captive colonies by leading the alates, when their swarms occur, to containers provided with moist soil and organic matter like dead leaves and twigs. This had to be planned on the basis of the time of the year alates were likely to emerge. Hence, surveys were done for three consecutive years to learn about the seasons during which alates emerge in the region where the authors work (Puducherry, India), and the frequency, the area-wise distribution, the intensity (in terms of estimated number of alates per swarm), and the duration of swarming of different termite species (Abbasi et al., 2014a; Harit et al., 2014). Also, unlike the studies of several other authors which had studied swarming behaviour of elates with the end objective of finding ways to kill the emerged alates to control termites in their regions – the goal of these authors was to capture alates without injuring or stressing them so that the alates can then be introduced in pre-arranged boxes to facilitate mating and consequent initiation of termite colonies. To this end, novel traps were designed with which alates can be captured efficiently as well as non-destructively (Abbasi et al., 2014a). A survey of prior art had indicated that in the tropics the swarming is usually confined to the rainy season, while in temperate regions warmer summer months are preferred (Nutting 1969; Mitchell 2008). In our surveys (Abbasi et al., 2014a; Harit et al., 2014), swarming was seen to be induced by a combination of three environmental factors – rainfall, temperature, and humidity – which have been identified as the prime triggers of alate swarming (Neoh & Lee 2009a,b). It appears likely that alates emerge when it has rained to benefit from the wet and softened soil created by the rains, which makes it easier for the alates to dig in and fashion their nuptial chambers. Wet soil also provides moisture to the early and sensitive stages of the incipient colony. The rain-induced wetting of large areas of upper levels of the soil, perhaps, causes a pattern of lowering of soil temperature and an increase in soil moisture, which signal the start of the alate flight (Nutting, 1969).

18.8.1.2 Prior Art on Generating Captive Termite Colonies

A survey of the prior art revealed that beginning with the work of Ausat et al. (1962), 33 reports have appeared on attempts to develop captive termite colonies within the confines of a laboratory (Harit et al., 2016). But, as per more than half of the reports, incipient colonies perished within less than 400 days of initiation. In other cases, the colonies survived longer, yet perished much before reaching maturity. No report was found describing the development in captivity of a mature colony or a colony that lasted as long as the colonies of the corresponding species last in nature. Connetable et al. (2012) reported that they managed to maintain colonies of *Pseudacanthotermes spiniger* and *P. militaris* – both higher termites that make their nests in wood – for about 20 years. But they needed to feed the termites with the fungi *Termitomyces eurrhizus* and *T. aurantiacus* obtained from fungus combs harvested from nests existing in nature. Moreover, even at their largest size, the colonies were small enough to be housed in 175 l containers. Much larger colonies were needed, capable of supplying workers in numbers that were required to process significant quantities of solid waste, when the objective was to use termites in controlled biocomposting.

The only instance of human-mediated development of a full-fledged termite colony has been recorded by Leuthold et al. (2004). They initiated a colony of *M. jeanneli* in confinement, supplied it with small pieces of the fungus comb when the first workers appeared, made elaborate arrangements to maintain humidity in the termite dwelling, and shifted the colony to a bigger container as it grew. Eventually, they transferred the colony to the below-ground soil. In the process, even though a giant 6 m tall mound came up and was enveloped by the authors in a glass case (to make it a showpiece), the colony was not really a captive one. Moreover, of the termite species established in the region where the author's work is, none have been explored earlier for *ab initio* colony development. Additionally, while the previous attempts at termite colony development were largely aimed at finding ways to control those species of termites, the aim of these authors was to use the colonies as "bioengines" for the processing of specific biowaste.

18.8.2 Present Work on Formation of Captive Termite Colonies

18.8.2.1 Attempts at Colony Initiation for Progressive Scale-Up

After each swarming, the trapped alate pairs, whose sex was determined on the basis of sternum and styli arrangement (McMahan 1960; Pearce 1997; Hu & Forschler 2012), were allowed to shed their wings. Four to six pairs, species-wise, were transferred into each plastic box or Petri dish set for the colony initiation (Harit et al., 2017). The containers were of sizes similar to the ones used by previous authors. It was observed that in the containers the males pursued the females, in tandem pairing similar to the one reported for other termite species earlier, leading to the mating (Hanus et al., 2009; Husseneder & Simms, 2008; Kawatsu & Matsuura 2013).

Five substrates were used as media for termite culture: normal soil (red-ferralitic), mound soil, sawdust, a mixture of sawdust and mound soil, and filter paper. Each substrate was previously sieved and oven dried (except filter paper) to prevent contamination with fungi and other microorganisms as reported by Ferraz and Cancello (2004). The plastic boxes and Petri dishes were filled with respective substrates to about half of the container height (Harit et al., 2017). The different substrates were used in this manner to enable determining which substrate(s) facilitate colony development better than others for the given species.

The authors had hoped that the colonies would grow further by controlling their feed, ambient temperature, humidity, etc. It was also planned to shift to larger containers as the colonies grew. Reproduction did begin, and small incipient colonies did develop, but then they died off one after another despite valiant efforts to sustain them. The period of survival of those incipient colonies varied both from species to species and within species. The maximum survival period (9 months) was recorded for *M. convulsionarius*, and the minimum (2 months 18 days) for *M. cameroni*. Generally, mortality was triggered by a fungal infection or a mite attack. All-in-all, none of the present authors' efforts resulted in any greater success than the previous attempts of other authors had been, albeit targeted at the colony development of different species. Even as some useful information on termite biology was obtained in the course of these studies, as documented by Harit et al. (2017), it also became clear that termitaria are possibly based on too fine a tuning of food, architecture, humidity, and temperature to be amenable to simulation at bench scale in a laboratory.

18.8.2.2 Development of Captive Termite Colonies by Full-Scale Simulation of Termite Colony Formation in Nature

It was decided not to make any more attempts to externally control the termite diet and living conditions. Instead, it appeared prudent to rely on mimicking the way colony initiation as well as its development takes place in nature. To that end, it was decided to develop colonies of the two most abundant termite species in the region – *H. obscuriceps* and *M. convulsionaris* – by simulating colony formation and growth of the two species as occurs in nature. It was aimed that the colonies would be of the same size as in nature, but held captive by directing them towards containers that are large enough to house full-scale colonies of those species. It was found that the colonies of the two species existing in the study area were 18 ± 2 m^3 in size. Hence, concrete chambers of slightly larger volume, of 21.6 m^3 ($3.0 \times 2.4 \times 3.0\,m^3$), were constructed. Each was filled with soil containing grass, twigs, wood pieces, leaves, etc., as they occur in the open where termite mounds of these species are located. Some soil taken from defunct mounds and from near active mounds was also used. This further ensured simulation of the conditions under which the concerned species of termites built colonies in nature.

All containers were periodically sprinkled with water to provide moisture. Layers of leaf litter of assorted plant species were placed at the top, as is seen in nature at the top of the soil, where termites dwell. This helped in creating the ambience of temperature and humidity very similar to that which exists in nature.

For nearly three months, there was no sign of any activity. After that, tunnels and mound-like structures began to appear. In about a year, vigorous colonies seemed to have developed as evidenced by the establishment of fungal comb, extensive foraging, soil-working, and gallery formation activities in the chambers (Figures 18.7 and 18.8). At the time of writing this report, which is about 10 years after the colony formation was complete, the colonies are continuing to thrive. It can be said that the strategy adapted by the authors can enable the generation and maintenance of captive termite colonies, or BBTs.

18.8.3 Controlled Use of BBTs in Processing Biowaste

In order to test the ability of the captive colonies as BBTs to process specific lignious biowaste such as the woody parts of ipomoea and lantana, rectangular termireactors were used. They were designed in such a way that they were perforated from the bottom up to a height of 30 cm, with holes of 1 cm diameter created at one inch distance from each other (Figure 18.9). This was done to provide access from all sides to the termites so that they are led to the feed placed in the reactors.

The reactors were run in semi-continuous, or "pulse" mode, in triplicate. They were placed on the surface of the chambers containing the *H. obscuriceps* and the *M. convulsionarius* BBTs. Each termireactor was charged with 2 kg of the twigs of either ipomoea or lantana. All quantities of the substrates and the post-biocomposting residues were assessed on "dry weight basis" – the equivalent of fresh weight of the substrate oven dried at 105°C to constant weight. The reactors were assessed for substrate consumption every 30 days. In other words, they were pulse fed once every 30 days.

An illustrative set of the findings, pertaining to the biocomposting of the woody parts of ipomoea (Figure 18.10), is presented in Figure 18.11. The average and standard deviations of substrate utilization in each set of reactors are also represented. The performance of the replicates agreed with each other within a relative error

FIGURE 18.7 Termite activity on and around termite-worked soil signalled the formation of a full-fledged captive termite colony, or a BBT.

of 15%. Considering the heterogeneous nature of the reactor content, this extent of agreement can be deemed to be very good.

About 42% of the waste was utilized during the first 30 days. Thereafter, the rate rose to above 50%, averaging 55% (or 1.1 kg out of 2 kg feed per pulse of 30 days), during the uninterrupted reactor operation lasting 7 months. The <50% consumption in the initial pulse was perhaps due to the time taken before the termites prospected the food source in the termireactor and began using it in full strength.

As was earlier seen with the ISTs, the rate of substrate consumption depended upon the fraction left unconsumed. When that fraction approached, termites seemed to depute progressively fewer workers. Upon replenishing the substrate at the end of the 30-day span, the number of foragers apparently increased till the dwindling food source brought proportionately fewer foragers. It is perhaps due to this phenomenon – of termites deputing lesser and lesser workers (waste feeders) as the source becomes

FIGURE 18.8 Mound formation witnessed in captive termite colonies, or BBTs.

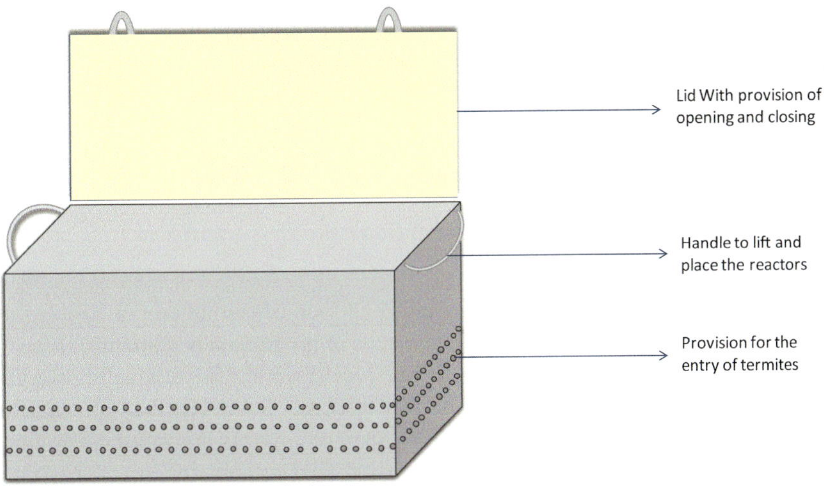

FIGURE 18.9 The schematic of the termireactors used in the BBTs.

FIGURE 18.10 Termigradation of ipomoea: *H. obscuriceps* at work under the protective reddish-brown cover created by them.

increasingly leaner – that the average rate of 2 kg food consumption per 30-day pulse remained more or less constant with time instead of progressively going up.

As the average substrate assimilation rate was close to 55% at a solids retention time of 30 days, it translates to 100% substrate assimilation in less than 60 days. This is significantly less than the 3–4 month time span required in conventional composting or vermicomposting. Also, unlike in conventional composting, no energy inputs are needed for periodic turning of the substrate, aeration, etc, in the course of the presently described termite-based biocomposting or "termigradation". Most significantly, the type of woody material termites were able to process cannot be composted or vermicomposted anyway.

The trend in the case of the BBTs based on *M. convulsionarius* was similar to the one observed with *H. obscuriceps,* with the difference that the foragers of

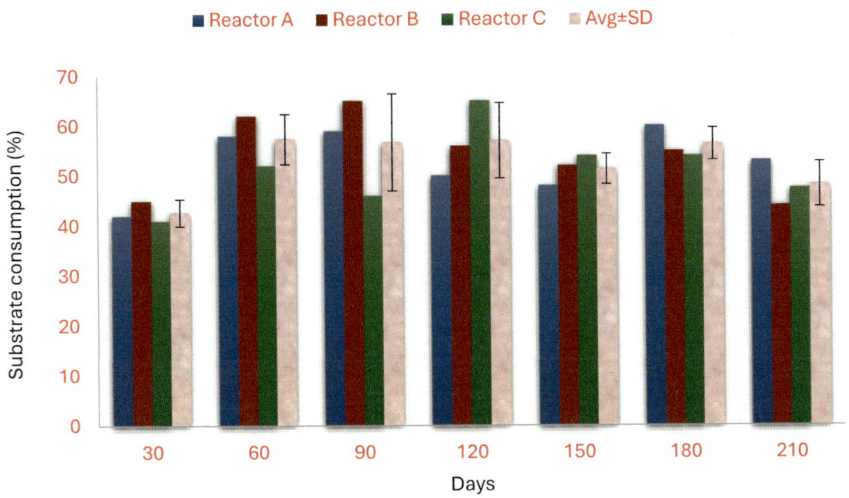

FIGURE 18.11 Biocompost of ligninous parts of the weed ipomoea occurring in the BBT served by *H. obscuriceps*.

M. convulsionarius kept the extent of consumption to about 30%, 28% for ipomoea, and 33% for lantana. This indicates that perhaps *M. convulsionarius* had a lesser taste for these weeds than *H. obscuriceps* had. Also, while *H. obscuriceps* had a slight preference for ipomoea over lantana, it was the other way for *M. convulsionarius*. More work is likely to bring out whether similar variation in feed preference occurs across other species also. In the meanwhile, an Indian patent, number 303,781, has been granted to the process and the reactors described above (Abbasi et al., 2014b) in 2019.

REFERENCES

Abbasi, S.A., Gajalakshmi, S., and Abbasi, T., 2014a. *Towards Opening a New Frontier in Bioprocess Technology*, Report BT/PR 11488/AGR/21/289/2008dt, Department of Biotechnology, Government of India; iv+339 pages.

Abbasi, S.A., Gajalakshmi, S., and Abbasi, T., 2014b. A novel process for eco-friendly treatment of ligninous biowaste. *Official Journal of the Patent Office*, *21*, p.14196.

Abensperg-Traun, M., 1993. A comparison of two methods for sampling assemblages of subterranean, wood-eating termites (Isoptera). *Australian Journal of Ecology*, *18*(3), pp.317–324.

Anantharaju, T., Kaur, G., Gajalakshmi, S. and Abbasi, S.A., 2017a. Termigradation of the weed *Solanum elaeagnifolium*. *International Journal of Advance Research in Science and Engineering*, *6*(10), pp.656–659.

Anantharaju, T., Kaur, G., Gajalakshmi, S. and Abbasi, S.A., 2017b. Controlled use of termites in assimilative disposal of the poisonous weed, *Parthenium hysterophorus*. In *6th International Conference on Emerging trends in Engineering, Technology, Science and Management* (IETE, Bengaluru), Institution of Electronics and telecommunication Engineers, Karnataka, India. ISBN: 978-93-86171-70–2.

Arora, C.B. and Pajni, H.R., 2008. Swarming and Foundation of the Colony in *Coptotermes heimi* (Wasmann) (Rhinotermitidae: Isoptera: Insecta). *Bugs 'R'All*, *16*, pp.9–11.

Ausat, A., Cheema, P.S., Koshi, T., Perti, S.L. and Ranganathan, S.K., 1962. Laboratory culturing of termites. In: M.L. Roonwal (ed), *Termites in the Humid Tropics: Proceedings of the New Delhi Symposium*, UNESCO, Paris, pp.121–125.

Banupriya, D., Abbasi, T. and Abbasi, S.A., 2022. Rapid, clean, and sustainable bioprocessing of toxic weeds into benign organic fertilizer. *Agriculture*, *12*(10), p.1511.

Carrijo, T.F., Brandao, D., de Oliveira, D.E., Costa, D.A. and Santos, T., 2009. Effects of pasture implantation on the termite (Isoptera) fauna in the Central Brazilian Savanna (Cerrado). *Journal of Insect Conservation*, *13*(6), pp.575–581.

Connetable, S., Robert, A. and Bordereau, C., 2012. Dispersal flight and colony development in the fungus-growing termites *Pseudacanthotermes spiniger* and *P. militaris*. *Insectes sociaux*, *59*, pp.269–277.

Donovan, S.E., Eggleton, P. and Bignell, D.E., 2001. Gut content analysis and a new feeding group classification of termites. *Ecological Entomology*, *26*(4), pp.356–366.

Dosso, K., Konaté, S., Aidara, D. and Linsenmair, K.E., 2010. Termite diversity and abundance across fire-induced habitat variability in a tropical moist savanna (Lamto, Central Côte d'Ivoire). *Journal of Tropical Ecology*, *26*(3), pp.323–334.

Ferraz, M.V. and Cancello, E.M., 2004. Strategies on the developmental biology of incipient colonies of *Coptotermes gestroi* (Isoptera: Rhinotermitidae) in different substrates. *Sociobiology*, *44*(1), pp.109–122.

Ganesh, P.S., Gajalakshmi, S. and Abbasi, S.A., 2007. Treatment of 'Hard'Biowastes with resource recovery by termite action–a new frontier in solid waste management. In Hilty, L., Edelmann, X., and Ruf, A. (Eds) *R'07 World Congress: Recovery of Materials and Energy for Resource Efficiency,* Davos, Switzerland, September 3–5, pp. 200–204.

Hanus, R., Luxová, A., Šobotník, J., Kalinov Á, B., Jiroš, P., KřeČek, J., Bourguignon, T. and Bordereau, C., 2009. Sexual communication in the termite *Prorhinotermes simplex* (Isoptera, Rhinotermitidae) mediated by a pheromone from female tergal glands. *Insectes sociaux*, *56*, pp.111–118.

Harit, A., Moger, H., Duprey, J.L., Gajalakshmi, S., Abbasi, S.A., Subramanian, S. and Jouquet, P., 2017. Termites can have greater influence on soil properties through the construction of soil sheetings than the production of above-ground mounds. *Insectes Sociaux*(Springer), *64*(2), pp.247–253.

Harit, A.K., Gajalakshmi, S. and Abbasi S.A., 2014. Swarming of the termite Coptotermesgestroi in north-eastern Puducherry. *Zoology and Ecology*, *24*(1), pp.62–69.

Harit, A.K., Gajalakshmi, S.A. and Abbasi, S.A., 2016. Studies on the development of captive termite colonies. *Zoology and Ecology*, *26*(4), pp.301–312.

Haritos, V.S., French, J.R. and Ahokas, J.T., 1994. Cytochrome P450 monooxygenase and glutathione S-transferase activity of two Australian termites: Mastotermes darwiniensis and Coptotermes acinaciformis. *Insect Biochemistry and Molecular Biology*, *24*(9), pp.929–935.

Hemachandra, I.I., Edirisinghe, J.P., Karunaratne, W.A.I.P. and Gunatilleke, C.V.S., 2010. Distinctiveness of termite assemblages in two fragmented forest types in Hantane hills in the Kandy District of Sri Lanka. *Ceylon Journal of Science* (Biological Sciences), *39*(1), pp.11–19.

Hu, J. and Forschler, B.T., 2012. Neotenic phenotype and sex ratios provide insight into developmental pathways in *Reticulitermes flavipes* (Isoptera: Rhinotermitidae). *Insects*, *3*(2), pp.538–552.

Husseneder, C. and Simms, D.M., 2008. Size and heterozygosity influence partner selection in the Formosan subterranean termite. *Behavioral Ecology*, *19*(4), pp.764–773.

Jones, D.T. and Eggleton, P., 2000. Sampling termite assemblages in tropical forests: testing a rapid biodiversity assessment protocol. *Journal of Applied Ecology*, *37*(1), pp.191–203.

Jones, D.T., Verkerk, R.H. and Eggleton, P., 2005. Methods for sampling termites. In: Leather, S. (Ed), *Insect Sampling in Forest Ecosystems*, Blackwell Publishing, Australia, pp.221–253.

Kaur, G., Anantharaju, T., Gajalakshmi, S., and Abbasi, S.A., 2016. Inventory of termite species in thickly vegetated region of Northern Puducherry, India. *International Journal of Biodiversity and Conservation*, *9*(8), pp.265–272.

Kaur, G., Gajalakshmi, S. and Abbasi, S.A., 2022. Rapid disposal of the weed lantana (*Lantana camara*) in the environment with controlled use of termites. *Ecology, Environment and Conservation*, *27*(4), pp.1896–1900.

Kaur, G., Gajalakshmi, S. and Abbasi, S.A., 2010. A survey of termi-fauna of Northeastern Puducherry for developing a Termigradation strategy, *International Conference on Energy, Environment and Development: from Stockholm to Copenhagen & beyond, P.G. Department of Environment Sciences*, Sambalpur, India.

Kaur, G., Gajalakshmi, S. and Abbasi, S.A., 2013. Termite biodiversity in Pondicherry University campus: a reappraisal on the basis of feeding preference studies, International *Journal of Chemical and Environmental Engineering System*, *4*(1), 30–38.

Kaur, G.T., Gajalakshmi, A.S. and Abbasi, S.A., 2018. Lab studies on trail following behavior of the termite *Hypotermes obscuriceps* towards 2-Phenoxyethanol. *International Journal of Entomological Research*, *6*(2), pp.49–61.

Kawatsu, K. and Matsuura, K., 2013. Preadaptation for parthenogenetic colony foundation in subterranean termites *Reticulitermes spp*. (Isoptera: Rhinotermitidae). *Journal of Ethology*, *31*, pp.123–128.

Lenz, M., Lee, C.Y., Lacey, M.J., Yoshimura, T. and Tsunoda, K., 2011. The potential and limits of termites (Isoptera) as decomposers of waste paper products. *Journal of Economic Entomology*, *104*(1), pp.232–242.

Leuthold, R.H., Triet, H., Schildger, B., Sa, I.H. and Cell, R., 2004. Husbandry and breeding of African giant termites (*Macrotermes jeanneli*) at Berne Animal Park. *Zoologische Garten*, *74*(1), pp.26–37.

McMahan, E.A., 1960. *External Sex Characteristics of Cryptotermes brevis (Walker) and Kalotermes immigrans Snyder.* Kalotermitidae, Isoptera.

Mitchell, J.D., 2008. Swarming flights of the fungus-growing termite, *Macrotermes natalensis* (Haviland) (Isoptera: Macrotermitinae), and the environmental factors affecting their timing and duration. *African Entomology*, *16*(2), pp.143–152.

Myles, T.G., 1995. The ecological importance of termites and the potential utilization of termites for the decomposition of lignocellulosic wastes. In Abe, T. (Ed.) *The Termite-Symbiont System: Proceedings of Kyoto Symposium, Center for Ecological Research, Kyoto University*, Kyoto (pp. 50–54).

Nasir, S. and Akhtar, M.S., 2011. Swarming of Termites (Isoptera) in the Mianwali District, Pakistan. *Sociobiology*, *58*(1), pp.151–164.

Neoh, K.B. and Lee, C.Y., 2009a. Flight activity of two sympatric termite species, *Macrotermes gilvus* and *Macrotermes carbonarius* (Termitidae: Macrotermitinae). *Environmental Entomology*, *38*(6), pp.1697–1706.

Neoh, K.B. and Lee, C.Y., 2009b. Flight activity and flight phenology of the Asian subterranean termite, *Coptotermes gestroi* (Blattodea: Rhinotermitidae). *Sociobiology*, *54*(2), pp.521–530.

Nutting, W.L., 1969. Flight and colony foundation. In: Krishna, K. and Weesner, F.M. (Eds.) *Biology of Termites*, Vol.1, pp.233–282.

Patnaik, P., Abbasi, S.A. and Tabassum-Abbasi 2022. Termigradation of un-compostible parts of major weeds prosopis (*Prosopis Juliflora*) and Ipomoea (*Ipomoea Carnea*). In: Siddiqui, N.A., Khan, F., Tauseef, S.M., Ghanem, W.S., Garaniya, V. (eds) *Advances in Behavioral Based Safety*. Springer, Singapore. https://doi.org/10.1007/978-981-16-8270-4_6.

Pearce, M.J., 1997. *Termites: Biology and Pest Management*. CABI Publishing, Wallingford, CAB International, xii + 172 pp + 32 colour plates. ISBN 0 85199 130.

Premalatha, M., Abbasi, T., Abbasi, T. and Abbasi, S.A., 2011. Energy-efficient food production to reduce global warming and ecodegradation: The use of edible insects. *Renewable and Sustainable Energy Reviews*, *15*(9), pp.4357–4360.

Severtson, D., 2006. Bioconversion of waste paper by termites: a landfill of opportunity (Doctoral dissertation, Thesis for the degree of Honours in the Bachelor of Science (Environmental Biology), Department of Environmental Biology, Curtin University of Technology, Perth, Western Australia).

Toninger, R.P., 1997. Termites as bioconversion organisms, effects of laboratory rearing conditions and food type on the vigor and consumption by Reticulitermes flavipes (Isoptera: Rhinotermitidae) (Doctoral dissertation).

Zeidler, J., Hanrahan, S. and Scholes, M., 2002. Termite species richness, composition and diversity on five farms in southern Kunene region, Namibia. *African Zoology*, *37*(1), pp.7–11.

Index

A

Abattoir waste, 305, 307–311, 313, 317, 324, 334, 347–349, 355, *See also* Food waste; ZOCOFA system
Abbasi, S.A., 19, 24, 26, 30, 33, 40, 51, 59, 60, 62, 103, 104, 110, 117–125, 129
Abelmoschus esculentus. See Ladies finger; Okra
Abrasion, in SEVS machine, 167–172
Acacia auriculiformis (acacia), 24, 277
Adult fly (BSF), 317, 322, 325, *See also* Black soldier fly (BSF)
  life cycle, 322–324, 348
Aeration
  in composting, 9, 13, 19–21, 23–24, 29
  in vermicomposting, 8, 26, 29–30, 34, 38, 60, 71–73, 76, 77, 97, 133, 139, 143, 145, 158, 162, 177, 188, 218, 244, 297, 353, 366, 392
Aerobic decomposition, 21, 24
Agriprotein, 334
Alates (termite reproductives), 363, 364, 379, 386–387, *See also* Termites
  capture of, 387
  swarming behavior, 363, 386–387
Algae-based sewage treatment, 293–294, 298
Allelochemicals
  fate during vermicomposting, 244–245, 287–288
  in Lantana, 244
  in Prosopis, 277, 282, 285
Allelopathy, 107, 110, 119–121, 134, 178, 179, 188, 190, 201, 221, 232, 241–242, 244, 259, 266, 278, 285, 290, 299, 305
Alternanthera sessilis, 296
Alternaria alternata, 187
Ammonium-nitrogen
  in LVC-amended soil, 200
  in vermicompost, 200
Anaerobic patches, avoidance of
  in FLUVTS, 143
  in HRVP design, 133
  in IPSCO, 158
  in MCVS, 162
Anaerobic pockets, 38, 105, 133, 143, *See also* Oxygen depletion
Anecic earthworms, 29, 34, 60–62, 278, *See also* Earthworms, ecological classification
Animal biomass. *See* Zoomass
Animal manure, 13, 41, 47, 58, 62, 63, 71, 90, 103, 122, 123, 134, 140, 167, 175, 176, 180, 188, 201, 202, 271, 277, 278, 282, 305, 360, *See also* Manure
  drawbacks of using, 122
  as primary feedstock, 103, 121
  supplementation with, 121–122
Animals, *See also* Zoocomposting
  as composers, 5, 10
  as scavengers, 3
Ants, comparison with termites, 359, 361–362, 366, 367
Anus (earthworm), 36, 37, 53
APSU (Automated Paper Soaking Unit), 144, 146–149
  components of, 147–149
  distinguishing features, 149
  flow chart, 147
  novelty of, 144
  prototype, 148
Appelhof, Mary Arlene, 47–50
Aquatic weeds, 107, 108, 112, 117–119, *See also* specific weeds
Architecture, termite-inspired, 367–370, 388, *See also* Mounds, termite
  Eastgate Centre, Harare, 369, 370
Aristotle, 42, 43

B

Batch size, effect on SEVS efficiency, 172–174
Bedding material
  jute cloth as, 133, 157, 162
  sand-gravel, replacement of, 133
Biodegradation resistance, 110, 121
Biodiesel. *See* Biofuels from BSF
Biodiversity. *See* Invertebrates
Bioengine-based termireactors (BBTs), 381, 386, 389–393, *See also* Termireactors
  colony initiation for, 386–389
  design and performance, 389–393
Biofuels from BSF, 310, 321, 355
Biogas production, 119, 120, 124, 278, 300, 334
Biomass, *See also* Phytomass; Zoomass
  distribution on Earth, 106
Bioreactor, earthworm as, 34–37
Biorefineries, 298–300
Biowaste, ligninous. *See* Ligninous waste

Black soldier fly (BSF) (*Hermetia illucens*), 8, 10, 11, 13, 30, 305, *See also* Zoocomposting
  adult, 317, 322, 325
  benefits and uniqueness of, 310, 317–321
  as biological control agent, 321
  compared to other fly species, 311, 317–318
  distribution of, 321
  frass. *See* Frass
  larvae. *See* Larvae (BSF)
  life cycle of, 322–325, 348–350
  name origin ("soldier"), 317
  pupa/pupae. *See* Pupa (BSF)
Blowflies, 310, 317, 318
  Lucilia sericata (green bottle fly), 11, 310, 311, 315, 376
Bolus (termite bricks), 368
Bouché, M.B., 60
Brazil, termite mound network in, 369, 370
Breeding (earthworms), 53–56, 68–70, 72, 73, 79, *See also* Life cycle (earthworms)
Brinjal (*Solanum melongena*),
  effect of PRVC on, 282–287
  nutrient content of, 283
Buccal cavity, 35–36
Bulk density, soil, 177, 183, 286
  effect of LVC on, 188, 198, 199,
  effect of PWVC on, 206, 209
Burrows (earthworm), 35, 51, 53, 58–61

C

C/N ratio, 33, 129, 130, 176–242
  desirable range for fertilizer, 241
  effect of earthworm species on, 259–260
  of parent weeds *vs.* vermicompost, 242–243, 290–291
  limitations as vermicomposting criterion, 130
CAD models, of BSF facility components, 324–328
Captive termite colonies, 378, 379, 381, 386–389, *See also* Bioengine-based termireactors (BBTs)
  development of, 388–389
  prior art on, 387–388
Carbon mineralization, 242, 259
Carbon-to-nitrogen (C/N) ratio, as completion criterion, 129, 130
Carotenoids, effect of vermicompost on, 217–220, 232
  effect of LVC on, 188, 196
  effect of PWVC on, 206, 208
  measurement method, 184
Cast. *See* Vermicast
Challenges in vermicomposting, 103–105
Chemical fertilizers. *See* Inorganic fertilizers
Chilli (*Capsicum annuum*), effect of PRVC on, 183, 282–283
Chlorophyll content
  effect of LVC on, 188, 194–195, 232, 285
  effect of PWVC on, 205, 208
  measurement method, 184
Chitin, from BSF pupae, 320, 323
Clitellum, 36, 53–54, 56
Cleopatra, 42, 43
Cluster bean (*Cyamopsis tetragonoloba*), 183
  effect of LVC *vs.* SCF, 220–221, 229–230
  effect of vermicompost on, 230–235
  flowering and disease incidence, 235
  germination value (GV) and percentage (GP), 233
  growth parameters, 230–231
  harvest index and yield attributes, 234
  plant growth and leaf pigments, 235
Cocoons (earthworm), 14, 53–55, 63, 270–273, 276, 280–282
Cold composting, 8, 9, 35, *See also* Composting, types of
Companies, BSF-based, 324, 334
Compost
  curing phase, 13, 24, 33
  defined, 103, 129
  finished, 24–25
  as vermicomposting feedstock, 13, 32–33
Composting, *See also* Curing phase; Hot composting; Thermophilic phase; Zoocomposting
  batch process nature of, 19, 26, 34, 131
  compared to vermicomposting, 8, 10, 19, 26, 29–30, 32–34, 38, 130, 131
  definition, 4
  distinguishing features, 26
  in history, 41–42
  human-controlled *vs.* natural, 3–5, 8
  pre-composting step, drawbacks of, 11, 141
  process steps, 19–26
Composting, types of
  cold, 8, 9, 35
  hot, 8–10, 19, 32–33
  hybrid, 8, 9
Constructed wetlands, 293
Continuous stirred tank reactor (CSTR), 34
Controlled composting. *See* Hot composting
Conventional vermicomposting systems/ technology
  drawbacks of, 103–105, 121–123, 139, 143
  limitations summarized, 123
  vs. FLUVTS, 143
Coprophagous organisms, 5, 7, 310, 318
Coptotermes acinaciformis, 377
Coptotermes formosanus, 378
Cow dung (CD), 179–181, 211, 213, 278–280, 282, *See also* Manure

comparison with CD vermicompost (CDVC), 222–224,
comparison with LVC, 231–232
comparison with PWVC, 202–210
nutrient comparison with PRVC, 285, 290
as supplement for paper waste, 11, 141, 144, 180
Cucumber (*Cucumis sativus*), 179, 181, 187
effect of IVC on, 190–192
effect of LVC on, 188–189, 191, 193–201
effect of PRVC on, 282–283, 288
effect of PVC on, 192–193
effect of SVC on, 197–198
Curing phase, 13, 20, 24, 33
Cui et al. (2024) system, 100

D

Darwin, Charles, 42, 44–47
Date palm phytowaste, 116, 118
Design. *See* Design of BSF facilities, Vermireactor design
Design of BSF facilities, 323–334, *See also* ZOCOFA system
approaches overview, 324
at different scales, 323
general considerations, 323–324
large-scale systems, 324–325, 329, 331–334
stackable tray systems, 330, 339
table of systems, 335–339
Digestion (earthworm), 29, 34–37, 59, 60, *See also* Gut microflora
Dimapur, India, BSF facility in, 340
Disease incidence. *See* Pest and pathogen resistance
Disease vectors, flies as, 311, 318
BSF as non-vector, 318
Drawida willsi, 61, 263, 264, 270, 274, 275
C:N ratio of vermicompost, 259–260
performance on salvinia and ipomoea, 271–273
vermicomposting of prosopis, 278–279
DSC (differential scanning calorimetry), 242
of ipomoea and vermicompost, 246–247
of salvinia and vermicompost, 246
use in assessing mineralization, 246–247, 263–264, 288, 291
Duckweed (*Lemna minor*), 107, 109, 114, 294–295

E

Earias vittella (pod borer), 187
Earthworms, 319, 347, 359, 367, 375, 378, *See also* Anecic earthworms; Endogeic earthworms; Epigeic earthworms
adaptation to single-weed diets, 270–276
anatomy, 35–37, 52–53
anecic, 124
breeding in captivity, 14, 63
as composers, 5
contact with feed, 105
ecological classification, 29, 60–62
epigeic, 124
feeding behavior/habits, 58–59, 104, 105
fecundity across generations, 271–273, 276
gut microflora, 5, 29, 33, 37, 38
high stocking density in HRVP, 133
history of appreciation, 41–50
life cycle, 53–56
manure as preferred feedstock for, 175–176
morphology, 52–53
mortality, 105
movement in substrate, 105
processing time, 103
regeneration ability, 52
sensitivity to moisture, 56
sensitivity to temperature, 29, 56
sensory organs, 57
separation from vermicast. *See* SEVS machines
social behaviour, 57, 103
species for vermicomposting, 61–62, 242, 259, 278
zoomass as poultry feed, 145
Eastgate Centre, Harare, 369, 370
Eggs, 14, 53, 54, 68
BSF, 318, 322, 325, 329, 330, 349, 351, 354, 355
laying in ZOCOFA, 349, 351, 354, 355
termite, 363, 365, 366, 386
Eisenia andrei, 122, 175, 264, 265, 270, 274, 275, 291
C:N ratio of vermicompost, 259–260
performance on salvinia and ipomoea, 271–273
vermicomposting of prosopis, 278–281
Eisenia fetida, 34, 125, 126, 175, 177, 179, 181, 259
in HRVP studies, 242
vermicomposting of prosopis, 278–279, 291
Endogeic earthworms, 29, 59, 60–62, *See also* Earthworms, ecological classification
Endophytes, 3
End product specification, 103, 123, 129
Enzymes
in composting process, 3
in earthworm digestion, 5, 29, 36–38, 59,
Epigeic earthworms, 29, 34, 41, 60–62, 124, 175, 176, 259, 278, *See also* Earthworms, ecological classification
Eremotermes paradoxalis, 379, 380
Eriksson, Ismail, 324–328, 332, 333

Eudrilus eugeniae, 34, 175, 176, 259, 278–280, 291
Excreta. *See* Vermicast
Exoskeleton (BSF), chitin from, 323

F

Face fly (*Musca autumnalis*), 12, 310, 315, 376
Facultative scavengers, 3, 5
Fecundity (earthworm), 31, 33, 54, 61, 62, 269–271
Feed, animal. *See* Poultry feed
Feedstock, 76, 84, 129, 167, 175, 298
  animal manure as primary, 103, 121
  for composting, 19, 21, 24
  phytomass as, 104, 106–123
  pre-composting of, 32–33
  for vermicomposting, 8, 13, 26, 29, 32–33, 63, 103, 121
Fertilizer, *See also* Frass
  from BSF frass, 319–320, 324, 334, 349, 355
  compost as, 4, 19, 24
  field assessment of, 349
  value of vermicompost, 177–210, 230–235, 282–287
  vermicast as, 4, 5, 19, 32, 33, 59, 68
Fertilizer value, *See also* specific plants and vermicomposts
  beneficial *vs.* excessive application, 177–178, 188, 202
  comparison of CDVC and LVC, 222–224
  comparison of LVC and SCF, 180, 220–221, 229–230
  of paper waste vermicompost (PWVC), 180, 202–210
  of phytomass vermicomposts, 179–201
Flesh fly (*Sarcophaga carnaria*), 12, 310, 316, 376
FLUVTS (Flappable Units Vermireactor Train System), 11, 140–144, 162
  comparison with conventional designs, 143
  distinguishing features, 141–143
  flow chart, 142
  need for, 11, 141
Fly species for waste processing, 310–311, 375–377, *See also* specific flies
  comparative drawbacks, 311, 317
Food waste, *See also* Abattoir waste; Black soldier fly (BSF); ZOCOFA system
  global generation of, 312
  insanitary conditions from, 309, 310
  management potential of insects, 310
Frank et al. (1983), 121
Frass (BSF), 319–320, 322–324, 329–331, 348, *See also* ZOCOFA system
  as fertilizer, 319–320, 324, 334, 349, 355
  harvesting of, 320, 330, 331, 349, 355
FTIR (Fourier transform infrared spectroscopy)
  of ipomoea and vermicompost, 262–263
  of lantana and vermicompost, 244–245
  of salvinia and vermicompost, 260–261
  use in tracking allelochemical degradation, 244–245, 260–261, 287, 290
Fungus gardens, termite, 386, 389

G

Gajalakshmi, S., 19, 24, 26, 30, 40, 51, 59, 60, 62, 64, 131, 132, 259, 269, 270
GC-MS (gas chromatography-mass spectrometry)
  chemical profiles of ipomoea and VC, 248–249
  chemical profiles of lantana and VC, 247–248, 251–253
  chemical profiles of parthenium and VC, 249–250, 255–256
  chemical profiles of salvinia and VC, 250, 257
  molecular rearrangements revealed by, 247, 291
Germination success
  effect of CDVC and LVC on, 222, 231
  effect of IVC on, 190, 203
  effect of LVC on, 187–189, 222, 231
  effect of PRVC on, 282–284
  effect of PWVC on, 202, 206–207
  effect of PVC on, 192–193, 203
  effect of SVC on, 197, 204
  effect of weed VCs in field studies, 208–212
  suppression by prosopis leaves, 282, 284, 289
Gizzard (earthworm), 29, 36, 37, 244, 259, 265
Global warming, 119, 122, 145
Grape marc, 122
Green bottle fly (*Lucilia sericata*), 11, 310, 311, 315, 376
Green gram (*Vigna radiata*), 179, 181, 187
  effect of IVC on, 190–192
  effect of LVC on, 188–189, 191, 193–201
  effect of PVC on, 192–193
  effect of SVC on, 197–198
Gut microflora, termite, 361, 365, 378

H

Harvest index, of cluster bean, 234
Harvesting
  of BSF larvae/pupae, 319, 324–326, 329–331, 334, 349
  of BSF frass, 320, 330, 331, 349, 355
  of compost, 24
  of vermicast, 26, 34, 35, 40, 63–64, 66–99 *passim*
  in ZOCOFA system, 349, 355
Helical transport system, 100
Hermetia illucens. *See* Black soldier fly (BSF)

HEVSPAR (High-Efficiency Vermicomposting System for Paper Waste and Other Refuse), 144–151, 153, 162
APSU unit. *See* APSU
prototype, 151, 152
VSV unit. *See* VSV unit
HI (humification index)
of parent weeds *vs.* vermicompost, 243–244
significance of, 242
High-rate vermicomposting, 13, 40, 103, 129
High-rate vermicomposting paradigm (HRVP), 131–134, 137, 139, 162, 167, 179, 180, 241, 259, 299, 305
attributes of, 133
premises of, 131–132
validation of, 134, 136
Homogenization of vermireactor, drawback of, 130
Horticultural waste, 42, 107, 116, 134, 176, *See also* Phytowaste
Hot composting, 8–10, 19, 32–35, *See also* Composting, types of; Thermophilic phase
House fly (*Musca domestica*), 12, 310, 314, 318, 321, 376
HRVP. *See* High-rate vermicomposting paradigm
Humification. *See* HI (humification index)
Humus, 3, 41, 42, 61, 96, 177, 244
Hybrid composting, 8, 9, *See also* Composting, types of
Hydrogen sulfide (H_2S), 105, 119
Hypotermes obscuriceps, 379, 380, 389, 392, 393

I

Inclined parallel stack continuously operable (IPSCO) vermireactor system, 151–159, 162
comparison with MCVS, 164–165
distinguishing features, 155
flow chart, 156
isometric and side views, 157
loading and unloading units, 158–159
novelty of, 156–158
Inorganic fertilizers
comparison with vermicompost, 230–231, 233–235, 283
equivalent to vermicompost, 233
In situ termireactors (ISTs), 375, 380–386, 390, *See also* Termireactors
design and embodiments, 381–385
performance evaluation, 385–386
with trails, 383, 385
Insects, *See also* specific insects
as food (entomophagous), 371–372
for waste management, 310, 375–377
Intestine (earthworm), 36, 37
Invasive species, 107, 110, 112, 117, 120, *See also* specific weeds
Invertebrates, 106, 175, 375–377, *See also* Zoocomposting
in composting process, 8, 10, 13
explored for waste degradation, 11–12
in nature, 3
Ipomoea (*Ipomoea carnea*), 107, 110, 111, 115, 124, 134, 179, 182, 385, 389, 392, 393
Ipomoea (Ipomoea crassipes), 107, 112, 113, 134, 293, 305
chemical profile (GC-MS), 248–249
DSC analysis, 246, 264
effect on earthworm fecundity, 271–273, 276
FTIR spectra, 262–263
TG analysis, 246, 263
vermicompost (IVC) chemical profile, 253–255
Ipomoea vermicompost (IVC), 180, 185, 199, 209, 210, *See also* Ipomoea
chemical profile, 253–255
effect on germination and growth, 190–192, 203
effect on soil properties, 199, 204
field study on ladies finger, 211–212, 214, 216, 218, 221, 222
IPSCO. *See* Inclined parallel stack continuously operable vermireactor system
IVC. *See* Ipomoea vermicompost

J

Juveniles, 14, 54, 55, 63, 270, earthworm production of, 271–273, 276, 280–281

L

Ladies finger (*Abelmoschus esculentus*), *See also* Okra
diseased leaves and pods, 226–227
effect of CDVC and LVC on, 222–224, 231–232
effect of IVC on, 190–192, 211–212, 214, 218, 222
effect of LVC on, 188–189, 191, 193–201, 208–213, 217, 221, 231–232
effect of PRVC on, 282–283, 288
effect of PVC on, 192–193, 211–212, 215, 219, 223
effect of PWVC on, 202, 206
effect of SVC on, 197–198, 211–212, 216, 220, 224
field-scale studies with weed VCs, 180, 185–187, 208–225
Lalander, Dr. Cesila, 324, 325, 328–331, 334
Lampito mauritii, 61, 259

Lantana (*Lantana camara*), 110, 111, 115, 134, 179, 181, 305, 385, 389, 393
chemical profile (GC-MS), 247–248
FTIR spectra, 244–245
vermicompost (LVC) chemical profile, 251–253
Lantana vermicompost (LVC)
chemical profile, 251–253
comparison with CD vermicompost (CDVC), 222–224, 231–232
comparison with synthetic fertilizers (SCF), 180, 220–221, 229–230
effect on germination and growth, 187–191, 193–201, 231
effect on soil properties, 188, 198–201
field study on ladies finger, 208–213, 217, 221
pest repellent properties, 213, 228
Larvae (BSF), 317–319, 322–324, *See also* Black soldier fly (BSF)
consumption of other larvae, 318
efficiency as waste processors, 319
harvesting of, 319, 324–326, 329–331, 334, 349
instars, 319, 322, 323
migration instinct, 319, 324, 325, 329, 331, 349
pathogen reduction by, 318–319
pre-pupa, 319, 320, 323, 324, 348
Leaf litter, 24, 58, 59, 103, 104, 107, 122, 134–137, 380, 389
Leaf miner infestation, 187, 284, 289
Leachate, 72–79, 83, 84, 86, 87, 90, 176, 331, 338
Leachate collection, in ZOCOFA, 350–353
Life cycle
BSF, 322–325, 348–350
earthworms, 41, 53–56
termite, 365, 366
Lignin degradation, 11, 14–15, 244, 245, 263, 264, 359, 360, 365, 367, 375
Ligninous waste, 11, 14, 360–361, 366, 378, 380, *See also* Termigradation
challenges in composting, 360
termite ability to degrade, 359, 360, 365, 367, 375
Liriomyza spp. (leaf miner), 187
Livestock zoomass, 106, 107
Lumbricus rubellus, 51, 56, 261, 263, 2 70–277
C:N ratio of vermicompost, 259–260
DSC analysis of vermicompost, 264
performance on salvinia and ipomoea, 270–273
vermicomposting of prosopis, 278–279, 291
Lumbricus terrestris, 52, 55, 58
LVC (lantana vermicompost). *See* Lantana

M

Macrotermes convulsionarius, 379, 380, 388, 389, 392–393
Macrotermes jeanneli (in text it is as "M. jeanneli"), 387
Manure, *See also* Cow dung
composting of, 41
as inoculant, 19
pivotal role in vermicomposting, 175–176
vermicomposting of, 13, 33, 62, 63
vermocompost of (CDVC), comparison with LVC, 222–224
Marloes et al. (2022) system, 100
Mating
BSF, 322, 325, 329–332, 335, 337, 348, 352, 354, 355,
termite, 362–363, 386–388
MCVS (multiple compartment vermicomposting system), 159–163
comparison with IPSCO, 164–165
distinguishing features, 161–162
flow chart, 163
schematic and drawer, 160–161
Mercaptans, 105, 143
Methane emissions, 105, 119, 122
Microbial biomass carbon
effect of IVC on, 204
effect of LVC on, 188, 201
effect of PVC on, 205
effect of SVC on, 205
measurement method, 184
Microchaetus rappi, 51
Microflora, termite gut, 361, 365, 378
Migration instinct, BSF larvae, 319, 324, 325, 329–331, 337, 349
in ZOCOFA system, 349–354
Mineralization, 1, 3–5, 19, 37, 177, 242, 246–247, 263–264, 288, 291
Moisture content/distribution, 104, 133, 172, 173
effect on SEVS efficiency, 172, 173
maintenance in HRVP designs, 133
Mounds, termite, 361, 362, 367–371, 380. 389
in Brazil, 369, 371
construction of, 368
strength and durability, 369, 371
Mouth (earthworm). *See* Peristomium
Multiple compartment vermicomposting system. *See* MCVS
Musca domestica. *See* House fly

N

Nagaland, India, BSF facility in, 340
Nayeem-Shah, M., 33, 40, 181, 270
Nests, termite. *See* Mounds, termite

Nitrate-nitrogen
in LVC-amended soil, 200
in vermicompost, 200
Nitrogen fixation, by termites, 359, 365
Nutrient overload, from excessive vermicompost, 178

O

Obligate scavengers, 3, 4
Odontotermes anamallensis, 379, 380
Oesophagus, 36, 37
Okra. *See* Ladies finger
Olivier, P.A., 329, 338–340
Oxygen. *See* Aeration
Oxygen depletion, 105, 118–119

P

Paper waste vermicompost (PWVC)
comparison with cow dung (CD), 202–210
fertilizer value of, 180, 202–210
Paper waste vermicomposting, 31, 135, 169, 170, 204, 222, 269, 378
with FLUVTS, 11, 140–144
with HEVSPAR, 144–151, 153, 162
need for nutrient supplementation, 11, 141, 180–181
Parthenium (*Parthenium hysterophorus*), 110, 111, 116, 117, 119–120, 134, 179, 182, 385
allelopathic compounds, 120
mammalian toxicity, 120
chemical profile (GC-MS), 249–250
vermicompost (PVC) chemical profile, 255–256
termigradation of, 385
Parthenium vermicompost (PVC)
effect on germination and growth, 192–193, 203
effect on soil properties, 199, 205
field study on ladies finger, 211–212, 215, 219, 223
Parthenin, 120
Particle density, soil, 177, 183, 286
effect of LVC on, 188, 198–199
effect of PWVC on, 206, 210
Patents, 321, 329–331, 334–339, 375, 393
Pathogen control, by vermicompost, 284
Pathogen reduction, by BSF larvae, 318–319
Pathogens, 24, 26, 175
killing in composting, 9, 10, 21, 32
killing in vermicomposting, 33, 38
Patnaik, P., 57, 103, 110, 120, 121, 124, 127, 277, 278, 290, 385
PDCOP (pseudo-discretized continuous reactor operation protocol)
definition and protocol, 270
use in prosopis vermicomposting, 279, 280
Perionyx excavatus, 52, 55, 259, 278–279, 291
Perionyx sansibaricus
C:N ratio of vermicompost, 259–260
FTIR spectra of vermicompost, 262–263
performance on salvinia and ipomoea, 271–273
vermicomposting of prosopis, 278–282, 291
Peristomium, 35, 59
Pest and pathogen resistance
mechanisms for, 216, 218
from weed-derived vermicomposts, 180, 187, 213–215, 228, 284
Pharynx, 36, 59
Phenolic compounds, inhibitory effects of, 178
Pheromone trails, termite, 383–385
Phytol, prevalence in vermicomposts, 247–258
Phytomass, 13, 59, 61, 103, 107, 277, 280, 300, 305, *See also* Weeds
annual production, 104, 106
defined, 106
direct vermicomposting with HRVP, 134, 241, 269, 299
as feedstock, 104, 106–123
vs. total biomass, 106
utilization challenges, 116
vermicompost as fertilizer, 179–201
vermicomposting attempts, 121–123
Phytowaste, 62, 106, 107, 116–121, 167, 179, 180, 222, 298, *See also* Weeds
adverse impacts of, 117–121
defined, 107
Pistia (*Pistia stratiotes*), 107, 109, 112, 114, 134, 293
Porosity, soil, 60, 157, 162, 175, 177, 183, 282, 286, 368
effect of LVC on, 188, 199
effect of PWVC on, 206, 211
Poultry feed
BSF pre-pupae as, 324, 348, 350
earthworm zoomass as, 145, 175
Pre-composting, 122, 123, 141, 181, 269, 277–278
Pre-pupa (BSF), 319, 320, 323, 324, 348, *See also* Larvae (BSF); Pupa (BSF)
harvesting of, 319, 324, 349, 355
as poultry feed, 348, 350
Process control
in composting, 9, 21, 23–24, 33
in vermicomposting, 13, 26, 29–31, 33, 63
Processing facilities. *See* Design of BSF facilities

Process monitoring, vermicast production as criterion for, 130, 131
Prosopis (*Prosopis juliflora*), 110, 112, 117, 120–121, 360, 385
allelopathy, 121
challenges in vermicomposting, 277–278
characteristics and problems, 277
charcoal from, 121
leaves, biodegradation of, 121
pre-treatment for vermicomposting, 278–279
PRVC (prosopis vermicompost)
comparison with CD and CDVC, 285, 290
as fertilizer, 282–287
as pest repellent, 284
transformations during production, 285, 287–288, 290–291
Protozoan, 5, 6
Puducherry, India, termite studies in, 379, 387
Pupa (BSF), 319, 321–323, 332, 333, 355, *See also* Pre-pupa (BSF)
chitin from, 321, 323
PVC (parthenium vermicompost). *See* Parthenium

Q

Queen, termite, 14, 362, 363, 365–367, 372
as food, 372

R

Ragn-Sells, 324, 332, 333
Ramasamy, E., 51, 62
Reactors. *See* Design of BSF facilities; Vermireactor design; Termireactors; ZOCOFA system
Retention time, 34, 103, 123, 139–140, 164, 247–257, 270, 279, 293, 300, 392
Reticulitermes speratus, 378

S

Salvinia (*Salvinia molesta*), 107, 108, 112, 113, 117–119, 134, 179, 183
biomass doubling time, 118
chemical profile (GC-MS), 250
DSC analysis, 246
environmental impact, 118–119
effect on earthworm fecundity, 271–273, 276
FTIR spectra, 260–261
TG analysis, 263
vermicompost (SVC) chemical profile, 257
in SHEFROL bioreactor, 295
Salvinia vermicompost (SVC)
chemical profile, 257
effect on germination and growth, 197–198, 204
effect on soil properties, 199, 205
field study on ladies finger, 211–212, 216, 220, 224
Scanning electron microscopy (SEM)
of ipomoea and vermicompost, 258, 264–265
surface characteristics revealed by, 258–259, 264–265, 288, 291
Scavengers, 3–5
Screw conveyor system, 100
Separators, for BSF pupae/larvae, 324, 325, 327, 328, 332–334
Setae, 36, 53
SEVS (Separator of Earthworms, Vermicompost, and Substrate) machines, 167–174
cam mechanism, 168, 170
effect of batch size, 172–174
effect of moisture, 172, 173
embodiments of, 167–169
testing efficacy of, 169–172
Sewage treatment, *See also* SHEFROL bioreactor
conventional technologies, 292
global and Indian scenario, 292
macrophyte/algae-based systems, 293–294
SHEFROL bioreactor
description and advantages, 297–298
development and validation, 294
pilot plants, 294–297
upgrading to biorefinery, 298–300
Soil engineers, 13–14, 44, 359, 366, 367
Soil health, improvement by vermicompost, 286
Soil properties, improvement by vermicompost
bulk density, 188, 198, 206, 209
microbial biomass, 188, 201, 204–205
particle density, 188, 198, 206, 210
porosity, 188, 199, 206, 211
water holding capacity, 183–184, 188, 199
Soldier (termite caste), 362, 363, 366
Soldier fly. *See* Black soldier fly (BSF)
Stable fly (*Stomoxys calcitrans*), 12, 310, 311, 316, 377
Stomach (earthworm), 37
Stratiomyidae, 317
Substrate, *See also* Manure; Paper waste; Phytomass; Weeds
aeration, 105
agitation, 105
moisture in, 104
separation from vermicast. *See* SEVS machines
Swedish University of Agricultural Sciences (SUAL), 324, 328–331
SVC (salvinia vermicompost). *See* Salvinia
Syntermes dirus, 369, 371
Synthetic chemical fertilizers (SCF), comparison with LVC, 180, 220–222, 229–231

T

Tabassum-Abbasi, 13, 51, 57, 62
Temperature
in composting, 9, 10, 19, 21, 23–25
earthworm sensitivity to, 29, 56
in vermicomposting, 26, 29, 38
Termigradation, 13, 357, 375–396, *See also* Bioengine-based termireactors (BBTs); *In situ* termireactors (ISTs); Termites; Zoocomposting
of ipomoea, 385, 389, 392, 393
prior attempts, 377–378
rate of, 385–386, 389–393
Termireactors, 375, 380, 381, 385, 389, 391, *See also* Bioengine-based termireactors (BBTs); *In situ* termireactors (ISTs)
Termites, 5, 13–15, 30, 359–374, *See also* Alates; Captive termite colonies; Termigradation
as architects, 367–370
castes, 362–363, 366
challenges in using, 375, 378–379
as delicacies (food), 371–372
distribution of, 361
gut microflora, 361, 365, 378
life cycle, 365, 366
lignin degradation by, 359, 360, 365, 367, 375
nitrogen fixation by, 359, 365
nests and mounds, 361, 362, 367–371
role and potential, 365–367
soil engineering by, 359, 366
species identification for study, 379–380
TG (thermogravimetric analysis), 242, 262
of ipomoea and vermicompost, 246
of salvinia and vermicompost, 263
use in assessing mineralization, 246, 263, 288, 290
Thermophilic phase
in composting, 8–10, 19, 21, 23, 26, 32–33, 38
earthworm intolerance to, 29
Tomato (*Solanum lycopersicum/Lycopersicon esculentum*)
effect of PRVC on, 282–283, 286–288
effect of PWVC on, 202, 204, 207–208
nutrient content of, 283
Total porosity. *See* Porosity, soil
Tray-based systems, problems of, 100
Trails, for termite attraction, 381, 383–385
Turning (in composting), 9, 19, 21, 24, 26, 30, 33, 46, 56, 392

V

Vermicast (vermicompost)
as aimed product, 103, 129–131
characteristics, 32, 59
deposition of, 105
as fertilizer, 4, 5, 19, 32, 33, 59
harvesting of, 26, 34, 35, 40, 63–64, 66–99 passim
production process, 29, 34–35, 37
separation from earthworms and substrate. *See* SEVS machines
Vermicompost (VC), *See also* CDVC; End product specification; Fertilizer value; IVC; LVC; PRVC; PVC; SVC;
advantages over compost, 130, 177
beneficial *vs.* inhibitory effects, 177–178, 202
comparison with inorganic fertilizer, 230–231, 233–235, 283
defined, 103
defined and reproducible product, need for, 176–177
effect on plant growth, 230–232
effect on soil properties, 286
as pathogen controller, 284
Vermicomposting, *See also* High-rate vermicomposting; High-rate vermicomposting paradigm (HRVP); Zoocomposting
completion, flawed criteria for judging, 130
vs. composting, 8, 10, 19, 26, 29–30, 32–34, 38, 130, 131
definition, 8, 10, 31–32
of phytomass (direct), 241, 269, 299
of pre-composted material, 32–33
process steps, 29–30
rate of, 26, 29, 34–35, 38–40
uniqueness as engineered process, 30–31
of xerophytes, 277, 299
Vermicomposting device (Marloes et al.), 100
Vermiculture, 46, 48, 49, 76–78, 95–96, 97, 99
Vermireactor design, 141–165, *See also* FLUVTS; HEVSPAR; IPSCO; MCVS; Process control
anaerobic pockets in, 38, 105, 133, 143
challenges, 26, 34–35, 38–40, 63
conventional systems overview, 63–99, 143, 294
depth of, 104
engineering of, 123
high-rate principles, 13, 34–35, 40
high surface area-to-volume ratio in HRVP, 133
low substrate height in HRVP, 133
operational modes, 34, 35
surface-to-volume ratio, 35
VFAs (volatile fatty acids), extraction from phytomass, 278, 300
VSV (vertical stacked vermireactor) unit, 144–145, 149–152
distinguishing features, 151
prototype, 152
working of, 150–151
Vultures, 3, 4

W

Waste, types of, *See also* Feedstock
abattoir/slaughter, 8, 11, 12, 33, 307–309, 375, 376
food/kitchen, 8, 33, 71, 89, 91, 338
ligninous, 11, 14, 291, 305, 359–361, 367, 375, 378, 380, 385, 389
phytomass/plant, 13, 59, 61, 62, 103–107, 116, 121, 134, 140, 167, 175, 266, 269, 298–300
Waste paper, termite processing of, 175, 176, 377–378, 384, 385
Water holding capacity, 218, 282, 286, 359
of soil amended with LVC, 188, 199
of vermicast, measurement method, 183–184
Water hyacinth (*Eichhornia crassipes*), 107, 108, 112, 113, 118, 293, 296
Weeds, 107–121, *See also* Aquatic weeds; *Ipomoea; Lantana; Parthenium; Salvinia*
control efforts, 110, 117
direct vermicomposting with HRVP, 134, 136
economic utilization of, 116, 119–121
environmental harm from, 117–121
termigradation of, 385, 389, 392, 393
toxic, vermicomposting of. *See* Ipomoea; Lantana; Parthenium; Salvinia
vermicompost as fertilizer, 179–201
Windrows
aeration of, 21, 23–24
setting up, 19, 21
static pile, 49, 64, 66
top-fed, 49, 64, 66
turning of, 21, 24
Workers (termite caste), 361–363, 366–368, 375, 378, 379, 383, 387

Z

ZOCOFA system, 347–356
bioreactor embodiments (S1, S2), 350–354
drawbacks overcome by, 348
novelty and features, 349–350
operation and production, 355–356
priming of reactor, 355
summary of invention, 348–349
Zoomass, 61, 63, 106–107, 145, 167, 322, 323, 334, 365, 386
defined, 106
livestock contribution, 106, 107
Zoocomposting, 309–396, *See also* Black soldier fly (BSF); Termigradation; Termites; Vermicomposting
defined, 310
definition and introduction of term, 1, 10
distinctive features, 10–12
processes covered in book, 13–15
rate compared to composting, 38
with termites (Termigradation), 357–396